ELECTRIC CIRCUITS AND MACHINES

ELECTRIC CIRCUITS AND MACHINES

Seventh Edition

Eugene C. Lister
Chief Electrical Engineer (Retired)
Stanley Consultants, Inc.

Robert J. Rusch
Chief Electrical Engineer
Stanley Consultants, Inc.

New York, New York   Columbus, Ohio   Mission Hills, California   Peoria, Illinois

The cover photograph is an interesting double image of a substation with an oil breaker as the main feature. A second photograph of the morning sunrise is superimposed for artistic impact. This contrived image is only for dramatic appeal. Courtesy: Comstock Inc./Comstock Inc.

Library of Congress Cataloging-in-Publication Data

Lister, Eugene C.
    Electric circuits and machines/Eugene C. Lister, Robert J. Rusch. —7th ed.
        p.   cm.
    Includes index.
    ISBN 0-02-801809-5
    1. Electric engineering.   I. Rusch, Robert J., Date
II. Title.
TK146.L54   1993
621.3—dc20   92-33286
                                              CIP

Electric Circuits and Machines, Seventh Edition

Imprint 1995
Copyright © 1993 by Glencoe/McGraw-Hill. All rights reserved. Originally copyrighted in 1984, 1975, 1968, 1960, 1952, and 1945 by McGraw-Hill, Inc. Except as permitted under the United States Copyright Act, no part of this publication may be reproduced or distributed in any form or by any means, or stored in a database or retrieval system, without prior written permission from the publisher.

Send all inquiries to:
Glencoe/McGraw-Hill
936 Eastwind Drive
Westerville, OH 43081

ISBN 0-02-801809-5

Printed in the United States of America.

3 4 5 6 7 8 9 10 11 12 13 14 15   RRDC/MC   03 02 01 00 99 98 97 96 95

Contents

14   ALTERNATING-CURRENT GENERATORS   234

15   DIRECT-CURRENT MOTORS AND CONTROLS   250

16 POLYPHASE INDUCTION MOTORS AND CONTROLS 281

17 SYNCHRONOUS MOTORS 309

21   INDUSTRIAL CONTROL   396

APPENDIXES   413

Preface

Electric Circuits and Machines, Seventh Edition, is a concise, practical, survey-type text covering the fundamentals of electric circuits and equipment. This new edition, like the earlier ones, is designed for use in technical institutes and community colleges; in home study, adult education, and job training courses; in training courses in the Armed Forces; and in survey courses for nonelectrical engineering students.

Essentially, the changes made for this edition update the text and illustrations along with adding new material to reflect changes in industry due to increasing computerization. Materials on obsolete or little-used equipment has, to some extent, been replaced by material on newer or more widely used equipment. However, in an increasingly solid-state electronic and computerized environment, the fundamental concepts behind electric circuits, machines, and their control need to be understood by the student. These newer devices can sometimes obscure the actual process and its fundamentals. Many of the components that promote this fundamental understanding have been eliminated by integration into electronic logic. Therefore, the concepts and descriptions of equipment that may not be in current common use have been retained where they illustrate these fundamental concepts.

The organization of *Electric Circuits and Machines*, Seventh Edition, has been changed in that the discussions of types of electric motors and their control have been integrated into the chapters describing motors rather than being scattered throughout the text. The chapter organization has also been revised. These changes provide a logical subject grouping and a useful order of presentation.

To achieve a broad yet concise text, detailed mathematical analysis of theories involved have been omitted. Mathematics is limited to simple algebra and to the use of a few trigonometric relationships. The simple trigonometry used is explained in Appendix G. For those students or instructors desiring to use the complex quantity notation for phasor representation, Appendix I includes a brief discussion of this notation.

Many numerical examples are worked out in detail in the text. In addition, problems are included in those chapters in which the student's solution of the problems will illustrate the application of the principles involved. Answers to the odd-numbered problems are included in the text. Review questions at the end of each chapter are valuable both in class discussion and as a study aid for individual students who wish to evaluate their grasp of the subject matter. An instructor's manual available separately from the publisher includes answers to the even-numbered problems and complete solutions for selected problems.

The comments and criticisms offered by users of earlier editions of this book are gratefully acknowledged. The authors are also indebted to the many companies and organizations that supplied information and photographs used in the preparation of this seventh edition.

Eugene C. Lister
Robert J. Rusch

1

Fundamental Units

While the exact nature of electricity is unknown, much is known about what it can do. By the mere closing of a switch, buildings are lighted, wheels are turned, ice is made, food is cooked, distant voices are heard, and countless other tasks—ordinary and extraordinary—are performed. Although many uses for electricity have been discovered and applied, electrical technology is not static. Electric machines and devices that have been used for many years are being improved and are finding wider applications. New uses for electricity are continually being found and applied. Much is still being learned about electricity and what it can do.

Electricity is a convenient form of energy. It is well known that when fuels such as coal, oil, and gas are burned, energy is released. A waterfall, whether it is constructed or natural, also possesses energy. Yet, to be of value, this energy must be made available at points where it can be used conveniently. Electricity furnishes the most practicable and convenient means yet devised for doing this. The energy of burning fuel or of falling water is changed to a more convenient form—electricity—by *electric machines*. It is transmitted to distant points over *electric circuits*. It is controlled by other electric machines. At points where it is to be used, it is converted into useful work by still other electric machines and devices.

Since electricity is a form of energy, the study of electricity is the study of energy, its conversions from one form to another, and its transmission from one point to another. Electric machines are energy-conversion devices, and electric circuits are energy-transmission devices.

Although no one knows precisely what electricity is, it has been possible to develop theories about electricity through experiment and by observation of its behavior. As a result, it is now believed that all matter is essentially electrical in nature.

In this introductory chapter, important ideas concerning electricity are presented together with a discussion of the basic units of measurement used in the study of electric circuits and machines.

1-1 THE STRUCTURE OF MATTER

Matter has been defined as anything that occupies space and has mass. Familiar examples of matter are water, air, copper, tin, smoke, iron, and salt. Matter is made up of extremely small particles called *molecules*. A molecule is the smallest particle into which a given substance may be divided without changing its identity. For instance, water is composed of two parts hydrogen and one part oxygen. If a drop of water were divided and subdivided into smaller and smaller particles, the smallest particle into which it could be divided and still contain two parts hydrogen and one part oxygen would be called a molecule of water. If this particle were subdivided further, the resulting parts would be hydrogen and oxygen, both of which are gases and obviously do not have the physical properties of water.

Small as they are, molecules may be divided by chemical means into *atoms*. There are as many kinds of molecules as there are kinds of matter, but there are only 109 different kinds of atoms corresponding to the number of known chemical elements. As a result of the work of such Nobel prize–winning scientists as Niels Bohr and Wolfgang Pauli, much is known about the structure of the atom.

The makeup of an atom may be described in a simplified way by likening it to a miniature solar system. The core of the atom, corresponding to the sun in the solar system, is called the *nucleus*. The nucleus is formed of subatomic particles called *protons* and *neutrons*. Surrounding the nucleus and in a state of continual motion similar to the orbiting of the planets about the sun are other subatomic particles called *electrons*. Atoms of different elements contain different numbers of electrons. However, the electrons of all atoms are identical.

It has been found that protons and electrons exert forces on one another over and above the forces of gravitational attraction between them. This force is attributed to a property of the protons and electrons known as *electric charge*. Protons exert forces of repulsion on other protons and likewise electrons exert forces of repulsion on other electrons. However, protons and electrons attract one another. There are, therefore, two kinds of electric charge, arbitrarily designated as *positive* and *negative*.

It has been found that all electrons have precisely the same negative charge and that all protons have precisely the same positive charge. Neutrons have no charge, or are electrically neutral. In the normal atom there are equal numbers of electrons and protons and therefore the atom is uncharged since the positive charges just balance the negative charges. Furthermore, a particular body made up of uncharged or neutral atoms is uncharged or neutral. However, nearly all atoms have some electrons that are loosely bound to their nuclei. These electrons are called *free electrons* and may be dislodged by one means or another and transferred from one atom to another.

Ans #3

If a body is caused to lose some of its electrons or negative charges, the body is said to be positively charged. Similarly, if a body is caused to gain electrons, it is said to be negatively charged. For example, a dry glass rod may be charged positively if it is rubbed vigorously with a silk cloth. Some of the electrons on the glass rod are transferred to the silk cloth, leaving the rod positively charged and causing the cloth to assume a negative charge.

The forces of attraction and repulsion of electrically charged bodies form the basis for a fundamental rule of electricity; that is, *similarly charged bodies repel one another and oppositely charged bodies attract one another.* The force of attraction or repulsion depends upon the distance between the charged bodies: the greater the separation, the weaker the force. If the distance between two charged bodies is doubled, the force is decreased to one-fourth as much. Stated in mathematical terms, the force is inversely proportional to the square of the distance between the charged bodies.

Ans #4

The atoms of some materials such as copper, zinc, silver, and aluminum have many free electrons. Such materials are called *conductors*. The atoms of nonmetallic materials such as glass, slate, mica, and porcelain have electrons that are held rigidly to their nuclei; that is, they have few free electrons. These materials are called nonconductors or *insulators*. Other materials such as germanium and silicon are neither good insulators nor good conductors. These materials are called *semiconductors*.

#6

#7

1-2 ELECTRIC CURRENT

In a conductor material, some of the free electrons are freely moving at random or "migrating" from atom to atom. When, in addition to this random motion, there is a drift or general movement of electrons along the conductor, this is called an *electric current*.

When a strip of zinc and a strip of copper are immersed in a solution of sulfuric acid to form a simple *electric cell,* the resulting chemical action causes the zinc strip to gain electrons from the solution, thereby causing it to become negatively charged. At the same time the copper strip has a tendency to lose some of its electrons, causing it to become positively charged. When the two plates are connected by a copper wire, which is a good conductor, the electrons on the zinc strip flow through the wire to the copper strip. This movement of the electrons from the zinc strip through the copper wire to the copper strip and back through the solution constitutes an electric current, and the entire path through which it flows is called a *circuit*. Thus, *an electric current is merely the movement of electrons or negative charges through a conductor.*

Early experimenters recognized the fact that an electric current was a movement of charges along a conductor. Since the direction of the flow of current was not known, unfortunately it was arbitrarily chosen to be from a positively charged body to a negatively charged body (positive to negative), and this convention has been so firmly established that it is still in use. Thus, many people believe the conventional direction or positive direction of current flow to be from positive to negative, even though it is now known that the direction of electron flow, which actually constitutes an electric current, is from negative to positive.

1-3 UNIT OF CURRENT: THE AMPERE

$1 \text{ coulomb} = 6.28 \times 10^{18}$ electrons

The smallest quantity of electricity is the electron. This particle of electricity is so small that the flow of one electron cannot be detected by normal current-measuring devices. A *coulomb* is a much larger unit of quantity of electricity or electric charge and is used for practical measurements. One coulomb (C) is equal to approximately 628×10^{16} electrons. Since a coulomb is a quantity of electricity, then a *rate of flow* of electricity may be specified in *coulombs per second*. In practice, the term *coulomb per second* is seldom used; a shorter term, *ampere* (symbol A), is used instead. One ampere is equal to the movement of one coulomb of electricity past a given point in one second.

For example, an ordinary 100-watt (W) lamp used on a house-lighting circuit takes a current of about 0.83 A. This means that electricity is passing through the lamp at the rate of 0.83 C every second, or more simply, the current is 0.83 A.

The flow of an electric current through a conductor is sometimes compared with the flow of water through a pipe. The quantity of electricity, measured in coulombs, is compared with the quantity of water, measured in gallons or cubic meters. The rate of flow of electricity measured in amperes is compared with the rate of flow of water measured in gallons per minute or cubic meters per second. This comparison is often helpful in learning the units of measurement of electricity.

1-4 RESISTANCE: THE OHM

It was pointed out in Sec. 1-1 that materials that have many free electrons are called conductors, while materials having few free electrons are called insulators. Conductors are said to offer a low *resistance* or opposition to the flow of an electric current, and insulators offer a high resistance to current flow. Electrical resistance, then, is defined as the opposition offered by a material to the flow of an electric current.

The practical unit of measurement of resistance is the *ohm*, which is represented by the symbol Ω (capital omega). As practical examples of resistance, an ordinary 100-watt electric lamp used on a 120-volt circuit has a resistance of about 144 Ω when hot. A 1000-ft length of No. 10 American Wire Gauge (AWG) copper wire (0.1 in. in diameter) has a resistance of about 1 Ω. A round copper wire with a cross-sectional area of 2 square millimeters (mm²) and a length of 1 kilometer (km) has a resistance of about 6.8 Ω. A 240-volt, 2500-watt electric heater has a resistance of about 23 Ω.

1-5 POTENTIAL DIFFERENCE AND ELECTROMOTIVE FORCE: THE VOLT

Just as there must be a difference in water pressure to cause water to flow between two points, so must there be a difference in electric pressure to cause an electric current to flow between two points in an electric circuit. This difference in electric pressure is called *potential difference,* and it is measured in *volts*. The volt (symbol V) is the amount of potential difference that will cause one ampere to flow through a resistance of one ohm.

Potential differences in common use vary from a few millionths of a volt to several million volts. The potential difference between the terminals of a common dry cell is about 1.5 V; between terminals of an automobile storage battery, about 12 V. Common potential differences applied to the terminals of electric motors are 115, 200, 230, 460, 575, and 2300 V. Potential differences between conductors on long power-transmission lines are as high as 765,000 V. Experimental lines have been operated with even higher potential differences.

If two bodies have different amounts of charge, a potential difference exists between the two bodies. Potential difference is then merely a difference in electric charge. When two points having a potential difference between them are joined by a conductor, a current flows along the conductor attempting to equalize the difference in charge on the two points. When the two charges are equalized, the flow of current stops. Therefore, if a current is to be maintained between two points, the potential difference between the points must be maintained.

A device that has the ability to maintain a potential difference or a difference in charge between two points, even though a current is flowing between those points, is said to develop an *electromotive force* (emf).

There are several ways in which an emf may be developed. The simple electric cell described in Sec. 1-2 develops an emf by *chemical* means. The electric generator, in which conductors are moved through magnetic fields, develops an emf by *mechanical* means. When a junction of two dissimilar metals is heated, a small emf is developed. This is the principle of the *thermocouple*. When certain substances are placed in contact with metals and the junction is illuminated, a small emf is developed. The *photoelectric cell* operates on this principle. If, when any of the above-mentioned devices is in operation, the terminals of the device are connected by a conductor, a continuous current flows through the completed circuit because the potential difference at the terminals is maintained.

Since the result of the action of an emf in a circuit is a potential difference measured in volts, emf is also expressed in volts.

Thus, a potential difference causes current to flow, and an emf maintains the potential difference. Since both are measured in volts, a common term, *voltage,* is used to indicate a measure of either. Although the terms potential difference, emf, and voltage do not mean exactly the same thing, they are often used interchangeably.

**1-6
MEASUREMENT
OF CURRENT,
VOLTAGE, AND
RESISTANCE**

A simple electric circuit consisting of a generator supplying a lamp is represented in Fig. 1-1. (For a table of symbols used in circuit diagrams refer to Appendix A.) The assumed or conventional direction of current flow is out of the positive terminal of the generator, through the lamp, and back to the negative terminal of the generator as shown by the arrows.

Electric current is measured by an *ammeter.* An ammeter measures the number of coulombs per second or amperes flowing in a circuit. To measure the current flowing through the lamp of Fig. 1-1, an ammeter is connected into the circuit as shown in Fig. 1-2. Note that the ammeter is *inserted into* or made a part of the circuit.

For the purpose at hand, electricity may be considered as a fluid that cannot be compressed. If electricity is thought of in this way, it should be easy to see that in the circuit of Fig. 1-2 the same current must flow through both the ammeter and the lamp. Because an ammeter is always

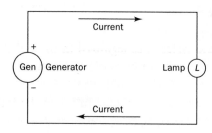

FIGURE 1-1 *A simple electric circuit consisting of a generator, a lamp, and the connecting conductors.*

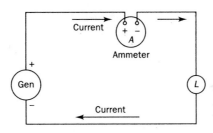

FIGURE 1-2 The ammeter is inserted into the circuit of Figure 1-1 to measure the electric current.

connected directly into the circuit, it must have a very low resistance so as not to hinder the flow of current. The ammeter is connected so that the current enters it through the positive $(+)$ terminal and leaves it through the negative $(-)$ terminal.

Potential difference or voltage is measured with a *voltmeter.* A voltmeter must be connected *between* or *across* the two points whose difference in potential is to be measured.

Figure 1-3 shows the method of connecting a voltmeter to measure the voltage across the lamp. Note that the current flowing through the lamp *does not* flow through the voltmeter; the voltmeter is merely *tapped* on the circuit.

Compare the methods of connecting ammeters and voltmeters in Figs. 1-2 and 1-3. The *ammeter* is *inserted into* the circuit, while the *voltmeter* is *connected across* the circuit. An ammeter should never be connected across a circuit, since it has such a low resistance that it would be ruined by the large rush of current through it. A voltmeter is not damaged by its connection across the circuit because it has a very high resistance.

Resistance may be measured by means of a Wheatstone bridge, an ohmmeter, or by the voltmeter-ammeter method, all of which are discussed in Chap. 20.

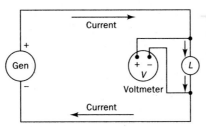

FIGURE 1-3 The voltmeter is connected to measure the voltage applied to the lamp.

1-7 THE INTERNATIONAL SYSTEM OF UNITS (SI)

The metric system was legalized in the United States by an act of Congress in 1866. The use of the metric system was not made obligatory, however. As a result, the firmly established U.S. Customary System (USCS) of units (pound, foot, etc.) continued in use in the United States. However, in 1975, the Metric Conversion Act was signed into law establishing a U.S. Metric

TABLE 1-1 BASIC SI UNITS

Quantity	Name	Symbol
Length	meter	m
Mass	kilogram	kg
Time	second	s
Electric current	ampere	A
Temperature	kelvin	K
Luminous intensity	candela	cd
Amount of substance	mole	mol

Board to coordinate the voluntary conversion activities of private industry. As a result, U.S. industries are gradually converting to the metric system on a voluntary basis.

Almost all major countries of the world, other than the United States, have officially adopted a modernized metric system of units called the International System of Units, officially abbreviated SI in all languages. In this system, all units used in the various technologies are derived from seven basic and arbitrarily defined units. These units and their symbols, which have been standardized by the General Conference on Weights and Measures (abbreviated CGPM from the official French name), are listed in Table 1-1. The units mole and candela are included in Table 1-1 for the sake of completeness and are not referred to further in this text.

In addition to the seven basic units, the CGPM has standardized many derived SI units, that is, units that have been derived from the basic units. All such units may be expressed algebraically in terms of the base SI units. As examples, the SI unit for area is the square meter (symbol m^2); the unit for volume is the cubic meter (symbol m^3); and the unit for velocity is the meter per second (symbol m/s).

One group of derived SI units have been given special names. The volt and ohm are examples of such units. Other derived units with special names that will be defined and used later in this text are the newton, watt, and joule, which are the units of force, power, and energy, respectively. Likewise, the unit of capacitance is called the farad; the unit of inductance, the henry; and the unit of frequency, the hertz.

1-8 SI UNITS The following is a brief description based on the official definitions of the five basic SI units that will be referred to in this text. Actually, it will be found that in electrical technology, SI or metric units rather than USCS units predominate and little conversion from one system to the other is required. However, important conversion factors for some of the basic units are included in the following.

The SI unit of length is the meter. One meter (m) is equal to 100 centimeters (cm), 1000 millimeters (mm), and 39.37 inches (in.). One inch is equal to 25.4 mm.

The unit of mass, the kilogram, is defined as the mass of a prototype platinum-iridium cylinder, which is kept in France. One kilogram (kg) is equal to 2.205 pounds mass, or one pound mass is equal to 0.4536 kilogram. Likewise, one ounce is equal to 0.02835 kilogram or 28.35 grams.

The unit of time, the second, is the same in SI as in USCS units and is approximately 1/86,400 of the mean solar day. The minute and the hour are not decimal multiples of the second and are, therefore, non-SI units, but they are so firmly established that their use is likely to continue indefinitely.

The SI temperature scale is the International Thermodynamic Temperature Scale and the SI unit is the kelvin. On this scale, the temperature of the freezing point of water is 273.15 K and the kelvin is defined as the fraction 1/273.15 of this temperature. A temperature change of one kelvin is the same as a temperature change of one degree Celsius (symbol °C), which is the unit of the Celsius scale (formerly called the Centigrade scale). Wide use is made, particularly in engineering areas, of the Celsius scale because of the equality of the kelvin and the degree Celsius. The relationship of the kelvin, the degree Celsius, and the degree Fahrenheit is shown in Table 1-2.

The ampere is defined as that constant current, which, if maintained in two parallel conductors spaced one meter apart, produces a specified force between the two conductors (due to their magnetic fields). The unit of current is the same in SI and USCS.

Two derived electrical SI units, the ohm and the volt, are so fundamental to electrical technology that their definitions are included here with the basic SI units. The volt is defined as the difference of electric potential between two points of a conductor carrying a constant current of one

TABLE 1-2 COMPARISON OF TEMPERATURE SCALE

	Kelvins	Degrees Celsius	Degrees Fahrenheit
Water boils	373.15	100	212
Water freezes	273.15	0	32
Absolute zero	0	−273.15	−459.57

ampere, when the power dissipated between these points is equal to one watt. The ohm is defined as the electric resistance between two points of a conductor when a constant difference of potential of one volt, applied between these two points, produces, in this conductor, a current of one ampere.

1-9 MULTIPLES AND SUBMULTIPLES OF SI UNITS

So that very large or very small numbers of SI units may be avoided, decimal multiples and submultiples of the basic units are formed by the use of standard prefixes. The standard prefixes used with SI units and the symbols used to represent them are given in Table 1-3. As an example of the use of these prefixes and symbols, one thousand volts (1000 V) is called one kilovolt (1 kV). Likewise, one million ohms (1,000,000 Ω) is called one megohm (1 MΩ) and one one-thousandth of one ampere (0.001 A) is called one milliampere (1 mA). The prefixes listed in Table 1-3 may be used with other electrical units as well as with the three basic units.

1-10 SUMMARY OF IMPORTANT IDEAS CONCERING ELECTRICITY

Electric current is the movement of *electric charges* along a conductor. An *electric circuit* is a path over which an *electric current* may flow. The *ampere* is the unit of measurement of *current* and is a *rate of flow* of *one coulomb* of electricity *per second*. *Resistance* is the *opposition* to the flow of an electric current. The *ohm* is the unit of measurement of *resistance*. *Potential difference* is that which causes current to flow, and its unit of measurement is the *volt*. *Electromotive force*, or *emf*, is required to maintain a potential difference between two points when a current flows between the two points.

TABLE 1-3 PREFIXES USED WITH ELECTRICAL UNITS

Prefix	Factor by Which the Unit Is Multiplied	Symbol
tera	$1,000,000,000,000 = 10^{12}$	T
giga	$1,000,000,000 = 10^{9}$	G
mega	$1,000,000 = 10^{6}$	M
kilo	$1,000 = 10^{3}$	k
hecto	$100 = 10^{2}$	h
deka	$10 = 10^{1}$	da
deci	$0.1 = 10^{-1}$	d
centi	$0.01 = 10^{-2}$	c
milli	$0.001 = 10^{-3}$	m
micro	$0.000\ 001 = 10^{-6}$	μ
nano	$0.000\ 000\ 001 = 10^{-9}$	n
pico	$0.000\ 000\ 000\ 001 = 10^{-12}$	p
femto	$0.000\ 000\ 000\ 000\ 001 = 10^{-15}$	f
atto	$0.000\ 000\ 000\ 000\ 000\ 001 = 10^{-18}$	a

TABLE 1-4 ELECTRICAL UNITS

Quantity and Symbol		Unit and Symbol		Measuring Device
Current	I	ampere	A	Ammeter
Electromotive force	E	volt	V	Voltmeter
Potential difference	V	volt	V	Voltmeter
Resistance	R	ohm	Ω	Wheatstone bridge Ohmmeter Voltmeter-ammeter

An emf may be produced *chemically, mechanically,* or by means of *light* or *heat. Voltage* is a common term used to indicate a measure of either *potential, difference* or *emf. Current* may be measured by means of a low-resistance *ammeter* that is *inserted into* the circuit. *Voltage* may be measured by means of a high-resistance *voltmeter* that is tapped *across* the circuit. *Resistance* may be measured by means of a *Wheatstone bridge,* an *ohmmeter,* or by the *voltmeter-ammeter method.* The *International System of Units (SI)* is based on seven base units: the *meter,* the *kilogram,* the *second,* the *ampere,* the *kelvin,* the *mole,* and the *candela.*

The *ampere* is a base SI unit, and the *volt* and the *ohm* are both derived SI units. The basic electrical units are listed in tabular form in Table 1-4. Electrical instruments and measurements are discussed in Chap. 20.

REVIEW QUESTIONS

1 Why is electricity a convenient form of energy?
2 What is matter? Give examples.
3 What is a molecule? What is an atom?
4 The nucleus of an atom is associated with what kind of charge?
5 What are the negative charges in an atom called?
6 What is the relation between the positive and negative charges in an uncharged or neutral atom?
7 Give the rules for attraction and repulsion of electric charges.
8 What is a conductor? Give examples.
9 What is an insulator? Give examples.
10 Describe the nature of an electric current.
11 What is an electric circuit?
12 Is the assumed or conventional flow of current in the same direction as the electron flow in a circuit?
13 What is the unit of electric charge used in practical measurements? How is this unit related to the ampere?
14 What is electrical resistance? What is its unit of measurement?

15 What is potential difference? What is its unit of measurement?

16 How is a potential difference maintained between two points, even though a current is flowing between the points?

17 What is the common term used to indicate a measure of either potential difference or emf?

18 How are ammeter and voltmeters connected in an electric circuit?

19 What precaution is necessary in connecting voltmeters and ammeters?

20 As a summary, write the units of measurement of electric charge, current, voltage, and resistance.

21 What is a simpler way of expressing the term 0.000 001 A?

22 Name the seven basic SI units.

23 How many millivolts are there in 1 V?

24 How does a derived SI unit differ from a base unit?

2

Direct-Current Circuits

The basic rules governing direct-current circuits and the interrelationship of voltage, current, and resistance in those circuits are discussed in this chapter.

There are two basic ways of connecting two or more pieces of electric apparatus: they may be connected in *series* or in *parallel*.

When electric devices are connected end to end or in tandem to form a single continuous circuit, they are said to be connected in *series*. The three resistances R_1, R_2, and R_3 in Fig. 2-1 are connected in series. Note that in Fig. 2-1 there is only one path over which current may flow.

When the apparatus is connected so that there is a divided path over which current may flow—two or more alternative routes between two points in a circuit—the arrangement is called a *parallel* circuit. This combination, shown in Fig. 2-2, is also known as the multiple or shunt connection.

Other connections are combinations or variations of the two basic circuits. For example, in Fig. 2-3, the parallel combination of R_1 and R_2

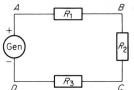

FIGURE 2-1 *The resistances R_1, R_2, and R_3 are connected in series.*

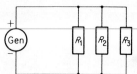

FIGURE 2-2 ◆ The resistances R_1, R_2, and R_3 are connected in parallel.

is in series with R_3. In Fig. 2-4, the series combination of R_1 and R_2 is in parallel with R_3. As still more apparatus is connected into a circuit, the combinations become more complex.

2-1 SERIES CIRCUITS: CURRENT RELATIONS
The same amount of current must flow in every part of a series circuit. This should be evident from a diagram of a series circuit. For example, in Fig. 2-1 the current that flows out of the positive terminal of the generator must flow successively through resistances R_1, R_2, and R_3 before returning to the negative terminal of the generator since there is only one path over which the current may flow. This is true regardless of the values of the several resistances in series. This may be easily verified experimentally by inserting an ammeter at several different points in a series circuit, such as at the points *A*, *B*, *C*, or *D* in Fig. 2-1. It will be found that the amount of current flowing at any of these points is the same.

The current is the same in all parts of a series circuit.

This rule may be stated mathematically by the expression

$$I_t = I_1 = I_2 = I_3 = \cdots \tag{2-1}$$

where I_t is the total current supplied by the generator and I_1, I_2, and I_3 are the currents in the several parts of the circuit.

This does not imply that the current of a series circuit cannot be changed by altering the circuit. A change in either the applied voltage or the resistance of the circuit will change the value of the current flowing. However, for any given value of circuit resistance and applied voltage, the same current must flow in every part of the circuit.

2-2 SERIES CIRCUITS: VOLTAGE
Water pressure is required to cause water to flow through a pipe. More pressure is required to force water to flow at a given rate through a small pipe than through a large pipe since more resistance is offered to the flow in the small pipe.

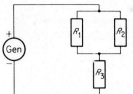

FIGURE 2-3 ◆ The parallel combination of R_1 and R_2 is connected in series with R_3.

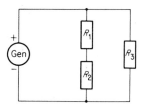

FIGURE 2-4   The series combination of R_1 and R_2 is connected in parallel with R_3.

Likewise, in an electric circuit an electric pressure or voltage is necessary to cause a current to flow. The greater the resistance of a circuit, the greater the voltage must be to cause a given current to flow through that circuit.

In the circuit shown in Fig. 2-5, three resistors R_1, R_2, and R_3 with resistances of 6, 18, and 12 Ω, respectively, are connected in series to a generator. Voltmeter V is connected to indicate the voltage or potential difference of 72 V maintained at the generator terminals and applied to the entire circuit. Voltmeters V_1, V_2, and V_3 are connected across R_1, R_2, and R_3 as shown. Ammeter A is connected in series with the three resistors to indicate the total circuit current of 2 A.

Voltmeter V_1 in Fig. 2-5 indicates a voltage of 12 V. This indicates that 12 V is required to cause the current of 2 A to flow through R_1. Stated another way, there is a potential difference or a *voltage drop* of 12 V across R_1. Voltmeter V_2 indicates 36 V, or three times the voltage required across R_1. This is to be expected since R_2 has three times the resistance of R_1 and the same current of 2 A is being caused to flow through both resistors. The voltage across R_3 is only 24 V since R_3 is only twice as large as R_1. In general, it may be stated that the voltage required to cause a current to flow in a dc circuit is directly proportional to the resistance of the circuit; that is, the higher the resistance, the higher the voltage must be.

Note in Fig. 2-5 that the sum of the voltage drops across the three resistors is equal to the voltage applied to the circuit

$$12 + 36 + 24 = 72$$

The entire applied voltage is used in causing the current to flow through the three resistors.

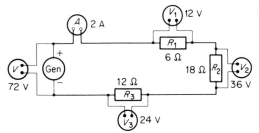

FIGURE 2-5   The sum of the voltages as indicated by voltmeters V_1, V_2, and V_3 is equal to the voltage indicated by voltmeter V.

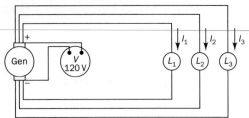

FIGURE 2-6   The three lamps are connected as three separate circuits to the same generator.

In a series circuit, the sum of the voltages across the several parts is equal to the total voltage applied to the circuit.
Stated mathematically,

$$V_t = V_1 + V_2 + V_3 + \cdots \qquad (2\text{-}2)$$

where V_t is voltage applied to the circuit and V_1, V_2, and V_3 are the voltages across the series components.

2-3 PARALLEL CIRCUITS: VOLTAGE RELATIONS

Figure 2-6 shows three lamps L_1, L_2, and L_3 connected as three independent circuits to a 120-V generator. Obviously, a voltage of 120 V is applied to each of the three lamps since each lamp is connected directly to the same generator. Now, if instead of making three separate circuit connections as in Fig. 2-6 the connecting wires are replaced by one outgoing and one returning wire, the same condition exists. That is, there is still a voltage of 120 V applied to each of the three lamps. The connecting wires of Fig. 2-7 may be considered to be merely extensions of the generator terminals so that the same voltage is supplied to each lamp. The connection shown in Fig. 2-7 is called a *parallel connection,* and the rule for voltages in a parallel circuit may be stated as follows:

The voltage across each branch of a parallel combination is the same as the voltage applied to the entire combination.
Stated mathematically,

$$V_t = V_1 = V_2 = V_3 = \ldots \qquad (2\text{-}3)$$

where V_t is the voltage across the combination and V_1, V_2, and V_3 are the voltages across the several branches.

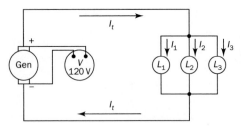

FIGURE 2-7   The three lamp circuits of Fig. 2-6 are combined to form a parallel circuit.

2-4 PARALLEL CIRCUITS: CURRENT RELATIONS

Since in Fig. 2-6 the generator supplies current to each of the three independent circuits, the total current supplied must be the sum of the currents in the three circuits. Then since Fig. 2-6 is equivalent to the parallel circuit shown in Fig. 2-7, the rule for currents in a parallel circuit may be stated as follows:

The total current supplied to a parallel circuit is the sum of the branch currents.

Mathematically,

$$I_t = I_1 + I_2 + I_3 + \cdots \tag{2-4}$$

where I_t is the total current and I_1, I_2, and I_3 are the currents in the several branches.

2-5 OHM'S LAW

#12

Ohm's law states that the amount of steady current flowing in a circuit is equal to the applied voltage divided by the resistance of the circuit. This law, experimentally determined by Georg Simon Ohm in 1826, is of the utmost importance as it forms the basis for nearly all circuit calculations. The law may be written briefly

$$\text{Current} = \frac{\text{voltage}}{\text{resistance}}$$

or

$$\text{Amperes} = \frac{\text{volts}}{\text{ohms}}$$

Written mathematically it is

#12

$$I = \frac{V}{R}$$

where I = current, amperes
V = voltage, volts
R = resistance, ohms

EXAMPLE 2-1

How much current will flow through a 12-Ω electric heater when a voltage of 120 V is applied?

$$I = \frac{V}{R}$$

$$I = \frac{120}{12} = 10 \text{ A}$$

Equation (2-5) may be transposed and written

$$V = IR \qquad (2\text{-}6)$$

That is, the voltage across a circuit is equal to the product of the current in amperes and the resistance in ohms.

EXAMPLE 2-2   What voltage is required to cause a current of 2 A to flow through a resistance of 55 Ω?

$$V = IR$$

$$V = 2 \times 55 = 110 \text{ V}$$

Since $V = IR$, the difference in voltage between two points is often referred to as the "IR drop" between the two points.

The third form of Ohm's law, by again transposing Eq. (2-5), is

$$R = \frac{V}{I} \qquad (2\text{-}7)$$

That is, the resistance of a circuit is equal to the voltage across the circuit divided by the current through the circuit.

EXAMPLE 2-3   What is the resistance of a lamp that draws 1.5 A from a 120-V line?

$$R = \frac{V}{I}$$

$$R = \frac{120}{1.5} = 80 \text{ }\Omega$$

The three forms of Ohm's law are summarized in Table 2-1. It is important that you become familiar with all three forms of Ohm's law, since you will be using them as long as you work with electricity.

2-6 APPLYING OHM'S LAW   Mistakes in electrical calculations are frequently made because Ohm's law is not used properly. The law may be applied to an *entire circuit* or to *any part of a circuit*. When used for the entire circuit, values of current, voltage, and resistance must be used for the entire circuit. When used for a certain part of a circuit, values of current, voltage, and resistance must be used from only that part.

In the circuit represented in Fig. 2-8, find the current flowing in and the voltage across each of the two series resistances.

EXAMPLE 2-4

$$I\text{(entire circuit)} = \frac{V\text{(entire circuit)}}{R\text{(entire circuit)}}$$

or $$I_t = \frac{V_t}{R_t} = \frac{240}{(40 + 20)} = \frac{240}{60} = 4\,\text{A}$$

$$V\text{(across } R_1) = I\text{(through } R_1) \times R_1$$

or $$V_1 = I_1 \times R_1 = 4 \times 40 = 160\,\text{V}$$

$$V\text{(across } R_2) = I\text{(through } R_2) \times R_2$$

or $$V_2 = I_2 \times R_2 = 4 \times 20 = 80\,\text{V}$$

TABLE 2-1   OHM'S LAW

$I = \dfrac{V}{R}$	Current $= \dfrac{\text{voltage}}{\text{resistance}}$	Amperes $= \dfrac{\text{volts}}{\text{ohms}}$
$V = IR$	Voltage $=$ current $\times$ resistance	Volts $=$ amperes $\times$ ohms
$R = \dfrac{V}{I}$	Resistance $= \dfrac{\text{voltage}}{\text{current}}$	Ohms $= \dfrac{\text{volts}}{\text{amperes}}$

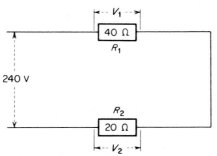

FIGURE 2-8   Circuit for Example 2-4. The symbol Ω is used to represent the word "ohm."

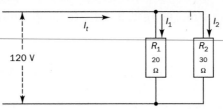

FIGURE 2-9   Circuit for Example 2-5.

EXAMPLE 2-5   Find the current in each branch and the line current in the parallel circuit shown in Fig. 2-9. What is the combined resistance of two branches?

$$I(\text{branch 1}) = \frac{V(\text{branch 1})}{R(\text{branch 1})}$$

or

$$I_1 = \frac{V_1}{R_1} = \frac{120}{20} = 6 \text{ A}$$

$$I(\text{branch 2}) = \frac{V(\text{branch 2})}{R(\text{branch 2})}$$

or

$$I_2 = \frac{V_2}{R_2} = \frac{120}{30} = 4 \text{ A}$$

$$\text{Line current} = I_1 + I_2 = 6 + 4 = 10 \text{ A}$$

$$R(\text{entire circuit}) = \frac{V(\text{entire circuit})}{I(\text{entire circuit})}$$

or

$$R_t = \frac{V_t}{I_t} = \frac{120}{10} = 12 \ \Omega$$

Notice that the combined resistance of R_1 and R_2 in parallel in Example 2-5 is less than the value of either R_1 or R_2. Resistances in parallel are discussed in Sec. 2-8.

2-7 RESISTANCE OF A SERIES CIRCUIT

It was shown in Sec. 2-1 that the same current flows through every part of a series circuit. Resistance is the opposition offered by a circuit to the flow of current. All parts of a series circuit have resistance, and each part of the circuit offers a part of the total resistance to the flow of current. Thus, the following rule may be stated for combining resistances in series:

The total or combined resistance of a series circuit is the sum of the resistances of the several parts.

Stated mathematically,

$$R_t = R_1 + R_2 + R_3 + \cdots \tag{2-8}$$

where R_t is the total resistance and R_1, R_2, and R_3 are the resistances of the several parts of a series circuit.

2-8 RESISTANCE OF A PARALLEL CIRCUIT

When one 120-Ω lamp is connected to a 120-V generator, 1 A of current flows. When another lamp of the same size is added in parallel, 1 A flows through the second lamp and the generator must supply a total of 2 A. Applying Ohm's law to the entire circuit with two lamps in parallel,

$$R_t = \frac{V}{I_t} = \frac{120}{2} = 60 \ \Omega$$

where R_t is the resistance of the entire circuit. Thus the combined resistance of resistances in parallel is *less* than the resistance of either branch. This is true since, as more paths are added through which current may flow, the easier it is for current to flow, or in other words, the lower is the combined resistance of the circuit.

The equation for combining resistances in parallel may be developed by means of Ohm's law and the current and voltage rules for parallel circuits. Applying Ohm's law to the entire circuit of Fig. 2-10,

$$I_t = \frac{V}{R_t}$$

where R_t is the resistance of the entire parallel combination. When Ohm's law is applied to the individual branches, the branch currents are

$$I_1 = \frac{V}{R_1} \quad I_2 = \frac{V}{R_2} \quad \text{and} \quad I_3 = \frac{V}{R_3}$$

From the current rule for parallel circuits given by Eq. (2-4)

$$I_t = I_1 + I_2 + I_3$$

Substituting the Ohm's law values of the currents in Eq. (2-4) results in

$$\frac{V}{R_t} = \frac{V}{R_1} + \frac{V}{R_2} + \frac{V}{R_3}$$

and, canceling out V from each term,

$$\frac{1}{R_t} = \frac{1}{R_1} + \frac{1}{R_2} + \frac{1}{R_3} \qquad (2-9)$$

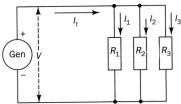

FIGURE 2-10   The combined resistance of a parallel circuit is less than the resistance of any one branch.

Equation (2-9) is the rule for combining resistances in parallel. It may be extended for any number of parallel branches.

EXAMPLE 2-6   What is the combined resistance of a circuit that has resistances of 20, 10, 40, and 80 Ω all connected in parallel?

$$\frac{1}{R_t} = \frac{1}{R_1} + \frac{1}{R_2} + \frac{1}{R_3} + \frac{1}{R_4}$$

$$\frac{1}{R_t} = \frac{1}{20} + \frac{1}{10} + \frac{1}{40} + \frac{1}{80}$$

$$\frac{1}{R_t} = \frac{4 + 8 + 2 + 1}{80} = \frac{15}{80}$$

$$15R_t = 80 \quad \text{or} \quad R_t = \frac{80}{15} = 5.33 \ \Omega$$

Note in Example 2-6 that the combined resistance is less than the resistance of any one of the branch resistances. This is true in any parallel circuit.

The resistance of two resistances in parallel may be found by simplifying Eq. (2-9) to

$$R_t = \frac{R_1 R_2}{R_1 + R_2} \tag{2-10}$$

That is, the combined resistance of any two resistances in parallel is equal to their product divided by their sum. This is true only for *two* resistances.

EXAMPLE 2-7   What is the combined resistance of a circuit composed of a 20-Ω resistor and a 30-Ω resistor connected in parallel?

$$R_t = \frac{R_1 R_2}{R_1 + R_2} = \frac{20 \times 30}{20 + 30} = \frac{600}{50} = 12 \ \Omega$$

The application of Eq. (2-9) or (2-10) to a circuit composed of two equal resistances in parallel shows that the combined resistance is one-half that of either resistance. Likewise, the combined resistance of three equal resistances in parallel is one-third of any one branch; the combined resistance of four equal resistances in parallel is one-fourth that of any one branch, and so on. In general, the combined resistance of any number of equal resistances in parallel is equal to the resistance of one branch divided by the number of branches. Stated mathematically, the combined resistance R_t of n paralleled equal resistances each of R Ω is

$$R_t = \frac{R}{n} \tag{2-11}$$

EXAMPLE 2-8

What is the combined resistance of three 30-Ω resistors connected in parallel?

$$R_t = \frac{R}{n} = \frac{30}{3} = 10\,\Omega$$

The relationships of current, voltage, and resistance in both series and parallel circuits are summarized in Table 2-2.

2-9 SERIES-PARALLEL CIRCUITS

Thus far, only simple series and simple parallel circuits have been considered. Practical electric circuits very often consist of combinations of series and parallel resistances. Such circuits may be solved by the proper application of Ohm's law and the rules for series and parallel circuits to the various parts of the complex circuit. There is no definite procedure to be followed in solving complex circuits, the solution depending on the known facts concerning the circuit and the quantities one desires to find. One simple rule may usually be followed, however—reduce the parallel branches to an equivalent series branch and then solve the circuit as a simple series circuit.

The following examples will illustrate the solution of typical series-parallel circuits.

EXAMPLE 2-9

A resistor of 30 Ω is connected in parallel with a resistor of 60 Ω. In series with the parallel combination is a 20-Ω resistor as shown in Fig. 2-11a (see p. 25). What is the resistance of the entire circuit?

The resistance between points B and C is

$$R_{BC} = \frac{R_2 R_3}{R_2 + R_3} = \frac{30 \times 60}{30 + 60} = \frac{1800}{90} = 20\ \Omega$$

Since R_2 and R_3 in parallel are equivalent to one 20-Ω resistance, the circuit may be reduced to the equivalent circuit shown in Fig. 2-11b. The total circuit resistance is then

$$R_t = R_1 + R_{BC} = 20 + 20 = 40\ \Omega$$

TABLE 2-2   SUMMARY OF THE LAWS OF SERIES AND PARALLEL CIRCUITS

	Series Circuits	*Parallel Circuits*
Diagram	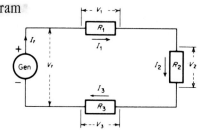	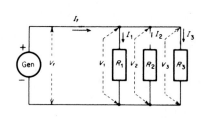
Current	The current is the same in all parts of the ciruit $$I_t = I_1 = I_2 = I_3 = \cdots$$	The total current supplied to the circuit equals the sum of the currents through the several branches $$I_t = I_1 + I_2 + I_3 + \cdots$$
Voltage	The total voltage equals the sum of the voltages across the different parts of the circuit $$V_t = V_1 + V_2 + V_3 + \cdots$$	The voltage across a parallel combination is the same as the voltage across each branch $$V_t = V_1 = V_2 = V_3 = \cdots$$
Resistance	The total resistance equals the sum of the resistances of the separate parts $$R_t = R_1 + R_2 + R_3 + \cdots$$	The reciprocal of the equivalent or combined resistance equals the sum of the reciprocals of the resistances of the individual branches $$\frac{1}{R_t} = \frac{1}{R_1} + \frac{1}{R_2} + \frac{1}{R_3} + \cdots$$

Anr
#20

EXAMPLE 2-10   Two resistors of 40 and 60 Ω in parallel are connected in series with two 0.5-Ω resistors as shown in Fig. 2-12*a* (see p. 26). Find the voltage across the series resistors and across the parallel resistors when 125 V is applied to the entire circuit.

EXAMPLE 2-10
(CONTINUED)

The resistance between B and C is

$$R_{BC} = \frac{R_2 R_3}{R_2 + R_3} = \frac{40 \times 60}{40 + 60} = \frac{2400}{100} = 24 \ \Omega$$

The circuit may now be represented as shown in Fig. 2-12b from which the total circuit resistance is

$$R_t = R_1 + R_{BC} + R_4$$
$$= 0.5 + 24 + 0.5 = 25 \ \Omega$$

Then $\quad I_t = \dfrac{V_t}{R_t} = \dfrac{125}{25} = 5A$

The voltage across R_1 is

$$V_1 = I_t R_1 = 5 \times 0.5 = 2.5 \ V$$

and the voltage across R_4 is

$$V_4 = I_t R_4 = 5 \times 0.5 = 2.5 \ V$$

The voltage across the parallel combination is then

$$V_{BC} = 125 - (2.5 + 2.5) = 120 \ V$$

or $\quad V_{BC} = I_t R_{BC} = 5 \times 24 = 120 \ V \ (check)$

2-10 LINE DROP

Thus far in dealing with electric circuits, the resistance of the connecting-line wires between the source and the load has been neglected. When connecting lines are short, this is permissible since the resistance is so small it can usually be ignored. However, a long line may have considerable resistance so that a certain amount of the applied voltage is used in overcoming this line resistance. This voltage is called *line drop,*

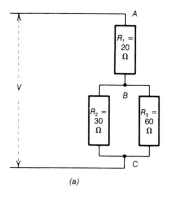

(a)

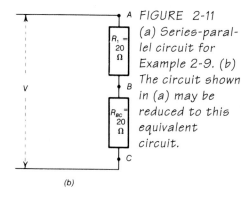

(b)

FIGURE 2-11 (a) Series-parallel circuit for Example 2-9. (b) The circuit shown in (a) may be reduced to this equivalent circuit.

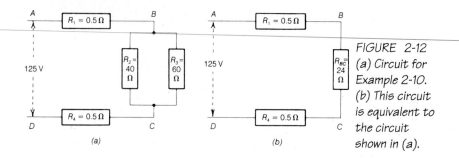

FIGURE 2-12 (a) Circuit for Example 2-10. (b) This circuit is equivalent to the circuit shown in (a).

and its value may be calculated by Ohm's law. Since $V = IR$, the voltage drop in the line wires is called the *IR drop* in the lines. Line drop or *IR* drop is often expressed as a percentage of the voltage applied to the load.

While line resistance is distributed throughout the line wires, it is convenient when dealing with line resistance to consider the resistance of the line wires to be lumped into two resistances, one representing the resistance of each line, with these two resistances connected in series with the load resistance by connecting wires having negligible resistance. The following example will illustrate the calculation of line drop.

EXAMPLE 2-11

Figure 2-13 represents a simple lighting circuit with a source voltage of 120 V. Each lamp requires a current of 0.6 A, and each connecting wire has a resistance of 1.5 Ω. It is desired to find the line drop in volts, the voltage at the lamps, and the percentage voltage drop. Note that this problem is very similar to Example 2-10.

The line current is $0.6 \times 2 = 1.2$ A. Applying Ohm's law to each line wire

$$V = IR = 1.2 \times 1.5 = 1.8 \text{ V}$$

The voltage at the lamps is equal to the source voltage minus the drop line in both lines, or

$$V = 120 - (1.8 + 1.8)$$
$$= 120 - 3.6 = 116.4 \text{ V}$$

The percentage voltage drop in the lines is

$$\frac{3.6 \times 100}{116.4} = 3.09 \text{ percent}$$

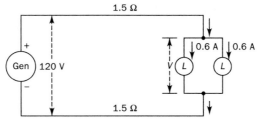

What is the voltage across each group of lamps in the lighting circuit of Fig. 2-14? Lines *AB* and *EF* have a resistance of 0.6 Ω each, and lines *BC* and *ED* have a resistance of 0.3 Ω each. Assume that each lamp takes a current of 1.5 A.

EXAMPLE 2-12

Current in lines *BC* and *DE* is $1.5 \times 2 = 3$ A. Current in lines *AB* and *EF* is 6 A. Line drop in *AB* and *EF* is

$$V = IR = 6 \times 0.6 = 3.6 \text{ V}$$

Line drop in *BC* and *DE* is

$$V = IR = 3 \times 0.3 = 0.9 \text{ V}$$

Voltage across lamps at *BE* = voltage at source − line drop in *AB* and *EF*

$$V_{BE} = 120 - (3.6 + 3.6)$$
$$= 120 - 7.2 = 112.8 \text{ V}$$

Voltage across lamps at *CD* = voltage at *BE* − line drop in *BC* and *DE*

$$V_{CD} = 112.8 - (0.9 + 0.9)$$
$$= 112.8 - 1.8 = 111 \text{ V}$$

**2-11
KIRCHHOFF'S
LAWS**

Complex electric circuits may be solved with the aid of two simple rules known as Kirchoff's laws.

Kirchhoff's Current Law. The sum of the currents flowing away from any point in an electric circuit must equal the sum of the currents flowing toward the point.

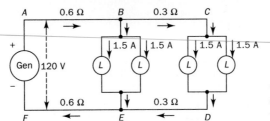

FIGURE 2-14 Lighting circuit with lamps at two locations. See Example 2-12.

Kirchhoff's Voltage Law. Around any closed path in an electric circuit, the sum of the potential drops is equal to the sum of the impressed emfs. This is equivalent to saying that the voltage drops must equal the voltage rises in any closed path, where a voltage rise is any voltage increase such as a generated emf.

The current law should be evident since this is the principle involved in the current rule for a parallel circuit; that is, the sum of the branch currents must equal the line current. Using Fig. 2-14 as an example, the current flowing into point B is 6 A. The current in line BC is 3 A and in BE it is 3 A, or a total of 6 A flowing away from point B.

The voltage law may also be illustrated by Fig. 2-14. Considering the closed path *ABEFA,* the impressed emfs or voltage rises must equal the voltage drops, or

$$\text{Rise through generator} = \text{drop in } AB + \text{drop in } BE + \text{drop in } EF$$
$$120 = 3.6 + 112.8 + 3.6$$
$$120 = 120 \text{ (check)}$$

Likewise in the closed path *ABCDEFA*

$$\text{Rises} = \text{drops}$$
$$120 = 3.6 + 0.9 + 111 + 0.9 + 3.6$$
$$120 = 120 \text{ (check)}$$

Kirchhoff's current and voltage laws restate, in concise and usable form, principles already discussed in connection with the rules for series and parallel circuits. By systematically applying these laws, the current and voltage values of the various parts of complex circuits may be found. One simple application will be illustrated in the solution of three-wire circuit problems in Sec. 2-12.

2-12 THREE-WIRE DISTRIBUTION CIRCUITS To obtain the advantage of a higher voltage-distribution circuit and yet provide 120 V for the operation of standard incandescent lights, Edison developed the 120/240-V three-wire system of power distribution. For a fixed amount of line loss, power can be transmitted by means of the three-wire system with only three-eighths the line conductor material required for a

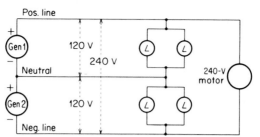

FIGURE 2-15 Three-wire distribution circuit.

two-wire 120-V system. The system has the further advantage of making available either 120 or 240 V for the operation of electric equipment.

The three-wire dc distribution system has been superseded generally by the three-wire ac (alternating current) system. However, since the principles of operation are essentially the same for the ac and dc systems, a description of the dc system is included here. Transformer connections used in supplying the three-wire ac system are discussed in Sec. 12-16.

Figure 2-15 shows one method of connecting the supply to a three-wire system. Two 120-V dc generators are connected in series, and three lines connect the generators to the system. The system positive line is connected to the positive terminal of one of the generators and the system negative line is connected to the negative terminal of the other generator. The third line is connected to the common point between the two generators and is called the neutral line or, more commonly, the *neutral*. With this arrangement, 120-V lamps and appliances may be connected between the neutral and either the positive or negative line, while 240-V equipment, such as motors and electric heaters, may be connected between the positive and negative lines, as shown in Fig. 2-15.

When the 120-V loads are balanced on each side of the neutral as in Fig. 2-16, the neutral carries no current. If a heavier load is connected between the positive line and neutral as in Fig. 2-17, the neutral carries toward the generators the difference in current between the upper and lower lines. If the heavier load is connected between the negative line and neutral as in Fig. 2-18 (see p. 31), the neutral carries the unbalanced current from the generators toward the load. That is, the current flowing in the neutral

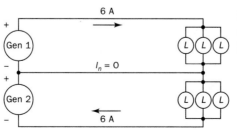

FIGURE 2-16 Three-wire circuit with balanced loads. The neutral line carries no current.

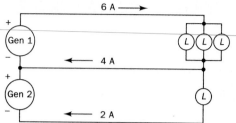

FIGURE 2-17 Three-wire circuit with heavier load between positive line and neutral. Neutral line carries current toward generators.

line is always the difference between the currents in the positive and negative lines. It is desirable to keep the loads as nearly balanced as possible to keep the flow of neutral current to a minimum.

When load is connected to only one side of a three-wire system, the neutral carries the full-load current. For this reason, the neutral conductor is usually of the same size as the positive and negative conductors.

An accidental opening of the neutral when an unbalanced load is being supplied results in badly unbalanced voltages across the loads. For this reason, the neutral conductor is connected solidly from the generator to the load and no fuses or other overcurrent devices are installed in it. For further protection against an open neutral conductor and for protection against lightning surges, the neutral is connected to earth at one or more points.

It is often necessary to find the voltage across the loads connected to a three-wire system when the line resistance and the load currents are known. The following example will illustrate this calculation for the simple case of two loads connected to the system at the same point.

EXAMPLE 2-13   Two loads located 1000 ft from a source are to be supplied from a three-wire system using three copper wires each having a resistance of 0.1 Ω. One load requires 30 A and the other 20 A. Find the voltage across each load. Generator voltages are 120 V.

SOLUTION: When the loads are connected as shown in Fig. 2-19, the neutral carries 10 A toward the source. The voltage drop in each line is

Line A	$IR = 30 \times 0.1 = 3\,V$
Neutral	$IR = 10 \times 0.1 = 1\,V$
Line B	$IR = 20 \times 0.1 = 2\,V$

Writing an equation for the upper half of the circuit in accordance with Kirchhoff's voltage law,

EXAMPLE 2-13
(CONTINUED)

Voltage rises = voltage drops

$$120 = 3 + V_1 + 1$$

$$V_1 = 120 - 3 - 1 = 116 \text{ V}$$

Kirchhoff's voltage equation for the entire outside loop is

$$120 + 120 = 3 + 116 + V_2 + 2$$

$$240 = 121 + V_2$$

$$V_2 = 240 - 121 = 119 \text{ V}$$

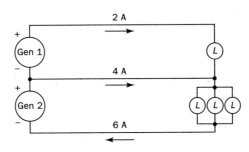

FIGURE 2-18   *Three-wire circuit with heavier load between negative and neutral lines. Neutral line carries current toward load.*

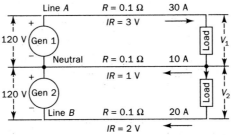

FIGURE 2-19   *Unbalanced three-wire system for Example 2-13.*

REVIEW QUESTIONS

1 What are the two basic ways in which electric apparatus may be connected?

2 What is the relation between the values of current flowing in different parts of a series circuit?

3 If the voltages across the several parts of a series circuit are known, how may the total circuit voltage be determined?

4 A voltage of 20 V is required to cause 2 A to flow through a resistance of 10 Ω. What voltage is necessary to cause the same current to flow through 40 Ω?

5 Give the rule for determining the voltage across parallel branches of a circuit when the applied voltage is known.

6 State Ohm's law in its three forms. Which form is most convenient for finding (*a*) the resistance, (*b*) the current, or (*c*) the voltage of a circuit?

7 What precautions should be observed in applying Ohm's law to a part of a circuit?

8 Give the rule for finding the combined resistance of several resistances (*a*) in a series and (*b*) in parallel.

9 Give the rule for finding the combined resistance of two resistances in parallel.

10 What is the general rule for finding the combined resistance of several equal resistances in parallel?

11 In reducing series-parallel circuits to simpler form, what is the general procedure?

12 What is meant by line drop? How is it calculated?

13 State Kirchhoff's current and voltage laws.

14 What are some advantages of the three-wire distribution system?

15 What determines the neutral current in a three-wire distribution system?

PROBLEMS

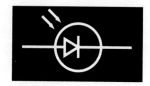

2-1 Two lamps are connected in series to a 120-V line. The voltage across one lamp is 70 V. What is the voltage across the second lamp?

2-2 Eight lamps of the same size are connected in series to a 120-V circuit. What is the voltage across each lamp?

2-3 Three lamps of equal size are connected in parallel to a lighting circuit. Each lamp draws a current of 1.5 A. What is the total line current?

2-4 Twenty lamps of equal size are connected in parallel to a 120-V lighting circuit. If the total current drawn by the 20 lamps is 15 A, what is the current through each lamp? If the lamps are reconnected in series in another circuit, so that the current through one of the lamps is 1 A, what will be the current through the other 19 lamps?

2-5 A 220-ohm lamp is connected to a 110-V circuit. What current does it take?

2-6 The field resistance of a shunt motor is 93 Ω. What will be the field current if 240 V is applied to the field terminals?

2-7 An accidental short circuit is made on a 240-V circuit by placing a 0.15 Ω resistance across the line. What will be the momentary short-circuit current?

2-8 A telephone receiver has a resistance of 1200 Ω. If a current of 0.02 A is to flow through the receiver, what voltage must be used?

2-9 What is the voltage drop across a rheostat if its resistance is 5 Ω and the current through it is 26 A?

2-10 Find the voltage drop across an electric furnace that has a resistance of 5 Ω and draws 43 A.

2-11 The voltage across the terminals of a motor field is 220 V, and the field current is 1 A. What is the field-circuit resistance?

2-12 A relay coil draws a current of 0.03 A when connected to a 12-V battery. What is its resistance?

2-13 Find the combined resistance of a 5-Ω rheostat in series with a 56-Ω shunt field.

2-14 What is the total resistance of a string of eight Christmas-tree lights connected in series, if the resistance of each lamp is 30 Ω?

2-15 How much voltage is required to cause a current of 0.4 A to flow through a 70-Ω, a 90-Ω, and a 120-Ω lamp all connected in series?

2-16 Three lamps are connected in series across a 120-V supply and take a current of 1.25 A. If the resistance of two of the lamps is 25 Ω each, what is the resistance of the third?

2-17 How much resistance must be added in series with a 60-Ω generator field if the field current desired is 2 A and the voltage applied to the field circuit is 230 V?

2-18 A certain series street-lighting system of 60 lamps operates at a constant current of 6.6 A. How much voltage must be applied to the circuit if each lamp has a resistance of 5 Ω and the total line resistance is 10 Ω?

2-19 Four lamps of equal resistance are connected in parallel across a 120-V circuit. If the total current supplied the lamps is 3 A, what is the resistance of each lamp?

2-20 Three resistances A, B, and C are connected in parallel across a 120-V line. If $A = 10$ Ω, $B = 30$ Ω, and $C = 40$ Ω, find (*a*) the voltage across each resistance, (*b*) the current through each resistance, (*c*) the total line current, and (*d*) the combined resistance of the circuit.

2-21 The combined resistance of two resistors connected in parallel is 8 Ω. The resistance of one resistor is 14 Ω. What would be their combined resistance if they were reconnected in series?

2-22 A trolley wire is paralleled for 3 miles by a heavy copper cable. The trolley-wire resistance is 0.3 Ω per mile, and the cable resistance is 0.10 Ω per mile. When the sum of the two currents through the lines is 150 A, find the current in each conductor.

2-23 A parallel circuit has three branches of 12, 4, and 16 Ω, respectively. If a current of 4 A flows in the branch containing 12 Ω, what current will flow in each of the others?

2-24 Three resistances of 3, 4, and 6 Ω are connected in parallel. In series with the group is an unknown resistance R. When the circuit is connected to a 12-V battery, a current of 4 A flows. What is the resistance of R?

2-25 A resistance of 12 Ω is connected in parallel with a series circuit of a 15-Ω and a 5-Ω resistance. A 40-Ω resistance is connected in series with the combination. What is the resistance of the entire circuit?

2-26 Three resistances of 4, 8, and 10 Ω are connected so that each resistance forms the side of a triangle. Point A is the connection between the 4- and 8-Ω sides; point B, between the 8- and 10-Ω sides; and point C, between the 4- and 10-Ω sides. Find the resistance between points A and B, B and C, and C and A.

2-27 Four resistors of 6 Ω each form the sides of a diamond. A fifth resistor of 10 Ω forms the long diagonal of the diamond. What is the resistance of the combination between the ends of the long diagonal? What is the resistance between one end of the long diagonal and one end of the short diagonal?

2-28 A dc motor is supplied by a dc generator. Each line connecting the two machines has a resistance of 0.15 Ω. What is the terminal voltage across the motor when the motor current is 50 A and the generator terminal voltage is 230 V?

2-29 A dc motor is located 2000 ft from a dc generator. The motor requires 30 A to operate a given load. The generator voltage is 220 V. Each line connecting the generator and motor is No. 0 cable, which has a resistance of 0.1 Ω per 1000 ft. Find the voltage at the motor.

2-30 A 240-V generator supplies 10 A to each of two motors connected in parallel. The resistance of each line wire between the generator and the first motor is 0.3 Ω, and the resistance of each line wire between the first and second motor is 0.4 Ω. What is the voltage across each motor?

2-31 Each wire of a two-wire distribution circuit has a resistance of 0.2 Ω per 1000 ft. Connected to this circuit at a point 500 ft from the generator is a motor that requires a current of 15 A and has a terminal voltage of 230 V. A second motor that requires 10 A is located on the same circuit but 750 ft from the generator. (*a*) What is the voltage at the second motor? (*b*) What is the generator voltage?

2-32 The voltage drop across an electric heater is 225 V. If the voltage at the source is 230 V and the heater draws 15 A, what is the resistance of each connecting line wire?

2-33 If each lamp in Fig. 2-20 requires a current of 1 A, find (*a*) the current and its direction in each of the three lines, (*b*) the *IR* drop in each line, and (*c*) the voltages V_1 and V_2.

2-34 Repeat the preceding problem if one lamp in the upper bank of lamps burns out.

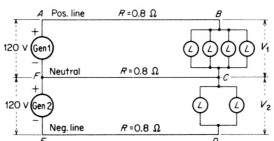

FIGURE 2-20   Three-wire circuit for Problems 2-33 and 2-34.

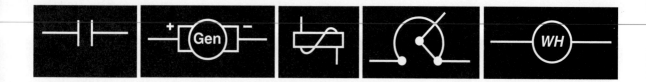

3

Electric Conductors

In this chapter, conductor materials, wire measurements, resistivity, and temperature coefficient of resistance, all items relating to electric conductors, are discussed.

**3-1
CONDUCTOR
MATERIALS**

Conductor materials are those materials that have many free electrons. Such materials are said to present a low resistance to the flow of electric current or to have a high *conductance*. A good electric conductor therefore has a high conductance and a low resistance.

In addition to having a high conductance, a good conductor material should have a high tensile strength, a relatively light weight, and a low cost. Copper and aluminum possess these qualities and are used extensively as conductors in the generation, transmission, and utilization of electric energy. Combinations of steel with copper or aluminum in the form of copper-clad steel or steel-reinforced aluminum are often used as overhead conductors where an especially high tensile strength is required.

Silver and gold are excellent conductor materials but are too expensive for extensive use. Both are used, however, as plating materials for electric-contact surfaces where corrosion-free, low-resistance contact must be ensured.

**3-2 WIRE
MEASUREMENTS,
U.S. CUSTOMARY
SYSTEM**

Because the diameter of wires used as conductors of electricity may be very small, it is convenient when using the U.S. Customary System (USCS) of measurement to express wire diameters in thousandths of an inch. To avoid the use of decimals in expressing wire diameters, a unit of measurement known as the *mil* is in general use in the United States. *One mil is equal*

38a) { *to one one-thousandth* (0.001) *of one inch.* For example, instead of saying that a certain wire has a diameter of 0.051 or 51/1000 in., it is said to have a diameter of 51 mils.

EXAMPLE 3-1

1 in. = 1 × 1000 = 1000 mils

0.5 in. = 0.5 × 1000 = 500 mils

0.253 in. = 0.253 × 1000 = 253 mils

38c) { The cross-sectional area of round conductors is expressed in units called *circular mils* (cmil). A conductor with a diameter of 1 mil has a cross-sectional area of 1 cmil.

Since the area of a circle is proportional to the square of its diameter (area = $\pi D^2/4$), *the area of a circular conductor in circular mils is equal to the square of its diameter in mils.*

EXAMPLE 3-2

What is the circular-mil area of a wire (*a*) 0.1 in., (*b*) 0.2 in., and (*c*) 0.325 in. in diameter?

(*a*) Diameter in mils = 0.1 × 1000 = 100 mils
Area = diameter in mils, squared
= 100 × 100 = 10,000 cmil

(*b*) Diameter = 0.2 × 1000 = 200 mils
Area = 200 × 200 = 40,000 cmil

(*c*) Diameter = 0.325 × 1000 = 325 mils
Area = 325 × 325 = 105,625 cmil

3-3 THE AMERICAN WIRE GAUGE   In the United States it is common practice to indicate wire sizes by a gauge number, and the commonly used gauge is the American Wire Gauge. American Wire Gauge (AWG) sizes range from No. 0000, the largest size, to No. 40, the smallest size. This system is based on a constant ratio between any wire diameter and the diameter of the next smaller wire. Appendix B gives the dimensions and properties of the more commonly used AWG sizes of solid copper wire.

Following are several approximate rules concerning the AWG tables. If these simple facts are remembered, approximate values concerning many wire sizes may be calculated without referring to the tables.

1. Easily remembered approximate data concerning No. 10 AWG solid copper wire are

$$\text{Diameter} = 0.1 \text{ in. or } 100 \text{ mils}$$
$$\text{Area} = 10,000 \text{ cmil}$$
$$\text{Resistance} = 1\,\Omega \text{ per } 1000 \text{ ft}$$
$$\text{Weight} = 32 \text{ lb per } 1000 \text{ ft}$$

2. An increase of three gauge numbers doubles the area and weight and halves the resistance.
3. An increase of 10 gauge numbers increases the area and weight by a factor of 10 and decreases the resistance by a factor of 10.

EXAMPLE 3-3   Find the approximate resistance of No. 0 AWG copper wire without referring to the tables.

Since No. 0 is 10 gauge numbers larger than No. 10, its resistance is $1.0/10 = 0.1\ \Omega$ per 1000 ft.

EXAMPLE 3-4   Find the approximate cross-sectional area of No. 4 AWG copper wire without referring to the tables.

Number 4 is six gauge sizes larger than No. 10; therefore its area is $10,000 \times 2 \times 2 = 40,000$ cmil.

3-4 WIRE MEASUREMENTS, SI UNITS   In metric or SI units, wire diameters are expressed in millimeters (mm) and cross-sectional areas in square millimeters (mm^2). Wire lengths are expressed in meters or kilometers (usually kilometers) and weights in kilograms per kilometer (kg/km). A table of dimensions, weights, and resistance of copper wire in SI units is included in Appendix D.

As an approximation, a 1-mm^2 copper conductor has a weight of about 9.2 kg/km and a resistance of about 17 Ω/km. Weights increase and resistances decrease in proportion to increases in cross-sectional area. Thus a 10-mm^2 copper conductor has a weight of approximately 92 kg/km and a resistance of approximately 1.7 Ω/km. Refer to Appendix D for more exact data.

It is often necessary to convert from one system of wire measurement to another. To convert from circular mils to square millimeters, it is necessary to multiply the number of circular mils by 506.7×10^{-6} (0.0005067). Similarly, to convert from square millimeters to circular mils, it is necessary to multiply the number of square millimeters by 1973.

EXAMPLE 3-5

A round solid No. 0000 AWG conductor is 0.46 in. in diameter. What is its cross-sectional area in circular mils and in square millimeters?

$$\text{Diameter} = 0.46 \text{ in. or } 460 \text{ mils}$$

$$\text{Cross-sectional area} = (460)^2 = 211,600 \text{ cmil}$$

$$\text{cmil} \times 506.7 \times 10^{-6} = \text{mm}^2$$

$$211,600 \times 506.7 \times 10^{-6} = 107.2 \text{ mm}^2$$

$$\text{mm}^2 \times 1973 = \text{cmil}$$

$$107.2 \times 1973 = 211,600 \text{ cmil (check)}$$

The following are helpful approximate factors for converting AWG to metric or metric to AWG sizes and for visualizing relative wire sizes in the two systems:

$$1 \text{ mm}^2 = (\text{approx}) \ 2000 \text{ cmil}$$

$$\text{No. 10 AWG} = (\text{approx}) \ 10,000 \text{ cmil}$$

$$6 \text{ mm}^2 = (\text{approx}) \ 12,000 \text{ cmil}$$

$$100 \text{ mm}^2 = (\text{approx}) \ \text{No. 0000 AWG}$$

3-5 STRANDED WIRE

To obtain the flexibility required for easy handling, the larger sizes of wire are made in the form of a cable consisting of several small strands. Standard numbers of strands commonly used in cable are 7, 19, 37, 61, 91, and 127 strands, since by using these numbers the strands may be laid in concentric layers about a single central wire. Wires larger than No. 0000 AWG are nearly always stranded, and it is often desirable to use stranded wire in the smaller sizes, even as small as No. 20 AWG or 0.5 mm^2 SI.

The separate strands of a stranded wire are not necessarily standard wire sizes since the total cross-sectional area is equal to a standard AWG or metric size, the area of each strand being the total area divided by the number of strands.

Appendix C gives the dimensions and properties of standard AWG stranded copper conductors. Note that conductors larger than AWG No. 0000 are listed by their circular-mil area. Appendix D gives dimensions and properties of stranded metric-size copper conductors.

**3-6
RESISTIVITY**

As previously indicated, materials classified as conductors have many free electrons and offer a low resistance to the flow of electric current. Not all conductor materials, however, offer the same resistance to the flow of current since different materials have different numbers of free electrons. For example, copper offers a lower resistance to the flow of current than does aluminum.

Resistance depends not only on the material used for the conductor but also on the size and temperature of the conductor. In a conductor of large cross-sectional area, the number of electrons free to move is larger than in a wire of small cross-sectional area. Thus, the larger the conductor, the lower its resistance.

Resistance also depends on the length of a conductor. A conductor 20 ft long has twice as much resistance as a conductor of the same material and size that is only 10 ft long.

It has also been found that temperature affects the resistance of most materials. The resistance of most materials increases with an increase in temperature. For example, a tungsten lamp filament has a much lower resistance when it is cold than it has when it is red hot.

Summarizing, the resistance of a conductor depends on the material, the cross-sectional area, the length, and the temperature. Expressed as an equation, the resistance of a conductor at any given temperature is

$$R = \frac{\rho l}{A} \qquad (3\text{-}1)$$

where R = resistance in ohms

l = length

A = cross-sectional area

ρ (rho) is a constant called the *resistivity* of the conductor material.

Resistivity is defined as the resistance of a conductor of unit length and unit cross-section. In USCS units, resistivity is usually expressed in ohms per circular mil-foot. A circular mil-foot is a conductor having a length of one foot and a cross-sectional area of one circular mil.

In SI units, the unit of resistivity is the ohm-meter and is the resistance in ohms of a conductor one meter long having a cross-sectional area of one square meter. It is generally more convenient in calculations to use the smaller units of microhms rather than ohms and millimeters rather than meters, resulting in the unit of resistivity called the microhm-millimeter. One microhm-millimeter is equal to 10^{-9} ohm-meter.

The resistivity of annealed copper at 20°C is 10.4 Ω per circular mil-foot and 17.2 microhm-millimeters in USCS and SI units, respectively. Likewise, aluminum has a resistivity of 17.1 Ω per circular mil-foot and 28.3 microhm-millimeters.

Equation (3-1) may be used to compute conductor resistance when the conductor resistivity and dimensions are known, as illustrated by the following examples.

EXAMPLE 3-6

What is the resistance of a 4110-cmil copper conductor 500 ft long, assuming a resistivity of 10.4 Ω per circular mil-foot?

$$R = \frac{\rho l}{A}$$

$$= \frac{10.4 \times 500}{4110}$$

$$= 1.26 \ \Omega$$

EXAMPLE 3-7

What is the resistance of a copper conductor (resistivity 17.2 microhm-millimeters) that has a length of 1000 m and a cross-sectional area of 2.5 mm²?

Since there are 1,000,000 square millimeters in one square meter, the conductor area A, is

$$A = \frac{2.5}{1,000,000} = 2.5 \times 10^{-6} \ \text{m}^2$$

Also, since one microhm-millimeter is equal to 10^{-9} ohm-meter

$$\rho = 17.2 \times 10^{-9} \ \text{ohm-meter}$$

$$R = \frac{\rho l}{A} = \frac{17.2 \times 10^{-9} \times 1000}{2.5 \times 10^{-6}} = 6.88 \ \Omega$$

#39

Nearly all conductors of electricity show a change in resistance with a change in temperature, as mentioned in the previous section. The resistance of a pure metal conductor increases with an increase in temperature and decreases with a decrease in temperature.

Conductor-resistance tables show resistances at a specific temperature. For example, Appendixes B, C, and D list the resistance of copper wires and cables at 20 degrees Celsius (20°C). (20 degrees Celsius is equivalent to 68 degrees Fahrenheit [68°F].) A true comparison of the resistance of conductors of different sizes or materials cannot be made unless the resistances are compared at the same temperature.

It has been found by experiment that for each degree of change in temperature above or below 20°C, the resistance changes by a given percentage of what it was at 20°C. This percentage change in resistance is called the *temperature coefficient of resistance*. This coefficient for copper at 20°C is 0.00393 and is nearly the same for other pure metals. Thus, for each degree that the temperature of a copper wire increases above 20°C, the resistance increases 0.393 of 1 percent of its value at 20°C. Likewise, for each degree the temperature falls below 20°C, the resistance decreases 0.393 of 1 percent of its value at 20°C.

The straight-line relationship between temperature and resistance holds throughout the range of temperatures normally encountered and may be expressed mathematically as follows:

$$R_2 = R[1 + a(t_2 - t)] \qquad (3\text{-}2)$$

where R_2 = resistance at temperature t_2
R = resistance at 20°C
t = 20°C
α (alpha) is the temperature coefficient of resistance at 20°C

The resistance of a length of copper wire is 3.60 Ω at 20°C. What is its resistance at 80°C?   **EXAMPLE 3-8**

$R_2 = R[1 + a(t_2 - t)]$

$= 3.60[1 + 0.00393(80 - 20)]$

$= 3.60 \times 1.236 = 4.45 \ \Omega$

1 What are the characteristics of a good metallic electric conductor?

2 One mil is equal to what part of 1 in.?

3 What unit of measurement is used for the cross-sectional area of conductors in the U.S. Customary System of measurement? In the metric system?

4 What is the relation between the cross-sectional area of No. 0 AWG and No. 6 AWG wire?

5 A 6-mm^2 conductor is slightly larger than what AWG size?

6 Define the terms resistivity and temperature coefficient of resistance.

7 Why are large conductors stranded?

PROBLEMS

3-1 What are the diameters in mils of wires that have diameters of 2.0, ¾, 0.675, 0.43, and ¼ in.?

3-2 What is the circular-mil area of each of the wires in Prob. 3-1?

3-3 What are the nearest AWG sizes of solid wires with areas of 5000, 10,000, 40,000, and 100,000 cmil?

3-4 What is the diameter in mils of a wire that has an area of 1600 cmil? What is the nearest AWG size?

3-5 If a 150-mm^2 copper wire has a weight of approximately 1375 kg/km, what is the weight of a 300-mm^2 copper wire 1000 m long?

3-6 A conductor with a resistivity of 28.3 × 10^{-9} ohm-meters has a cross-sectional area of 120 mm^2 and is 1 km long. What is its resistance?

3-7 What is the cross-sectional area in square millimeters of a No. 2 AWG solid conductor?

3-8 A conductor of certain specifications can carry safely 1 A for each 300 cmil of its cross-sectional area. If the current to be carried is 20 A, what is the AWG size of the wire required?

3-9 The resistance of a copper coil is 200 Ω at 20°C. What will be the resistance of the coil when it is heated to 90°C?

3-10 The resistance of a generator field coil wound with copper wire is 130 Ω at 80°C. What is its resistance at 20°C?

3-11 Find the temperature coefficient of resistance at 20°C of a conductor that increases resistance from 20 to 24 Ω when heated from 20 to 80°C.

3-12 A copper conductor has a resistance of 400 Ω at 40°C. What will its resistance be at 60°C? (*Hint*: First find its resistance at 20°C.)

3-13 The resistance of a copper generator coil is 0.10 Ω at 20°C. Following operation of the generator at full load, the same coil measures 0.114 Ω. What was the temperature increase in the coil?

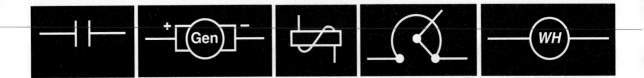

4

Energy and Power

As stated in Chap. 1, electricity is a convenient form of energy. The study of electricity, particularly electric power technology, therefore involves a study of energy and related subjects.

In this chapter, the concepts of weight, mass, force, work, energy, and power are introduced, together with a discussion of the units of measurements used for these quantities.

4-1 WEIGHT AND MASS

The weight of a body is a measure of the pull or force exerted on the body by the gravitational effect of the earth. The weight of a body will therefore vary with its position (altitude and latitude) on the earth in accordance with the earth's gravitational effect. However, this variation of weight at different locations on the earth is usually negligible in most engineering calculations. Although weight is actually a measure of gravitational force, the normally accepted units of measurement in commercial and everyday use are the pound (USGS) and the kilogram (SI).

The mass of a body is defined as the quantity of matter that it contains. Mass is usually determined by dividing the weight of a body, as determined by a weighing scale, by the *gravitational constant g,* or

$$\text{Mass} = \frac{\text{weight}}{g}$$

The unit of mass in the USGS is the pound (lb) and in the SI is the kilogram (kg). The gravitational constants at sea level on the earth are

USGS, $g = 32.2$ feet/sec/sec

SI, $g = 9.81$ meters/sec/sec

4-2 FORCE

Force cannot be defined in terms of more fundamental quantities. For this reason, the concept of force and the attempts at definition have undergone many changes as systems of scientific and engineering units have evolved. However, a presently well-accepted engineering definition is as follows: *A force is that which changes or tends to change the state of rest or motion of a body.*

According to Newton's second law of motion, a body, when acted upon by a force, will be accelerated in direct proportion to the magnitude of the force so acting. Thus, in whatever system of units being used, a unit force acting on a unit mass causes unit acceleration.

The USCS unit of force used in engineering calculations has been the *pound-force,* from which the suffix force was, in practice, usually dropped. This is the force that will give a mass of one pound an acceleration of one foot per second per second.

The SI unit of force is called the *newton* (N). One newton is the force required to impart an acceleration of one meter per second per second to a mass of one kilogram. Thus, explicitly distinct units are used in the SI for force and mass, namely, the newton and the kilogram.

4-3 WORK

Work is done when a force overcomes a resistance. In a mechanical sense, work is measured by the product of a force and the distance through which it acts. When a force of 1 lb acts through a distance of one foot, one foot pound (ft-lb) of work is done. If a force of 10 lb is required to lift a body 6 ft, the work is 10×6 or 60 ft-lb. Thus

$$\text{Work} = \text{force} \times \text{distance}$$

In SI units, the unit of work is the *joule* (J), which is defined as the work done when a force of one newton is exerted through a distance of one meter. If, for example, a force of 20 N is applied to move a body 30 m, the work done is 20×30 or 600 J. One joule is equal to 0.737 ft-lb.

Work is not done unless resistance is overcome. A building foundation exerts a tremendous force in holding up the building, but since there is no motion involved or no resistance being overcome, no work is done.

4-4 ENERGY

Energy is the ability to do work; energy is also stored work. A tank of air under pressure has energy because when it is released through an air hammer, it performs work. A coiled spring in a clock has energy since it can operate the clock for days or weeks. Water stored behind a dam also has energy since it may be used to operate turbine generators to generate electricity.

Since energy is stored work, energy is expended whenever work is done. The fact that 10,000 ft-lb of work is done in raising a 1000-lb object

10 ft means that 10,000 ft-lb of energy must be expended in raising the object. Thus the units of work and energy are the same.

Energy may exist in many forms—mechanical, electric, chemical, heat, and light—and its form is readily changed. For example, an electric generator changes mechanical energy to electric energy; a storage battery changes chemical energy to electric energy. When coal is burned, chemical energy is changed to heat energy, and so on. In accordance with the principle of the conservation of energy, energy may be transformed but it cannot be created, nor can it be destroyed. The electric generator does not create electric energy, it merely changes mechanical energy to electric energy. Nor is the energy supplied to an electric lamp destroyed; it is merely converted to light and heat energy.

4-5
MECHANICAL
POWER

According to the definition of work, if 3000 lb of cargo is lifted to a height of 40 ft, the work required is 3000×40 or 120,000 ft-lb. Nothing is said of the time required to raise the cargo, only that 120,000 ft-lb of work is required.

One motor driving the cargo winch may require 2 minutes (min) to raise the load, while a second motor may accomplish the same result in $^1/_2$ min. Work is done four times as fast by the second motor as by the first, or it is said that the second motor develops four times as much *power* as the first. Thus, power is defined as *the rate of doing work,* or

$$\text{Power} = \frac{\text{work}}{\text{time}}$$

When work is done at the rate of 33,000 ft-lb per min, the power in USCS units is 1 *horsepower* (hp). In the example of the cargo winch, the power required to raise the cargo in 2 min is

$$\frac{120,000}{2} = 60,000 \text{ ft-lb per min}$$

$$\frac{60,000}{33,000} = 1.82 \text{ hp}$$

The horsepower required in the second case is

$$\frac{120,000}{0.5} = 240,000 \text{ ft-lb per min}$$

$$\frac{240,000}{33,000} = 7.28 \text{ hp}$$

Likewise, in SI units, when work is done at the rate of one joule per second, the power is one *watt* (W), which is the SI unit for both mechanical and electric power.

The distinctions between work, energy, and power are important. *Work is the overcoming of resistance. Energy is the ability to do work. Power is the rate of doing work or the rate at which energy is expended.*

The commonly used mechanical units of energy, work, and power are summarized in the following:

The USCS unit of work or energy = foot pound (ft-lb).
The SI unit of work or energy = joule (J).

$$\text{One joule} = 0.737 \text{ foot pound}$$

The USCS unit of power = horsepower (hp).
The SI unit of power = watt (W).

$$\text{One horsepower} = 746 \text{ watts}$$

4-6 ELECTRIC POWER The unit of electric power in both the U.S. Customary and the metric systems is the *watt*. In SI units, one watt is defined as being equal to work being done at the rate of one joule per second. The watt is also defined as the energy expended or the work done per second by an unvarying current of one ampere flowing under a pressure of one volt.

In a dc circuit this is expressed as

$$P = IV \tag{4-1}$$

where P = power, watts
I = current, amperes
V = voltage, volts

What is the power used by an electric lamp that draws 2.5 A from a 120-V dc line?   **EXAMPLE 4-1**

$$P = IV$$

$$P = 2.5 \times 120 = 300 \text{ W}$$

The fact that the watt is a unit of power or a unit of a rate of doing work cannot be emphasized too strongly. Remember that current in amperes is a rate of flow of electricity or is equal to the number of coulombs per second. The power formula may then be written

$$\text{Power in watts} = \frac{\text{coulombs}}{\text{second}} \times \text{volts}$$

In other words, *the watt is a measure of how fast a quantity of electricity is being moved through a difference in potential.*

Since by Ohm's law $V = IR$, this value of V may be substituted in Eq. (4-1) to obtain another useful power formula:

$$P = IV = I \times IR$$

or
$$P = I^2R \tag{4-2}$$

EXAMPLE 4-2 What is the power used in a 60-Ω generator field rheostat when the field current is 2 A?

$$P = I^2R$$
$$P = (2)^2 60 = 4 \times 60 = 240 \text{ W}$$

A third power formula may be derived from the fact that $I = V/R$ by Ohm's law. Substituting in Eq. (4-1)

$$P = IV = \frac{V \times V}{R}$$

or
$$P = \frac{V^2}{R} \tag{4-3}$$

EXAMPLE 4-3 What is the power used by a 15-Ω electric heater when a voltage of 120 V is applied?

$$P = \frac{V^2}{R}$$

$$P = \frac{(120)^2}{15} = \frac{14,400}{15} = 960 \text{ W}$$

As illustrated in the above examples, Eq. (4-1) is used to find the power in a circuit when the current and voltage are known, Eq. (4-2) when the current and resistance are known, and Eq. (4-3) when the voltage and resistance are known.

Since the watt is a small unit, a larger unit, the kilowatt (kW), is often used instead. *One kilowatt is equal to 1000 watts.*

Calculations concerning electric machinery often involve both the electric unit of power (watt) and the mechanical unit (horsepower). *One horsepower is equal to 746 watts.* Therefore, to change power in watts to power in horsepower, it is necessary to divide the number of watts by 746.

For many purposes, the relation between the horsepower and the kilowatt may be taken as

$$1 \text{ hp} = {}^3/_4 \text{ kW (approximately)}$$

EXAMPLE 4-4

The input to a motor is 20 kW. What is the horsepower input?

$$20 \text{ kW} = 20 \times 1000 = 20{,}000 \text{ W}$$

$$\text{Horsepower} = \frac{\text{watts}}{746} = \frac{20{,}000}{746} = 26.8 \text{ hp}$$

4-7 ELECTRIC WORK AND ENERGY

Power is a measure of how fast work is being done or of how fast energy is being expended, that is,

$$\text{Power} = \frac{\text{work or energy}}{\text{time}}$$

Thus, the energy used by an electric device is the rate at which the energy is being used (the power) multiplied by the time during which the device is in use. When power is measured in watts and time in hours, then

$$\text{Power} \times \text{time} = \text{energy}$$

or

$$\text{Watts} \times \text{hours} = \text{watthours}$$

the *watthour* (Wh) being the energy expended when one watt is used for one hour.

The watthour is a relatively small unit and the *kilowatthour* (kWh) is used much more extensively in commercial measurements. One kilowatthour is equal to 1000 watthours.

EXAMPLE 4-5   How much energy is used by a 1500-W heater in 8 h?

$$\text{Energy} = \text{power} \times \text{time}$$
$$= 1500 \times 8 = 12,000 \text{ Wh}$$
$$= \frac{12,000}{1000} = 12 \text{ kWh}$$

If power is measured in watts and time in seconds, then

$$\text{Power} \times \text{time} = \text{energy}$$

and
$$\text{Watts} \times \text{seconds} = \text{watt-seconds}$$

The watt-second is called a *joule,* which is the SI unit for electric as well as mechanical energy. Since there are 3600 seconds in an hour and 1000 watts in a kilowatt, one kilowatthour is equal to 3,600,000 joules or 3.6 megajoules (3.6 MJ).

EXAMPLE 4-6   How much energy in joules, megajoules, and kilowatthours is used by a 100-W lamp in 12 h?

$$\text{Energy in joules} = \text{watts} \times \text{seconds}$$
$$= 100 \times 12 \times 3600$$
$$= 4,320,000 \text{ J}$$
$$= 4.32 \text{ MJ}$$
$$\text{Energy in kilowatthours} = \frac{\text{watts} \times \text{hours}}{1000}$$
$$= \frac{100 \times 12}{1000}$$
$$= 1.2 \text{ kWh}$$
$$1.2 \text{ kWh} \times 3.6 = 4.32 \text{ MJ (check)}$$

Power is the rate of expending energy just as speed is a rate of motion. If the average speed of an automobile is known for a given time, the distance traveled is the average speed multiplied by the time traveled. Likewise, if the average power required by an electric motor for a given time is known, the energy used by the motor is the average power multiplied by the time the motor is used. The reader should make sure that the difference between power and energy is understood. Power is the *rate* of expending energy or of doing work, just as speed is a *rate* of motion.

The commonly used electrical units of energy, work, and power are summarized in the following:

The USCS unit of work or energy = watthour (Wh).
The SI unit of work or energy = joule (J).

$$\text{One watthour} = 3600 \text{ joules}$$

$$\text{One kilowatthour} = 3.6 \text{ megajoules (MJ)}$$

The USCS unit of power = watt (W).
The SI unit of power = watt (W).

It should be noted that the SI units for work, energy, and power are the same whether the process is mechanical or electrical.

**4-8
MEASUREMENT
OF ELECTRIC
POWER AND
ENERGY**

The power being used by a lamp, motor, or other device in a dc circuit may be determined by measuring both the current and the voltage and calculating the power by means of the formula $P = IV$. Power may also be measured directly by means of a *wattmeter.* The wattmeter, which is described in Chap. 20, is an instrument with both a current and a voltage or potential element. The current element is connected in series and the potential element in parallel with the circuit in which the power is being measured, as shown in Fig. 4-1. The power in watts, being a product of current and voltage, is indicated directly on the scale of the wattmeter.

Electric energy is measured by means of the *watthour meter.* This is the familiar type of meter used in the home to determine the monthly electric bill. The connections of the watthour meter are the same as those of the wattmeter, a current element being connected in series and a potential element being connected in parallel with the circuit. However, the watthour meter has a rotating element, the speed of which at any time is proportional to the power being used at that time. The total number of revolutions over a period of time is then proportional to the energy used in that time.

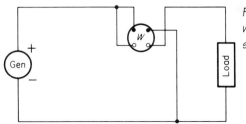

FIGURE 4-1 *Connection of a wattmeter to measure the power supplied to the load.*

The rotating element drives a gear train that records on a row of dials the number of kilowatthours used. The amount of energy, in kilowatthours, used during a certain period is then the dial reading at the beginning of the period subtracted from the dial reading at the end of the period.

The user of electricity then pays for the amount of *energy* or *kilowatthours* used and not the power in kilowatts. Power is the measure of how fast energy is being used at any given time as evidenced by the speed of the rotating element of the watthour meter at that time. The energy used, however, over a certain period of time is determined by the total number of revolutions made by the watthour meter in that time.

4-9 LINE LOSS   Whenever a current flows through a resistance, the resistance becomes heated, or it is said that electric energy is transformed into heat energy. The rate at which the electric energy is transformed into heat is called power, and the common way of finding this power is by the formula $P = I^2R$. In other words, the rate at which energy is expended in a resistance is proportional to the square of the current.

Because the conductors of a transmission or distribution circuit have resistance, the conductors become heated whenever a current flows through them. Since this heat is lost to the surrounding air, it is called a *line loss*. This loss of power is proportional to the square of the current flowing, and for this reason it is desirable to transmit power with as low a current value as possible to keep the line losses from being excessive. Since the power supplied to a load is a product of current and voltage, the current required to transmit a given amount of power may be made small by using higher voltages; that is, the higher the voltage, the lower the current for a given amount of power. The advantages obtained by increasing the voltage used to transmit a given amount of power will be illustrated by the following example.

EXAMPLE 4-7   A 12-kW load is supplied from a line that has a resistance of 0.1 Ω in each conductor. Find the line loss in watts when the load is supplied at (*a*) 120 V and (*b*) 240 V.

$$(a)\, I = \frac{P}{V} = \frac{12,000}{120} = 100 \text{ A}$$

Line loss (both lines) $P = I^2R$

$$P = (100)^2 \times 2 \times 0.1 = 2000 \ W$$

EXAMPLE 4-7
(CONTINUED)

$(b)\ I = \dfrac{P}{V} = \dfrac{12,000}{240} = 50\ \text{A}$

Line loss (both lines) $P = I^2R$

$P = (50)^2 \times 2 \times 0.1 = 500\ \text{W}$

Note in the above example that the same power may be transmitted at 240 V with one-fourth the line loss as at 120 V. Or the power could be transmitted at the same loss with conductors one-fourth as large at 240 V as at 120 V.

REVIEW QUESTIONS

1 Define the term *weight*. What are the USCS and SI units of measurement in everyday commercial use?
2 What is meant by the term *mass*? How is it determined for any body? What are the units?
3 Give a generally accepted definition of *force*.
4 What is the SI unit of force? How is it defined?
5 Define *work*. What are the USCS and SI units? What is the numerical relationship between the two units?
6 What is energy? Give some examples of different kinds of energy.
7 How is power defined? What are the USCS and SI units of mechanical power?
8 What is the unit of electric power?
9 Give the electric power formula in three forms.
10 One kilowatt is equal to how many watts?
11 One horsepower is equal to how many watts?
12 One horsepower is equal to approximately how many kilowatts?
13 How many horsepower are there in one kilowatt?
14 What is another name for the watt-second?
15 One kilowatthour is equal to how many megajoules?

PROBLEMS

4-1 The work done in moving a body through a distance of 40 m is 800 J. What is the value of the force?
4-2 A crane lifts an 8000 lb cargo to a height of 20 ft. What is the work done by the crane?
4-3 If the crane in the preceding problem raises the cargo in 30 s, what is the average power *(a)* in foot-pounds per minute; *(b)* in horsepower?

4-4 A force of 60 N is applied to a body to move it through a distance of 20 m in 8 s. What is the value of the power?

4-5 An electric heater has a resistance of 57 Ω. *(a)* Find the power in kilowatts when the heater is drawing a current of 3.8 A. *(b)* What is the energy used by the heater in 4 h in kilowatthours and in megajoules?

4-6 A 120-V lamp draws 0.8 A. How much power does it consume?

4-7 The energy used by an electric heater in 24 h is 1728 MJ. What was the energy used in kilowatthours? If the heater operated continuously for the 24 h, what was the average power used?

4-8 A current of 8 A flows through a resistance of 20 Ω. How much power is consumed *(a)* in watts; *(b)* in kilowatts?

4-9 A motor takes 11.2 kW when it is connected to a 220-V supply. How much current does the motor take?

4-10 The output of a motor is 1.2 kW. What is the horsepower output?

4-11 A dc motor is connected to a 110-V supply and draws 50 A. What is the horsepower input to the motor?

4-12 A dc shunt motor takes 60 A at 550 V. *(a)* Find the power consumed in kilowatts. *(b)* If the energy cost is $0.05 per kilowatthour, find the cost of operating this motor for 10 h.

4-13 An electric heater uses 20 kWh in 8 h. If the voltage at the heater is 240 V, what is the resistance of the heater?

4-14 Find the current taken by 40-, 60-, and 100-W lamps when connected to a 120-V power supply. Determine the time (in hours) needed for each of these lamps to consume 1 kWh.

4-15 The load voltage of a 12-kW load is 400 V. The resistance of each line wire connecting the load to the source is 0.1 Ω. *(a)* What is the voltage at the source? *(b)* What is the power dissipated as heat by the lines? *(c)* How much energy is used by the load in 24 h? *(d)* How much energy is lost in the lines in 24 h?

 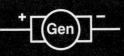  

5
Primary and Secondary Batteries

Devices in which chemical energy is changed to electric energy are called *electric cells*. When several cells are connected electrically, they form a *battery*.

The essentials of any cell are two *dissimilar* metals or conductors that are immersed in a conducting liquid.

#26

Cells or batteries may be classified into two general types: *primary* and *secondary*. In the primary cell it is necessary to renew the active materials from time to time or to discard the cell when electric energy can no longer be obtained from it. In the secondary, or storage, cell the active materials may be restored by passing a current through the cell in a direction opposite to that of discharge.

In this chapter, the principles of operation and the characteristics of the several types of both primary and secondary cells are discussed.

5-1 PRIMARY CELLS When two dissimilar conductors are placed in a conducting solution that will act chemically on one of the conductors, an electromotive force is developed between the two conductors. The conductors are called *plates* or *electrodes,* and the solution is called the *electrolyte.* When the electrodes are connected by a conductor, a current will flow from one electrode to the other through the conductor, the circuit being completed through the electrolyte. The electrode from which the current leaves the cell is called the *positive electrode,* and the other is called the *negative electrode.*

The emf developed in any cell depends on the materials used for electrodes and electrolyte. For example, if electrodes of zinc and lead are placed in a solution of sulfuric acid, the emf developed is about 0.5 V; zinc and copper in sulfuric acid, about 1.1 V; and zinc and mercuric oxide in potassium hydroxide, about 1.35 V. Many other combinations of materials have been used for electrodes and electrolytes of primary cells. The size and spacing of the electrodes do not affect the emf developed but do affect other characteristics of the cell.

Primary cell performance depends on the conditions of service such as temperature, rapidity of discharge, and on whether discharge is continuous or intermittent. While some cells are designed for general-purpose use, cells are often designed for specific duties. A cell may be designed and rated, for example, for high-pulse current drain as required for photo-flash service, or it may be designed for continuous low-current drain as in an electronic watch or hearing aid.

Early primary cells had spillable electrolytes and were therefore not readily portable. Because of its portability and compact size, the *dry cell* is now by far the most widely used primary cell.

5-2 DRY CELLS

There are six generally used types of dry cells: the general purpose carbon-zinc, the heavy-duty zinc-chloride, the alkaline, the mercuric oxide, the silver oxide, and the lithium cells. Actually, these cells are not dry but have an electrolyte that is combined with an absorbent inactive material. The cells are, however, sealed into liquid-tight containers so that they may be easily transported or operated in any position.

The most widely used and probably the most familiar type of dry cell is the *carbon-zinc cell,* often referred to as the general-purpose cell. The basic design of this cell has remained largely unchanged for over half a century. The construction of a typical cell of this type is shown in Fig. 5-1.

The liquid-tight zinc-can enclosure of the carbon-zinc cell also serves as the negative electrode. Inside the zinc can is an absorbent or gelatinous substance that is saturated with the electrolyte, which is zinc chloride and ammonium chloride. The positive electrode is a mixture of manganese dioxide and powdered carbon surrounding a carbon rod, which serves as a current collector. The mixture serves as a depolarizing agent, which prevents the formation of hydrogen bubbles on the carbon rod during discharge of the cell.

The open-circuit voltage of a new carbon-zinc dry cell is about 1.5 V. The capacity of the cell is dependent on the physical size of the cell and the rate at which energy is withdrawn: the lower the rate, the more efficiently

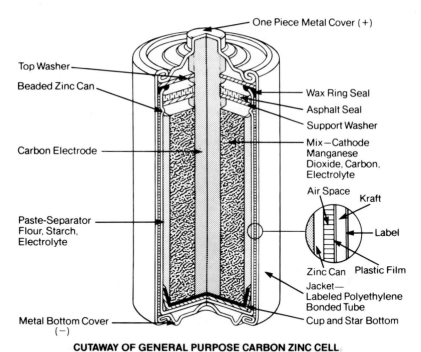

One Piece Metal Cover (+)

Top Washer

Beaded Zinc Can

Wax Ring Seal

Asphalt Seal

Support Washer

Carbon Electrode

Mix—Cathode Manganese Dioxide, Carbon, Electrolyte

Air Space

Kraft

Paste-Separator Flour, Starch, Electrolyte

Label

Zinc Can   Plastic Film

Jacket— Labeled Polyethylene Bonded Tube

Metal Bottom Cover (−)

Cup and Star Bottom

CUTAWAY OF GENERAL PURPOSE CARBON ZINC CELL

FIGURE 5-1   Cutaway view of a general-purpose carbon-zinc cell. (Eveready Battery Company, Inc.)

the cell is able to convert the available chemical energy into electric energy. For this reason, the best application of this cell is in general-purpose intermittent-, light-, or medium-drain service, such as in flashlights or portable radios.

The *zinc-chloride cell* is designated by some manufacturers as their heavy-duty cell. It has the same type of construction and the same open-circuit voltage as the carbon-zinc cell. The essential difference between the two cells is that all or most of the ammonium chloride is omitted from the electrolyte of the zinc-chloride cell. Because of its more concentrated electrolyte, the zinc chloride cell has a lower internal resistance. This gives the cell the ability to deliver higher rates of discharge for longer periods of time than the carbon-zinc cell. Both the carbon-zinc and zinc-chloride cells are recommended for intermittent rather than continuous service.

The alkaline-manganese cell, commonly called the *alkaline cell,* was developed to fill the need for a higher continuous-rate energy source than was available in the carbon-zinc and zinc-chloride cells. Alkaline cell applications include portable radios, cassette recorders and players, battery-powered toys, electronic flash photography, and similar heavy-drain applications.

The alkaline cell employs a powdered zinc negative electrode, a positive electrode of manganese dioxide, and an electrolyte of potassium hydroxide. It has an open-circuit voltage of 1.5 V and a low internal resistance, making it suitable for uses requiring heavy drain and long life. The alkaline cell may contain 50 to 100 percent more total energy than a carbon-zinc cell of the same size.

Mercuric oxide and *silver oxide cells* have similar low-rate, continuous-drain applications. They are usually designed for specific uses rather than for general use. Typical applications are hearing aids, watches, calculators, and cameras. Both are most often fabricated in a flat button or disc configuration. The cases of these cells are precision-made of nickel-plated steel and tightly sealed to prevent leakage.

Electrochemically, the mercuric oxide and silver oxide cells both use zinc negative electrodes with an electrolyte of either potassium hydroxide or sodium hydroxide. The mercuric oxide cell uses a mercuric oxide positive electrode and has an open-circuit voltage of 1.35 V. This cell has good storage life and is noted for its high capacity per unit volume.

The positive electrode of the silver oxide cell is silver oxide mixed with a small amount of manganese oxide, resulting in an open-circuit voltage of 1.6 V. The output voltage of both mercuric-oxide and silver oxide cells is fairly stable for their entire life.

The *lithium cell* is a relatively new type of dry cell developed in Japan during the early 1970s; it has a number of advantages over other dry cells. The use of lithium as the negative electrode in this cell results in a cell that is lightweight, and has long storage life, high energy density, and high voltage. Cell voltages range from 2.0 to 3.6 V, depending on the electrolyte and positive electrode materials used.

Manufacturers have used numerous combinations of materials for electrolytes and for the positive electrodes of the lithium cell. Materials used as the active positive electrode include manganese dioxide, copper sulfide, and carbon monofluoride. Several nonaqueous solvents have been used as the electrolyte, two of which are acetonitrile and propylene carbonate.

The lithium cell has become competitive in price with mercuric and silver oxide cells. Because of this and its many desirable characteristics, the lithium cell is expected to be used increasingly for applications where other types of primary cells have been used.

5-3 INTERNAL RESISTANCE OF A CELL

The entire resistance encountered by a current as it flows through a cell from the negative terminal to the positive terminal is called the internal resistance of the cell. As in any other conductor of electricity, the resistance of a cell depends on the materials, the cross-sectional area, the length of the current path, and the temperature. Thus both the area and spacing of the electrodes affect the internal resistance of a cell, even though the emf developed is independent of these factors.

The current delivered by a cell depends not only on the emf of the cell and the load resistance, but on the internal resistance of the cell as well. By Ohm's law the current delivered by a cell is

$$I = \frac{E}{R_i + R_l} \tag{5-1}$$

where I = current, amperes
E = emf, volts
R_i = internal resistance of cell, ohms
R_l = load resistance, ohms

On open circuit, the terminal voltage of a cell is equal to the emf developed. However, when a cell delivers current, some of the emf developed must be used in overcoming the resistance of the cell itself so that the terminal voltage is lower than the emf. The greater the current delivered, the greater the voltage drop, or IR drop, in the cell and the lower the terminal voltage becomes. The terminal voltage in any case is

$$V_t = E - IR_i \tag{5-2}$$

where V_t = terminal voltage, volts
E = emf, volts
I = current, amperes
R_i = internal resistance, ohms

When Eq. (5-2) is transposed, the following expression for internal resistance results:

$$R_i = \frac{E - V_t}{I} \tag{5-3}$$

The relation given in Eq. (5-3) suggests a method of measuring internal resistance. First, the open-circuit voltage of the cell is measured to get the value of E. A high-resistance voltmeter should be used so that the current drawn by the voltmeter is negligible. Then, with the cell delivering a current I, the terminal voltage V_t is measured. The internal resistance R_i may then be obtained by means of Eq. (5-3).

The following example will illustrate the effect of different values of internal resistance on the amount of current that a cell is able to deliver to a particular load.

EXAMPLE 5-1   A coil that has a resistance of 0.05 Ω is connected to a cell that develops an emf of 1.5 V. Find the current flowing if the internal resistance of the cell is *(a)* 0.1 Ω, and *(b)* 0.01 Ω.

$$(a) I = \frac{E}{R_i + R_l} = \frac{1.5}{0.1 + 0.005} = 10\,A$$

$$(b) I = \frac{E}{R_i + R_l} = \frac{1.5}{0.1 + 0.005} = 25\,A$$

5-4 SECONDARY, OR STORAGE, CELLS

A cell in which the physical state and chemical condition of the electrodes and electrolyte may be restored by charging is called a secondary, or storage, cell. Charging such a cell consists of passing current through the cell in a direction opposite to the direction of current flow on discharge. There are several types of storage cells in use, but the more common ones are the *lead-acid* and the *nickel-cadmium-alkaline* cells.

As in the primary cell, the emf of a secondary cell depends on the materials used for electrodes and the electrolyte. The average terminal voltage of a lead-acid cell is approximately 2.0 V, while that of the nickel-cadmium-alkaline cell is approximately 1.2 V.

The capacity of a cell, for a given plate thickness, depends on the total plate area; the greater the plate area, the greater the capacity.

5-5 THEORY OF THE LEAD-ACID CELL

The positive plate of a lead-acid storage cell is lead peroxide, PbO_2, and the negative plate is pure sponge lead, Pb. Dilute sulfuric acid, H_2SO_4, is used as the electrolyte. When the cell supplies current to a load or is discharging, the chemical action that takes place forms lead sulfate, $PbSO_4$, on both plates with water being formed in the electrolyte. After a certain amount of energy has been withdrawn from the cell, both plates have been transformed into the same material and the cell is no longer able to develop an emf.

To charge the cell, a current is caused to flow through the cell in the opposite direction; this reverses the chemical process and again forms a lead peroxide positive plate and a pure lead negative plate, at the same time restoring the electrolyte to its original condition. The chemical reaction may be represented as follows:

Positive plate		Negative plate		Electrolyte		Positive plate		Negative plate		Electrolyte
PbO_2	+	Pb	+	$2H_2SO_4$	$\rightleftharpoons$	$PbSO_4$	+	$PbSO_4$	+	$2H_2O$
		Cell charged						Cell discharged		

This equation represents the discharging action when read from left to right and the charging action when read from right to left.

5-6 SPECIFIC GRAVITY OF LEAD-ACID CELLS

The ratio of the weight of any liquid to the weight of an equal volume of water is called the *specific gravity* of the liquid. For example, the specific gravity of sulfuric acid is 1.840, which means that sulfuric acid is 1.840 times as heavy as water.

The electrolyte used in lead-acid batteries is a solution of sulfuric acid and water. In batteries used for stationary applications in which a large volume of electrolyte is used, the specific gravity in the fully charged state is about 1.210 to 1.225. In portable batteries, such as those used in automobile and aircraft service, where space and weight are limited, a smaller volume of electrolyte having a higher specific gravity is used. The specific gravity of portable batteries is about 1.250 to 1.270 when fully charged. The specific gravity of all batteries varies with temperature; the higher the temperature, the higher the specific gravity.

As a lead-acid battery is discharged, its specific gravity decreases. The amount of decrease from the fully charged state to the discharged state varies with the volume of the electrolyte. For batteries with large volumes of electrolyte, the drop may be as small as 0.030, and for batteries with smaller volumes, it may be as large as 0.145. If, however, the specific gravity of a given battery's electrolyte is known for both the fully charged and discharged conditions, specific gravity readings may be used as an indication of the state of charge of the battery.

For measuring the specific gravity of a cell, a simple instrument called a *syringe-type hydrometer* is commonly used. When a sample of the electrolyte is drawn from the cell into the glass barrel of the hydrometer, a calibrated weighted bulb floats in the electrolyte. The depth at which the bulb floats is a measure of the specific gravity of the electrolyte being tested.

To compare accurately specific gravity readings taken at different times, readings must be corrected to a normal reference temperature of 77°F (25°C). One point (0.001) of gravity should be added for each 3°F above 77°F or subtracted for each 3°F below 77°F.

5-7 CONSTRUCTION OF LEAD-ACID BATTERIES

In the pasted-plate battery, used in both stationary and motive power applications, the cell plates are formed by applying a special lead oxide paste to lead frames or grids. Grids are alloyed with either antimony or calcium to harden the lead and to facilitate the casting of the grids. After charging, the lead oxides become the active material of the plate. The grids serve to hold the active material in place and to distribute the current evenly over the surface of the plate.

Another type of lead-acid cell, known as the tubular-plate cell, is often used in stationary applications. The positive-plate active material in this type of cell is retained in a series of porous plastic or glass fiber tubes supported on a lead grid. The tubular construction is used only on the positive plate for this cell. The positive tubular and the negative pasted plates of a typical tubular-plate cell are shown in Fig. 5-2.

Plates of lead-acid cells are formed into positive and negative groups so that they may be nested together. The number of negative plates is always one more than the number of positive plates so that both sides of each positive plate will be acted upon chemically. This is necessary since the active material on the positive plate expands and contracts as the battery is charged and discharged, and this expansion and contraction must be kept the same on both sides of the plate to prevent buckling.

Separators of wood, rubber, or glass mat similar to those shown in Fig. 5-2 are placed between the positive and negative plates to prevent the plates from coming into contact with each other. Separators are grooved vertically on one side and are smooth on the other. The grooved side is placed against the positive plate to permit free circulation of the electrolyte around the positive plate where the greater chemical action takes place.

The well-known automobile storage battery consists of three or six cells (6- or 12-V nominal rating) connected in series and assembled into one case. The case is made of lightweight acid-proof material such as polypropylene and is divided into separate compartments to accommodate

FIGURE 5-2   Positive (left) and negative (right) plates with separators used in the tubular-plate cell construction. (Exide Corporation.)

each of the cells. Intercell connections are made with lead-alloy links, and the battery terminals are brought out through sealed openings on the side or top of the case.

Lead-acid cells intended for stationary use are usually contained in glass or plastic jars with hard-rubber covers. For the common application of stationary batteries as the source of control power in power plants or in emergency lighting systems, 60 cells are mounted on steel or wood racks and connected in series to form a nominal 125-V battery. The battery is maintained at full charge by "floating" it continuously on a battery charger at 129 V. A typical 60-cell installation is shown in Fig. 5-3.

5-8 LEAD-ACID BATTERY RATINGS

The capacity of a battery depends on the number, design, and dimensions of the plates and the quantity of the electrolyte. The amount of energy that any given fully charged battery can deliver also depends on several variables such as the discharge rate, the temperature, and the specific gravity of the electrolyte.

Because of the many variables involved, there are several methods of rating battery capacity. Generally, the capacity of any battery may be expressed in *ampere-hours* (Ah), which is simply the product of the discharge in amperes and the duration of the discharge in hours. However, since batteries are often adapted to a particular kind of service, the rating may be based on the requirements for that service.

Stationary batteries for comparatively low-rate discharge service, such as those used in telephone exchanges or as control power supplies in power

FIGURE 5-3 Typical installation of a 60-cell tubular-plate-type lead-acid battery. (Exide Corporation.)

plants, are rated in ampere-hours at an *8-h* rate. This is an ampere-hour rating based on a continuous discharge rate for 8 h under specified conditions, to a final voltage of 1.75 V per cell. Thus, a battery capable of an 8-h discharge rate of 25 A would have an 8-h capacity of 8×25, or 200 Ah.

Batteries used in automotive service are given two ratings to express the battery power characteristics. One rating is called the *cold cranking performance* rating. This rating is defined as the number of amperes that a battery can deliver at 0°F (-17.8°C) for 30 s and maintain a voltage of 1.2 V per cell, or higher. The second rating, called the *reserve capacity performance* rating, is a measure of the ability of a battery to supply an automobile electric load in the event of a failure of the charging system. This rating is defined as the number of minutes a battery at 80°F (26.7°C) can be discharged at 25 A and maintain a voltage of 1.75 V per cell, or higher. As an example, a six-cell, 12-V passenger automobile battery might have a cranking performance rating of 450 A and a reserve capacity performance rating of 138 min.

5-9 OPERATION AND MAINTENANCE OF LEAD-ACID BATTERIES

In the normal operation of a conventional lead-acid storage battery, a certain amount of water is lost from the electrolyte by evaporation and because of gassing. Gassing takes place during charging when the water in the electrolyte is decomposed to hydrogen and oxygen, each of which is given off in the form of gas. The level of the electrolyte should never be allowed to fall below the tops of the plates, and it should be kept at the proper level by adding pure distilled water at regular intervals. Acid is not added to a battery unless the electrolyte has been spilled.

Lead-acid batteries should not be discharged further after a terminal voltage of about 1.75 V per cell is reached and the specific gravity has dropped its normal amount. Further discharge produces an excess of lead sulfate on the plates.

A lead-acid battery should not be left in the discharged condition for any length of time since the lead sulfate on the plates crystallizes and proper reforming of the plates becomes difficult, if not impossible. If a battery is to be stored for any length of time, water should be added to the proper level, after which the battery should be fully charged. At intervals of 4 to 6 weeks the battery should be given a freshening charge to keep it in the fully charged state.

Batteries may be charged at almost any rate provided the cell temperature does not exceed 110°F and that excessive gassing does not take place. Usual practice is to charge at a tapered rate, that is, at a high rate at first but at a gradually reduced rate as the battery becomes nearly charged. Charging should continue until all cells are gassing freely and until the specific gravity of the electrolyte and the terminal voltage of the battery are constant for a period of 1 h. Although violent gassing or gassing for long periods should be avoided, a small amount of gassing for a short time at the end of a charge is desirable to ensure that no lead sulfate remains on the plates.

5-10 MAINTENANCE-FREE AUTOMOTIVE BATTERIES

As indicated in Sec. 5-9, a lead-acid battery produces both hydrogen and oxygen during the recharging process. This results in the loss of water in the battery electrolyte, particularly in an automobile battery that may often be subjected to overcharging. This has led to the development of the so-called maintenance-free battery for automotive use.

The maintenance-free battery has a modified design that limits the amount of gassing and thereby eliminates the need for adding water periodically. Grids of this battery are alloyed with calcium rather than antimony. Impurities in grids, electrolyte, lead, and other parts of the battery are reduced to a minimum. More space is also provided for the electrolyte. As a result of these design modifications, gassing is virtually eliminated.

Since there is no need to add water in the maintenance-free battery, no vent caps are provided. Thus the specific gravity of the battery cannot be measured by conventional means to indicate the state of charge. The state of charge must be determined from voltage readings taken during a load test prescribed by the battery manufacturer. Some batteries have a built-in hydrometer to give some measure of the battery state of charge.

In addition to the elimination of the need for the periodic addition of water, the maintenance-free battery has reduced terminal corrosion and long storage life, and, because of its low internal resistance, it has a good cold cranking performance rating. A typical heavy-duty maintenance-free 12-V automotive battery with a built-in hydrometer is shown in Fig. 5-4 (see p. 66). This battery has a cold cranking performance rating of 475 A and is recommended for use with large diesel or gasoline engines.

5-11 THE NICKEL-CADMIUM BATTERY

The nickel-cadmium-alkaline battery, usually referred to as the nickel-cadmium battery, has been used in Europe since about 1910 but did not find extensive acceptance in America until after about 1945. Experience indicates that the nickel-cadmium battery is extremely reliable and has a long life expectancy.

In the nickel-cadmium cell, the principal active material in the positive plate is nickel hydroxide; in the negative plate it is cadmium hydroxide. The electrolyte is potassium hydroxide. During charge or discharge there is practically no change in the specific gravity of the electrolyte. The sole function of the electrolyte is to act as a conductor for the transfer of

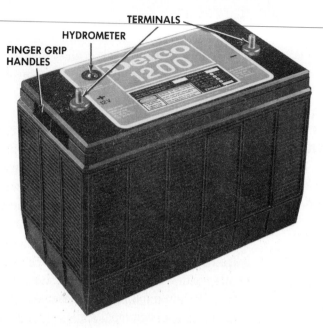

FIGURE 5-4 *Maintenance-free automotive battery.* (Delco-Remy Division, General Motors Corporation.)

hydroxyl ions (electrically charged particles) from one plate to the other depending on whether the cell is being charged or discharged.

Nickel-cadmium cells may be the vented type or the sealed type. Under certain charging conditions the vented type of cell will liberate gases, which are vented to the atmosphere through a valve in the top of the cell. In the sealed type, the design of the cell is such that the evolution of hydrogen gas is completely suppressed at the negative plate, and the oxygen that is evolved at the positive plate combines chemically with the cadmium of the negative plate.

The average discharge voltage per cell of the nickel-cadmium battery is 1.2 V. However, the voltage per cell may be as high as 1.40 to 1.44 V when the cell is being maintained in the fully charged state by a trickle charger. Ten cells make up a nominal 12-V battery, which is normally trickle-charged at about 14.0 V.

The nickel-cadmium battery is characterized by its low maintenance cost, long life, and reliability under severe operating conditions. The battery can be left idle for long periods of time in any state of charge without deteriorating. It will not freeze even in the completely discharged condition. Sealed cells and batteries have a life expectancy in excess of 300 to 500 cycles of charge and discharge under normal operating conditions.

Vented-cell life expectancy in cyclical operation is claimed to be more than 2000 cycles.

Two different basic constructions are used in the nickel-cadmium cell: the pocket-plate and the sintered-plate constructions.

5-12 THE POCKET-PLATE NICKEL-CADMIUM CELL

The term *pocket-plate cell* is derived from the manner in which the cell plates are constructed. In this type of cell, pockets are used to hold the active materials. As constructed by one manufacturer, the pockets are made with finely perforated nickel-plated steel strips formed into channels. A perforated steel cover strip is crimped into place to form a permanent envelope to hold the active materials. Pockets are interlocked, cut to length, and inserted into a steel frame. Alternate positive and negative plates, with insulation between, form plate groups that are assembled into translucent plastic containers. A number of cells may be enclosed in one container to form a battery of the required nominal voltage, or the battery may be formed by interconnecting a number of individual cells mounted on a rack. A cutaway view of a single pocket-plate cell is shown in Fig. 5-5, and an assembly of cells forming a 24-V battery is shown in Fig. 5-6 (see p. 68).

Because of their rugged construction and ability to withstand temperature extremes, pocket-plate batteries are often used for train-car lighting

FIGURE 5-5   Cutaway view of a pocket-plate nickel-cadmium cell. (McGraw-Edison Power Systems Group.)

FIGURE 5-6   A typical 24-V assembly of pocket-plate nickel-cadmium cells. (McGraw-Edison Power Systems Group.)

and diesel engine cranking service. Other applications include use as emergency power supplies and as control power batteries in power stations.

5-13 THE SINTERED-PLATE NICKEL-CADMIUM CELL

Sintered-plate cells may be either vented or sealed and are made in several physical configurations. The most widely used is the sealed cylindrical type, which is sometimes used as a direct replacement for primary cells in such devices as portable tape recorders, radios, or calculators. These cells are available in capacities ranging from about 1 to 7 Ah.

Plates of the sintered-plate cell are made from a nickel powder applied to a nickel-plated steel base and heated to a very high temperature. The resulting plate is a highly porous structure, and the pores are impregnated with the active material, forming a plate with a large surface area.

The plates in cylindrical sintered-plate cells are wound to form a compact coil and are isolated from each other by a porous separator, usually nylon. The separator is saturated with the potassium hydroxide electrolyte. A nickel-plated steel shell serves as the negative terminal, and the cell cover, which is insulated from the shell, serves as the positive terminal. Several sizes of cylindrical sealed nickel-cadmium cells are shown in Fig. 5-7.

In addition to the widely used cylindrical configuration, sealed sintered-plate cells are available in rectangular and oval shapes having greater capacities than the cylindrical types.

Applications of sealed nickel-cadmium cells, in addition to

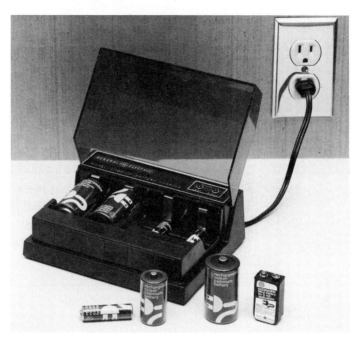

FIGURE 5-7 *Sealed nickel-cadmium batteries with charger. (General Electric Company.)*

replacements for primary cells, include their use as power supplies for cordless devices, as alternate battery power sources, and as emergency stand-by power sources.

5-14 APPLICATIONS OF STORAGE BATTERIES

A very important use of storage batteries is providing stand-by power for various electrical systems. In some electrical systems, storage batteries are connected in parallel with the generator and the load. When the generator is in operation, the battery draws enough current to keep it fully charged. When the generator is shut down, the battery supplies the load. Batteries are also an essential part of uninterruptable power supplies, described in Chap. 13.

Railway car lighting systems are supplied from axle-driven generators when the train is in motion, with batteries supplying the system when the train runs at slow speeds or is stopped.

Automobile electrical systems are similar to the above-mentioned system in that the generator, battery, and load are connected in parallel, the battery supplying power for starting and lighting when the generator is not in operation.

In ac generating plants, storage batteries are used to energize control apparatus and, during emergency shutdowns of the generators, are used to supply emergency lights. Hospitals and other places where a continuous source of power is absolutely essential often use batteries as an emergency supply.

Applications in which storage batteries supply the primary or normal current are in industrial-truck or mine-locomotive propulsion, portable lighting equipment, portable radios, and other applications in which continuous connection to a generator is impracticable.

1   What is the function of an electric cell?
2   What are the essential parts of an electric cell?
3   Upon what does the emf of a cell depend?
4   How do primary and secondary cells differ?
5   Upon what factors does the internal resistance of a cell depend?
6   Why does the terminal voltage of a cell drop with increases in current output?
7   How can internal resistance be determined?
8   Name several types of dry cells.
9   Describe the construction of the carbon-zinc dry cell.
10   What materials are used for electrodes in the mercuric oxide cell?
11   Name two types of storage cells.
12   Describe briefly the chemical action that takes place in the lead-acid cell.
13   What is meant by the term *specific gravity?* Why is it valuable to know the specific gravity of the electrolyte of a lead-acid cell? How is it measured?
14   How are lead-acid batteries rated as to capacity?
15   Describe briefly the construction of the lead-acid battery.
16   Why is it necessary to add water to the electrolyte of a lead-acid cell?
17   Why should lead-acid cells not be discharged below certain levels? Should a lead-acid battery be left standing in the discharged condition? Why?
18   What two things limit the charging rate of a lead-acid cell?
19   What are the active materials in the nickel-cadmium cell?
20   What are two types of construction used in the nickel-cadmium cell?
21   Give some of the more important characteristics of the nickel-cadmium battery.
22   Give several general applications of storage batteries.

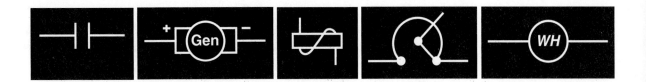

6

Magnetism

Magnetism is involved in the operation of a great number of electric devices, such as generators, motors, measuring instruments, and transformers. For this reason, a knowledge of the underlying principles of magnetism is essential before the operation of electric machinery may be understood. This chapter and the following one are intended to provide background information for the later discussions of electric machinery.

6-1 MAGNETS AND MAGNETIC MATERIALS

It has been known for centuries that certain materials have the ability to attract iron and steel. A body possessing this property is called a *magnet.* Magnets are found in a natural state in the form of the mineral called magnetite. However, natural magnets have little practical value, and commercial magnets are made artificially from iron and steel or alloy materials.

Magnets may be classified as *permanent* or *temporary,* depending on their ability to retain magnetism. Hardened steel and certain alloys of nickel and cobalt, when magnetized, retain their magnetism indefinitely and are called permanent magnets. However, when a piece of soft iron is magnetized, it retains only a small part of its magnetism after the magnetizing force is removed. The amount of magnetism retained by a magnet after the magnetizing force is removed is called *residual magnetism.*

Permanent magnets are used extensively in electric instruments and in meters, telephone receivers, and magnetos. In electric generators and motors where it is desirable to control the amount of magnetism present in the magnet, soft-iron temporary magnets are used.

Materials that are attracted or repelled by a magnet are called *magnetic materials*. Iron and steel are by far the most common magnetic materials. Nickel and cobalt and some of their alloys are also magnetic, the alloys being used in high-grade, permanent magnets. Ceramic magnetic materials are being used increasingly for permanent magnets because of their improved magnetic properties. The basic raw material used in these magnets is iron oxide, which is combined and compacted under pressure with either strontium or barium carbonate.

When a material is easy to magnetize, it is said to have a high *permeability*. Soft iron, being relatively easy to magnetize, has a high permeability. Steel is hard to magnetize and, therefore, has a much lower permeability than soft iron.

A straight bar of steel when magnetized is called a *bar magnet*. When a bar magnet is dipped into iron filings, it is found that the filings are attracted in great numbers at the ends of the bar while very few are attracted to the center of the magnet. The areas at the end of the magnet where the attractive force is the greatest are called the *poles* of the magnet.

6-2 MAGNETIC FIELDS

It has been stated that magnets have an attractive force for certain materials. With the aid of a compass, which is merely a freely suspended magnetized steel needle, the direction of the magnetic force may be determined at various points near a magnet. As shown in Fig. 6-1, the marked end of the compass needle always points away from one pole and toward the other. At the center of the magnet the needle points in a direction parallel to the magnet. The pole toward which the needle points is called the *south pole* of the magnet, and the other pole is called the *north pole*.

Another way of determining the polarity of a magnet is to suspend or pivot it at its center. The magnet will then come to rest in a north-south direction. The end of the magnet pointing north is called the *north pole* of the magnet, while the end pointing south is called the *south pole*.

When the directions to which the compass needle points as it is moved slowly from the north to the south pole of a magnet are plotted as in Fig. 6-1, the resulting figure shows that the magnetic force has definite direction at all points along a curved line from north to south. Such a line is called a *magnetic flux line*. While these lines are really imaginary lines surround-

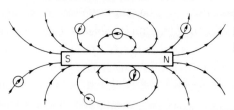

FIGURE 6-1   Direction of the magnetic force around a bar magnet as indicated by a compass needle.

ing a magnet, they are helpful in visualizing the nature of magnetism and magnetic fields.

The space around a magnet, or the space in which magnetic forces act, is called a _magnetic field_ and may be considered to be made up of many magnetic flux lines. The compass needle shows that the flux lines emerge from the north pole of the magnet, pass through the surrounding medium, and reenter the south pole. Inside the magnet, each flux line passes from the south pole to the north pole, forming a complete closed loop or magnetic circuit independently; that is, the lines do not cut across or merge into other flux lines.

An excellent graphical demonstration of the magnetic field pattern around a magnet may be made by placing a sheet of cardboard over a magnet and sifting fine iron filings over the cardboard. Since the iron filings are magnetic bodies, they arrange themselves in definite paths or lines between the poles. The pattern formed shows the shape of the field to be about as shown in Fig. 6-2.

To show that the magnetic field exists on all sides of the magnet, the magnet may be turned on edge. A pattern similar to that of Fig. 6-2 again results. The magnetic field exists in the entire space surrounding a magnet.

The magnetic field surrounding a magnet is called the _magnetic flux,_ and the SI unit of magnetic flux is the _weber_ (wb). The intensity of the magnetic flux or the flux per unit area is called the _flux density._ The SI unit of flux density is the _tesla_ (T) and is equal to a density of one weber per square meter.

The path in which magnetic flux is established is called a _magnetic circuit._ The magnetic circuit of a bar magnet consists of the path of the magnetic flux through the magnet and the surrounding space. The opposition offered to the establishment of magnetic flux in a magnetic circuit is called the _reluctance_ of the circuit. Air has a much higher reluctance than does iron or steel. For this reason, magnetic circuits such as those used in generators and motors are designed with very small air gaps, the greater part of the path followed by the flux being iron.

FIGURE 6-2   Field around a bar magnet.

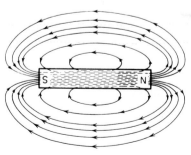

6-3 MAGNETIC ATTRACTION AND REPULSION

When two magnets are suspended freely with their north poles toward each other, the two magnets push each other apart. The same happens when two south poles are placed near each other. However, when a north pole of one magnet and a south pole of the other are placed near each other, the two magnets attract each other.

The rule of magnetic attraction and repulsion is, then, that *like magnetic poles repel and unlike magnetic poles attract one another.* A more accurate picture of a magnetic field is based on this rule. If a small north pole were free to move in a field about a magnet, it would be repelled by the north pole of the magnet and attracted by the south pole. The path that this small north pole would follow in moving from north to south is called a magnetic flux line.

6-4 NATURE OF MAGNETISM

If a bar magnet is broken into two parts, each part is in itself a complete magnet with both a north and a south pole. If each part is again broken, the resulting parts are magnets. If the process is continued, it is found that the smaller and smaller particles retain their magnetism. If it were physically possible to break a magnet into so many pieces that each piece was a molecule, it would be logical to assume that each of these molecules would be a magnet.

#62

Weber's theory of the nature of magnetism is based on the assumption that each of the molecules of a magnet is a tiny magnet. According to this theory, an unmagnetized bar of iron or steel is composed of these tiny molecular magnets haphazardly arranged so that the magnetism of each of the molecules is neutralized by an adjacent molecule. However, when a magnetizing force is applied to the iron or steel, the molecules become arranged in a definite pattern with their north and south poles pointing in opposite directions. The magnetism of each of the molecules acting in the same direction establishes the north and south poles of the magnet.

After a magnetizing force is removed from a piece of hardened steel, the molecules of the steel retain their positions indefinitely, while the molecules of a soft-iron bar tend to return to their original position.

6-5 FIELD AROUND A CURRENT-CARRYING CONDUCTOR

In 1819, Oersted found that a definite relation existed between electricity and magnetism. This discovery and later experiments by Henry and Faraday laid the foundation for the development of modern electric machinery.

When a compass is brought into the vicinity of a current-carrying conductor, the needle sets itself at right angles to the conductor, indicating the presence of a magnetic field. If a conductor is passed through a hole in a sheet of cardboard as shown in Fig. 6-3 and a current is passed through the conductor, the shape and direction of the field may be determined by setting the compass at various points on the cardboard and noting its

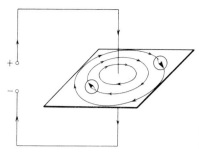

FIGURE 6-3   *Experiment for exploring the field around a conductor.*

deflection. This experiment shows that the magnetic field exists in concentric circles around the conductor. When the current is flowing downward as shown in Fig. 6-3, the field direction is clockwise. However, if the supply polarity is reversed so that current flows upward, the field is found to be counterclockwise. A simple rule, called the right-hand rule for a conductor, for relating the directions of current and field has been established.

Imagine the conductor being grasped in the right hand with the thumb pointing in the direction of the current flow. The fingers then point in the direction of the field around the conductor.

The symbol ⊙ is used in diagrams to denote a cross-sectional view of a conductor carrying current toward the reader, while the symbol ⊕ is used to indicate current flowing away from the reader. These symbols may be thought of as views of an arrow pointing in the direction of current flow; in the former the arrow is approaching, and in the latter the arrow is going away from the reader. Figure 6-4 (see p. 76) illustrates the use of these symbols.

6-6 FIELD AROUND A COIL

The magnetism associated with a current-carrying conductor can be intensified by forming the conductor into a *coil* or *solenoid*.

It may be shown how the field is established around a coil by first considering two parallel conductors carrying current in the same direction as in Fig. 6-5. Magnetic flux lines pass around each conductor in the same direction, resulting in a field that entirely surrounds the two conductors. Likewise, the field established by a belt of several conductors, all carrying current in the same direction, completely envelops the conductors as shown in Fig. 6-6*a* (see p. 76). When the current is reversed, the direction of the field is reversed, as in Fig. 6-6*b*.

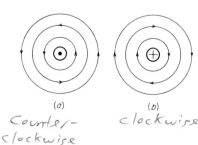

(a)   (b)

FIGURE 6-4   *Representation of the field around a conductor carrying current (a) toward the reader and (b) away from the reader.*

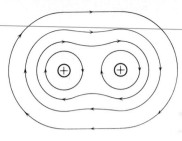

FIGURE 6-5   Field around two parallel conductors.

Figure 6-7*a* represents a coil formed by wrapping a conductor on a hollow fiber or cardboard tube. Note that when current flows through the coil in the direction shown, current is flowing away from the observer in the upper part of each turn (at points 1, 2, 3, 4, and 5) and toward the observer in the lower part of each turn (at points 6, 7, 8, and 9). This is further illustrated in the cross-sectional view of the coil in Fig. 6-7*b*. As indicated by the right-hand rule for a conductor, the field around conductors 1, 2, 3, 4, and 5 is in a clockwise direction and in a counterclockwise direction around conductors 6, 7, 8, and 9. Thus the field established by the coil is similar to that of a bar magnet, with flux emerging from one end of the coil and entering the other. The end of the coil from which the flux emerges is called the north pole of the coil.

The polarity of any coil may be found by means of the right-hand rule for a coil, which may be stated as follows:

Imagine the coil being grasped in the right hand with the fingers pointing in the direction of the current in the coil; the thumb then points toward the north pole of the coil.

6-7 MAGNETOMOTIVE FORCE OR MMF   A measure of the ability of a coil to produce flux is called *magnetomotive force* (abbreviated mmf). Magnetomotive force corresponds to emf in an electric circuit and may be considered to be a magnetic pressure, just as

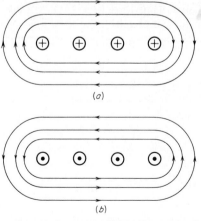

FIGURE 6-6   Field around several conductors, all carrying current (a) away from the reader and (b) toward the reader.

(a)

(b)

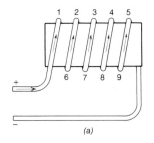
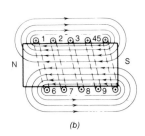

FIGURE 6-7
(a) Coil wound on a hollow tube.
(b) Cross-sectional view showing the field produced by the coil.

emf is considered an electric pressure. The mmf of a coil varies directly with the current flowing in the coil and the number of turns on the coil. The product of the current in amperes and the number of turns is called the *ampere-turns* of the coil. The ampere-turn is taken as a practical unit of mmf.

6-8 ELECTRO-MAGNETS

A coil with a given amount of mmf is able to produce a much greater amount of flux when an iron core is inserted into the coil, since the permeability of iron is so much greater than that of air. Very powerful magnets called *electromagnets* may be made by placing a coil around an iron core.

The strength of an electromagnet depends on the number of ampere-turns of the exciting coil and on the permeability of the core. Soft iron is the material usually used for the core of an electromagnet because of its high permeability. The strength of an electromagnet with a given number of turns on the exciting coil may be varied by varying the amount of current through the coil. This is the method of varying the amount of flux, and hence the amount of generated emf, in a generator.

Electromagnets have a multitude of applications in electric machinery. One important application, as mentioned above, is in the generator. The magnetic circuit of a two-pole dc generator is shown in Fig. 6-8. A strong magnetic field is produced by the two field coils that are wound around the iron pole cores. As the armature is turned through the magnetic field, emf is generated in the armature conductors.

The *relay* shown in Fig. 6-9 illustrates an important application of electromagnetism. A relay is a switch that is operated by an electromagnet.

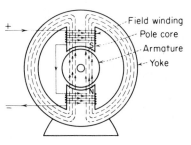

Field winding
Pole core
Armature
Yoke

FIGURE 6-8 Magnetic circuit of a two-pole generator.

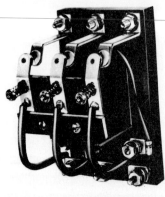

FIGURE 6-9 Typical three-pole relay. (Automatic Switch Co.)

The relay electromagnet consists of a coil and a stationary iron core. The relay contacts are mounted on a hinged iron bar called the *armature*. When the electromagnet is energized by a current passing through the coil, the armature is attracted to the electromagnet core, closing the relay contacts. When the coil is deenergized, the contacts are opened by a spring.

A very small amount of current is required to energize a relay electromagnet and close the relay contacts. Relay contacts can be designed to carry large currents. For example, the relay shown in Fig. 6-9 requires only 0.24 A for operation, while the contacts are rated to carry 25 A. Relays, therefore, are used when it is desired to control a device with a high current rating, such as an electric heater or similar device, from a remote point. It is not necessary to run the main supply lines for the device to the point of control. The remote-control wires need to be large enough to carry only the relay coil current.

Another application of electromagnetism is in the solenoid-operated brake. The brake shown in Fig. 6-10 is called a spring-set, electrically

FIGURE 6-10 Solenoid-operated brake. (General Electric Company.)

released brake. As the name indicates, this brake is set by the coil spring on the top of the brake and is released electrically by energizing the solenoid on the left-hand side of the mechanism. Brakes of this type are used as holding brakes on cranes, hoists, conveyors, machine tools, or similar machines. They may also be used as stopping brakes when they are applied within their heat-dissipating ability.

Other typical applications of electromagnets are in lifting magnets for moving scrap iron, solenoid-operated valves, magnetic clutches, and magnetic separators used for removing tramp iron from coal before coal is fed into pulverizers.

6-9 SATURATION

In a coil with air as the core, the amount of flux produced is directly proportional to the coil mmf in ampere-turns. This is approximately true for a coil having a core of iron or other magnetic material up to a certain stage of magnetization of the core. Above this point, increases in mmf produce smaller and smaller increases in flux in the core, and the core is said to be *saturated.*

The saturation of iron (and other magnetic materials) may be explained by Weber's theory of magnetism. According to this theory, the molecules of an unmagnetized piece of iron (or other magnetic material) are not arranged in any definite order but exist in a disorganized state. When the iron is magnetized by passing current through a coil placed around the iron, the molecules become arranged in a definite order. To arrange the greater part of the molecules in a definite order or to magnetize the iron up to a certain point requires relatively few ampere-turns of applied mmf. In this stage of magnetization, the amount of flux established in the iron increases almost directly with increases in the ampere-turns applied. However, above this point, which is called the *saturation point,* it becomes increasingly difficult to magnetize the iron further, since the unmagnetized molecules become fewer and fewer. Above the saturation point, when much larger increases in ampere-turns are required for corresponding increases in flux in the iron, the iron is said to be saturated. The fact that the iron has become saturated does not mean that a further increase in magnetism is impossible; the increases in magnetism merely require very much larger increases in applied ampere-turns than before the iron became saturated.

The concept of the saturation of magnetic materials is illustrated in Fig. 6-11 (see p. 80). As shown, increases in magnetizing force produce decreasing amounts of flux above the saturation point.

The effects of the saturation of magnetic materials are further discussed in Chaps. 8 and 12.

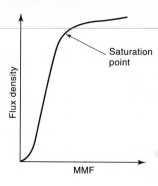

FIGURE 6-11 Curve showing saturation of a magnetic material.

6-10 RESIDUAL MAGNETISM

After the magnetizing force has been removed from an iron-core coil, the iron core displays some remaining magnetization called *residual magnetism*.

As described in Sec. 6-9, the molecules of a magnetized iron core are arranged in a definite order. When the magnetizing force is removed, most of the molecules return to their disorganized condition instantaneously. The remaining molecules stay in their previous order for an indefinite time period and maintain some of the magnetic flux in the coil, but in a weaker state. Over time, as more and more of the molecules return to their disorganized state, the magnetic field disappears entirely. The time it takes for the field to disappear depends totally on the material involved and the strength of the mmf applied when the iron core was magnetized.

The fact that magnetic materials are able to retain even a small amount of magnetism for an indefinite time plays an important part in the operation of generators, as will be shown in Chap. 8.

REVIEW QUESTIONS

1 What is a magnet? Into what two classifications may magnets be divided?
2 What is the magnetism called that is retained by a magnet after its magnetizing force is removed?
3 What distinguishes magnetic materials from nonmagnetic materials?
4 How can the north and south poles of a magnet be identified?
5 What is a magnetic flux line? What is a magnetic field?
6 What is the SI unit of magnetic flux?
7 What is meant by the term flux density? What is the SI unit of flux density?
8 Give the rule of magnetic attraction and repulsion.
9 Describe, briefly, Weber's theory of magnetism.
10 How may it be shown that there is a magnetic field around a current-carrying conductor? Give the rule that relates the current direction and the field direction.

11 What determines the polarity of a coil? State the rule for finding the polarity of a coil.

12 Define magnetomotive force. What is the practical unit?

13 In an electric circuit, an emf causes a current to flow through a resistance in accordance with Ohm's law. What are the corresponding quantities in a magnetic circuit?

14 Upon what does the strength of an air-cored coil depend? An iron-cored coil? How is the strength of an electromagnet varied?

15 Give some applications of electromagnets.

16 When is a bar of iron said to be saturated?

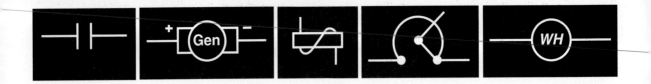

7

Electromagnetic Induction

It can be shown experimentally that an electromotive force can be produced in a conductor by moving the conductor through a magnetic field. The discovery of this principle was announced by Faraday in 1831. It has been called the most important in the history of electricity, since it has led to the development of the electric generator, the transformer, the telephone, and numerous other electric devices.

7-1 INDUCED EMF When the ends of a conductor are connected to a low-reading voltmeter and the conductor is moved into the magnetic field of a magnet as shown in Fig. 7-1, a momentary reading will be noted on the voltmeter. When the conductor is withdrawn from the field, the meter will deflect momentarily in the opposite direction. If the conductor is held stationary while the magnet is moved so that its field is moved across the conductor, the same results are obtained. The phenomenon whereby an emf is induced in a conductor when the conductor is moved through a magnetic field (or when the magnetic field is moved across the conductor) is called *electromagnetic induction* and the emf so induced is called an *induced electromotive force*. The current caused to flow in the conductor by the induced emf is called an *induced current*.

The principle of inducing an emf in a conductor by moving the conductor through a magnetic field is used in the dc generator (see Chap. 8). Stationary electromagnets establish a field through which conductors are moved, causing an emf to be induced in them.

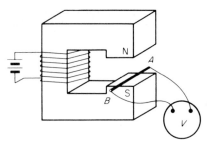

FIGURE 7-1 *An emf is induced in the conductor AB when it is moved through the magnetic field.*

In ac generators, because of insulation and mechanical design problems, the conductors are stationary, while the electromagnets are revolved (see Chap. 14). The principle is the same in both dc and ac generators, however, since there is relative motion between a magnetic field and a conductor in either case.

7-2 FACTORS AFFECTING THE AMOUNT OF INDUCED EMF When the conductor of Fig. 7-1 is replaced by a coil of several turns as in Fig. 7-2 and the experiment is repeated, it will be found that the voltmeter deflection is greater than when the single conductor was used, other conditions being the same. Each turn of the coil now has an emf induced in it, and since the turns are in series, the total emf of the coil is the sum of the emfs of each of the turns. If a coil of still more turns is used, correspondingly greater emfs will be induced, the amount of emf increasing directly with the number of turns on the coil.

When the speed at which a given coil is moved into a magnetic field is increased, the emf induced is increased since the emf induced is directly proportional to the rate at which the magnetic flux lines are cut across. The strength of the field is also a factor since, at a given speed, more flux lines are cut across in a strong magnetic field than in a weak one. Other factors involved in inducing an emf are the angle at which the field is cut and the length of the section of conductor that is being moved through the field.

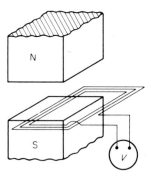

FIGURE 7-2 *Electromotive force is induced in each turn of the coil as it is moved through the magnetic field.*

7-3 FARADAY'S LAW OF INDUCTION

All of the previously mentioned factors involved in the induction of an emf concern the *rate* at which a magnetic field is being cut across by a conductor or the rate at which the number of lines of magnetic flux through a coil is changing. In general:

Whenever the number of magnetic flux lines threading through a coil is changed, an emf is induced in that coil. The amount of emf induced is proportional to the rate at which the number of flux lines through the coil is changing.

This is known as Faraday's law of induction.

7-4 NUMERICAL VALUE OF INDUCED EMF

It has been shown that when flux is changing through a coil, an emf is induced. When the flux through a one-turn coil is changing at the rate of one weber per second, one volt is induced in the coil.

7-5 DIRECTION OF INDUCED EMF

As the conductor in Fig. 7-1 (see p. 83) is moved into the magnetic field from right to left, the voltmeter indicates that an emf is induced from A to B. When the conductor is withdrawn from the field, the direction of the emf is reversed or is induced from B to A. If the polarity of the electromagnet is reversed, emf is induced from B to A as the conductor is moved into the field and from A to B when the conductor is withdrawn. The relation between the directions of motion, field, and induced emf is given by *Fleming's right-hand rule:*

Extend the thumb, forefinger, and middle finger of the right hand at right angles to one another as in Fig. 7-3. Point the forefinger in the direction of the magnetic field and the thumb in the direction of motion of the conductor. The middle finger then points in the direction of the induced emf.

Fleming's right-hand rule is established for a *stationary field* and a *moving conductor*. However, it may also be applied in the case of a moving field and a stationary conductor if the *relative* motion of the conductor is considered. If the conductor AB were stationary in Fig. 7-1 and the field moved toward the right, emf would be induced from A to B just as if the conductor were moved to the left into a stationary field. That is, the motion

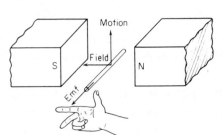

FIGURE 7-3   Fleming's right-hand rule for determining the direction of an induced emf.

of the conductor relative to the field is to the left in either case. Therefore, it must be kept in mind when using Fleming's right-hand rule that the thumb must point in the direction of relative motion of the conductor.

7-6 LENZ'S LAW OF INDUCTION

According to Faraday's law, whenever the flux threading through or linking with a coil is changed, an emf is induced in the coil. If this coil is a part of a closed circuit, the induced emf will cause a current to flow in that circuit. The direction of the resulting current has a definite relation to the variation of the field that produces it. This relation is formalized by Lenz's law of induction, which may be stated as follows:

An induced emf will cause a current to flow in a closed circuit in such a direction that its magnetic effect will oppose the change that produces it.

This rule follows directly from the law of the conservation of energy; that is, to cause an induced current to flow requires the expenditure of energy. In the case of a generator, for example, when the induced current is caused to flow through a load connected to the generator, electric energy is expended. The field produced by the current is always in a direction so that it reacts with the main generator field to oppose the turning action of the prime mover driving the generator. Thus, the greater the electric energy supplied to the load, the greater is the reaction and, in turn, the greater is the mechanical energy required from the prime mover. Energy must be supplied to the generator at the same rate that it is being taken from the generator.

7-7 SELF-INDUCED EMF: INDUCTANCE

When a conductor is carrying a current, there is a magnetic field around the conductor. When the current ceases to flow, the field ceases to exist. When the current changes in intensity, the field likewise changes in intensity. It has been shown that when the amount of flux threading through, or linking with, a circuit is changed, an emf is induced in that circuit. When due to a change of current in a circuit itself, the induced emf is called the *emf of self-induction*.

Since, according to Lenz's law, any induced emf acts to oppose the change that produces it, a self-induced emf is always in such a direction as to oppose the *change* of current in the circuit in which it is induced. When a coil or other electric circuit has the property of opposing any change of current in the circuit, it is said to possess *self-inductance* or simply *inductance*.

The unit of inductance is the *henry*. *A circuit has an inductance of one henry if an emf of one volt is induced in the circuit when the current changes at the rate of one ampere per second.*

7-8 SELF-INDUCTION IN A COIL

A circuit consisting of a straight wire contains a certain amount of inductance because a change of current in the wire produces a change in the flux enveloping or linking the circuit. However, a circuit containing a coil has a much higher value of inductance since a change in current in such a circuit produces much more of a change in the flux linking the circuit.

Figure 7-4 represents a vertical cross-section of a coil. When the switch S is closed, current will flow from the battery into the coil at A. As the current increases in value, the magnetic lines expand from the center of each turn of the coil and cut across adjacent turns. For example, the lines set up by the current in turn A expand and cut across turn B, which is the equivalent of moving B into the field of A. Fleming's right-hand rule shows the resulting induced emf to be toward the reader at B or *against* the flow of current. Other turns of the coil react in the same way. The increase of current in the coil is thus opposed.

When the switch is opened, the collapsing field of the coil cuts across the coil in a direction opposite to that of the increasing flux. This induces an emf in the reverse direction or in the *same* direction as the current flow, tending to maintain the current flow or to oppose its decrease.

Thus, a circuit that has inductance opposes any *change* in current through the circuit. In the coil of Fig. 7-4, the self-induced emf opposes the increase in current as the switch is closed and opposes the decrease in current as the switch is opened. When a current is unchanging in value, the inductance of the circuit has no effects on its flow. A steady current is opposed only by the resistance of the circuit.

When a highly inductive circuit (such as a generator field circuit) is opened, the self-induced emf may be high enough to damage the insulation of the coil or to endanger the life of the person opening the switch. For this reason, a discharge resistor is often connected across the coil at the same instant that the switch blade is opened. The induced emf then causes a current to flow through the resistor and permits the field to collapse gradually, thus limiting the value of the self-induced emf.

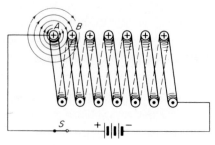

FIGURE 7-4 As the current increases in the coil, the expanding flux lines from each turn of the coil move across adjacent turns, inducing an emf in them.

7-9 MUTUAL INDUCTION

In Sec. 7-1 it was shown how an emf may be induced by moving either a conductor or the magnets that produce the field, as in the case of electric generators. Electromotive forces may also be induced by devices in which neither the conductor nor the magnets move but in which the magnetic field is made to vary in strength or direction.

If two coils are placed adjacent to one another as in Fig. 7-5, a part of the flux produced by coil *A* passes through or links with coil *B*. If the field strength of coil *A* is increased or decreased, there is a corresponding increase or decrease in the field strength inside coil *B*. Since, according to Faraday's law, an emf is induced whenever the flux is changing through a coil, each change of field strength through coil *B* causes an emf to be induced in that coil. Electromotive force is induced in one direction in coil *B* as the field expands and in the other direction as the field collapses, the direction in each case being that indicated by Fleming's right-hand rule or by Lenz's law.

Coil *A,* the coil that is connected to the supply and that produces the original flux, is called the *primary coil*. Coil *B* is called the *secondary coil*. If the two coils are wound on the same iron core, the flux established by the primary coil is not only greater, but a larger part of it is made to cut or link with the secondary coil as it expands or collapses. An emf induced in a secondary coil by a change of current in the primary coil is said to be induced by *mutual induction*.

7-10 MUTUAL-INDUCTION DEVICES

The field established by the primary coil, and which links the secondary coil of a mutual-induction device, may be made to vary by one of three methods: by periodically reversing the primary current, by varying the intensity of the primary current, or by interrupting or "making and breaking" the primary current.

The transformer is a device used for changing the value of alternating voltages. The magnetic field in this mutual-induction device is varied by the periodic reversal of the primary current. Its primary and secondary windings are wound on the same iron core. When an alternating voltage is applied to the primary winding, an alternating current, which is a current

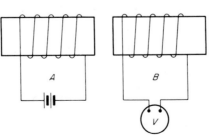

FIGURE 7-5   A change in the amount of current flowing in coil A induces an emf in coil B.

that reverses itself periodically, flows in the primary winding. The flux produced in the core increases, decreases, and reverses with the primary current. This induces an alternating emf in the secondary winding that is directly proportional to the number of turns on the secondary winding. Transformers are discussed in detail in Chap. 12.

The conventional telephone is an example of the use of a varying primary current to induce an emf in the secondary circuit. Although this type of telephone is still providing reliable service, it is being replaced gradually by the more-efficient electronic telephone.

In the conventional telephone system, a battery, the telephone transmitter, and the primary winding of an induction coil are connected in series as shown in Fig. 7-6a. The sound waves of a speaker's voice cause the

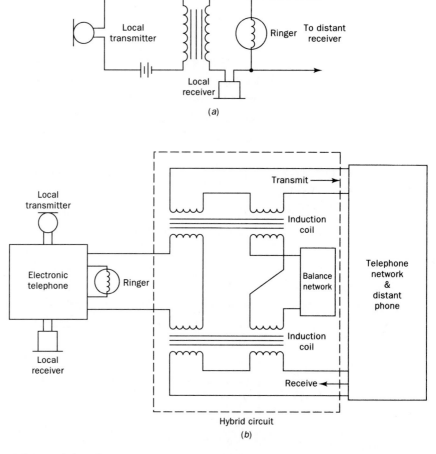

FIGURE 7-6   Telephone circuits illustrating use of induction coils. (a) In conventional telephone systems. (b) In the hybrid circuit of an electronic telephone system.

transmitter diaphragm to vibrate, which, in turn, varies the resistance of the transmitter. The primary current increases and decreases in value with the variations of the transmitter resistance, causing a variation in the amount of flux linking the secondary winding of the induction coil. The varying emf so induced in the secondary circuit causes a varying current to flow over the line and through a distant receiver.

The electronic telephone system uses a device called a *hybrid* to couple the telephone to the network, permitting simultaneous transmission and reception of signals. The hybrid has multiple primary and secondary coils wound on a common core and operates in a manner similar to the conventional telephone induction coil. The simplified diagram in Fig. 7-6*b* illustrates the concept. In this system, electronic devices and circuits replace the transmitter and receiver of the conventional telephone set. The ringer in both telephone systems uses magnetic induction from the telephone line to ring the telephone.

The method of interrupting the primary-coil current to induce an emf in the secondary coil is used in internal-combustion-engine ignition systems. Figure 7-7 shows a simplified circuit diagram of an automobile engine-ignition system.

The ignition coil consists of a primary coil of a few turns of relatively heavy wire wound on an iron core and a secondary coil of many turns of fine wire wound directly around, but insulated from, the primary coil and the core. Primary current is supplied from a 12-V storage battery.

In automobile ignition systems used prior to about 1975, contacts

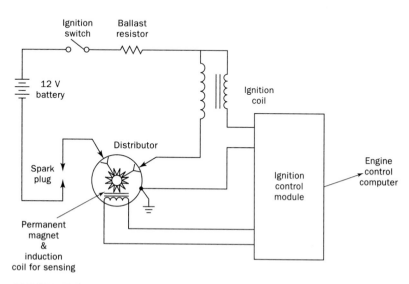

FIGURE 7-7 *Simplified diagram of a solid-state ignition system for an internal combustion engine.*

called breaker points were connected in series with the primary coil and the battery and were alternately opened and closed by a mechanically driven cam. When the points were closed, current from the battery flowed through the primary coil, establishing a flux that linked both primary and secondary coils. At the proper instant the cam "broke" the points, opening the primary circuit. At the same time the distributor arm connected the secondary coil to the proper spark plug. As the field collapsed around the secondary coil, the high voltage induced was applied across the spark-plug gap, causing an arc between the spark-plug points, which ignited the mixture in the combustion chamber. The sequence of events was then repeated when the distributor arm advanced to fire the next spark plug as the cam again broke the points.

In the newer electronic-type ignition systems, the ignition coil is used to perform the same function as in the older breaker-point systems. In the electronic system, however, the alternate opening and closing of the primary-coil circuit is performed by a solid-state switching device instead of by breaker points. The solid-state switching device or ignition control module receives its input signal from a magnetic induction device on the distributor shaft.

This magnetic device consists of a permanent magnet and a pickup or sensing coil. The permanent magnet is shaped to induce pulses of current in the sensing coil as it rotates on the distributor shaft like the cam in older ignition systems. The current pulses act as a "trigger" for the ignition module electronics. This input signal is combined with signals from the engine control computer to determine the pulse timing to the ignition coil. The distributor then directs the high voltage to the correct spark plug.

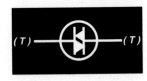

REVIEW QUESTIONS

1 What is an induced emf?
2 What factors affect the amount of emf induced in a coil?
3 State the general rule that covers all the variable factors involved in inducing an emf.
4 Numerically, what rate of change of flux is necessary to induce 1 V?
5 State Fleming's right-hand rule. How may it be applied to a moving field and a stationary conductor?
6 State Lenz's law of induction. Upon what fundamental law is it based?
7 What is an emf of self-induction? Under what conditions of current flow is this emf induced? What is its direction in relation to any change in current?
8 In an inductive circuit, does the inductance have any effect on the flow of a steady current? Why?

9 Why is a coil said to have a higher value of inductance than a straight wire?

10 What is the unit of inductance?

11 What is the purpose of a discharge resistor used in conjunction with opening a highly inductive circuit?

12 Show how a change of current may be made to induce an emf in an adjacent circuit. Using the relation given by Lenz's law, determine the direction of the flux established by an induced secondary-coil current in relation to an increasing primary flux.

13 Give three ways in which flux that mutually links two circuits may be varied. Which of these methods is used in *(a)* the telephone, *(b)* the transformer, *(c)* the ignition system?

14 What would be the effect of connecting a transformer to a dc supply?

15 Why is the secondary coil of an ignition coil wound with many turns of wire?

8

Direct-Current Generators

A dc machine may be used either to convert mechanical energy to electric energy or to convert electric energy to mechanical energy. When a dc machine is driven mechanically by a prime mover such as a steam turbine, hydraulic turbine, or diesel engine and delivers electric energy to electric lights or machines, it is called a *generator.* If electric energy is supplied to the machine and its output is used to drive mechanical devices such as conveyors or machine tools, it is called a *motor.* Generators are rated as to the *kilowatts* they can deliver without overheating at a rated voltage and speed. Motors are rated as to the *horsepower* (USCS) or *kilowatts* (SI) they can deliver without overheating at their rated voltage and speed. A summary of generator and motor characteristics is included in Appendix E.

This chapter deals with dc machines used as generators, although much of what is said concerning generators is equally applicable to motors. Motors are discussed in Chap. 15.

8-1 GENERATOR CONSTRUCTION

Flux produced by the field windings of a generator is established in the field yoke, pole cores, air gap, and armature core, all of which form what is known as the *magnetic circuit* of a generator. The magnetic circuit of a four-pole generator is shown in Fig. 8-1. The *field yoke,* or frame, made of cast steel or rolled steel, serves as a mechanical support for the pole cores as well as serving as part of the magnetic circuit. *End bells,* which support the brush assembly and which in all but the very large machines also support the armature bearings, are also attached to the yoke.

Generator *pole cores* are made of sheet-steel laminations that are insulated from each other and riveted together, the core then being bolted

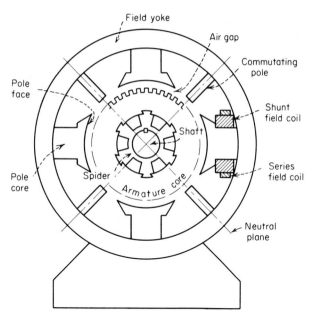

FIGURE 8-1 The parts of the magnetic circuit of a four-pole generator.

to the field yoke. The pole face, which is the surface of the core next to the air gap, is larger than the main body of the core. This is to reduce the reluctance of the air gap and to provide a means of support for the field coils.

An assembled *armature core* and commutator, without the armature windings, are shown in Fig. 8-2. The core is made of sheet-steel laminations that are keyed to the shaft or to a spider, which, in turn, is keyed to the shaft. The outer surface of the core is slotted to provide a means of securing the armature coils.

FIGURE 8-2 Armature core with assembled commutator in the foreground. (General Electric Company.)

The *air gap* is the space between the armature surface and the pole face and varies in length with the size of the machine but is of the order of $1/16$ to $1/4$ in.

The *electric circuits* of a dc generator are made up of the armature winding, commutator, brushes, and field windings. Except on small armatures, armature windings consist of coils that are wound to their correct shape and size on a form, after which they are completely insulated. After the formed coils are slipped into their proper places in the armature slots, they are securely wedged into place and the coil ends are connected to the proper commutator segments. On small armatures, the windings are not form-wound but are wound by hand or by machine directly into the slots of the armature core.

The *commutator* consists of a number of copper segments assembled into a cylinder, which is secured to, but insulated from, the shaft as shown in Fig. 8-2. The segments are well insulated from each other, mica being the insulating material commonly used. To these commutator segments are soldered the ends of the armature coils.

Brushes that rest on the face of the commutator form the sliding electrical connection between the armature coils and the external circuit. Brushes are made of carbon of varying degrees of hardness and in some cases are made of a mixture of carbon and metallic copper. The brushes are held in place under spring pressure by brush holders, the electric connection between the brush and brush holder being made by a flexible copper conductor called a *pigtail*.

The *field coils* are placed around the pole cores as shown in Fig. 8-1. The coils of each of the poles are connected in series to form the field circuit. Field circuits may be designed to be connected either in series or in parallel with the armature circuit. Parallel- or shunt-field coils have many turns of wire of small cross-section and a relatively high resistance, while series-field coils have few turns of wire of large cross-section and relatively low resistance. In order to obtain special operating characteristics, some generators (compound generators) are equipped with both shunt- and series-field coils placed on the same pole core, as indicated in Fig. 8-1.

Most generators are equipped with small poles called *interpoles* or *commutating poles*, which are placed midway between the main poles, as shown in Fig. 8-1. Flux is established in these poles only when current flows in the armature circuit, the purpose of the flux being to improve commutation, as explained in Sec. 8-8.

A dismantled small generator is shown in Fig. 8-3. Note that the brush assembly and the bearings are supported by the end bells. A completed

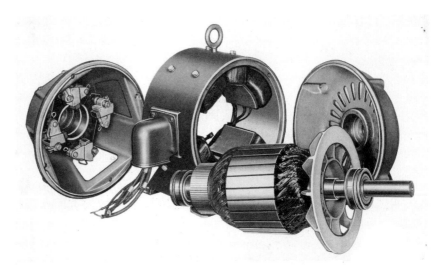

FIGURE 8-3 *Generator dismantled to show the various parts. (Robbins & Myers, Inc.)*

FIGURE 8-4 *DC generator for direct engine drive. (Kato Engineering/Division of Reliance Electric.)*

larger generator designed for direct connection to an engine is shown in Fig. 8-4.

8-2 THE SINGLE-COIL GENERATOR

Figure 8-5 represents a simple two-pole, single-coil generator with the armature core and commutator omitted for the sake of simplicity. The single armature coil can be rotated about its axis 0-0 in a uniform magnetic field produced by the two poles. As the coil is rotated at a constant rotational speed by some mechanical means not shown, the number of magnetic flux lines through the coil changes continually. In Chap. 7 it was shown that when the number of flux lines through a coil is changed, an emf is induced in that coil. The amount of emf induced depends on the rate at

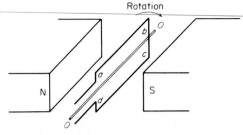

FIGURE 8-5 *Two-pole single-coil generator with the armature coil positioned in the neutral plane.*

which the number of flux lines is changing through the coil, and its direction is determined by Fleming's right-hand rule. In a single-coil generator with a constant field strength, the amount of emf induced at any instant depends on how fast the coil is moving across the field at that instant. An emf produced by the relative movement of a coil and a magnetic field, as in a generator, is called a *generated emf.*

At the instant that the coil is in the vertical position shown in Fig. 8-5, the coil edges are moving parallel to the flux lines so that the flux enclosed by the coil is not changing. Thus no emf is generated when the coil is in this position and the coil is said to be in the *neutral plane.*

As the coil is moved in a clockwise direction from the neutral plane at a constant speed, the coil edges begin to move across the flux lines slowly at first but at a gradually increasing rate. The instantaneous values of emf therefore gradually increase as the coil is moved around to the position shown in Fig. 8-6, which is 90° from the position shown in Fig. 8-5. As determined by Fleming's right-hand rule, the direction of the emf generated as the coil is moved from the neutral plane to the 90° position is from b to a and from d to c. In the 90° position, the coil edges are moving at right angles to the field and are therefore moving across the field at a maximum rate; consequently, the instantaneous value of the emf generated at this point is at a maximum.

As the coil is turned on past the 90° position, the rate at which the coil conductors move across the field gradually decreases, causing the

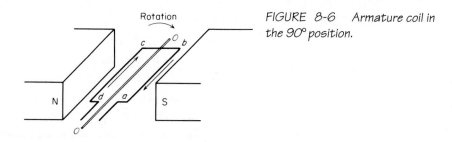

FIGURE 8-6 *Armature coil in the 90° position.*

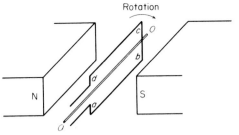

FIGURE 8-7 *Armature coil in the 180° position.*

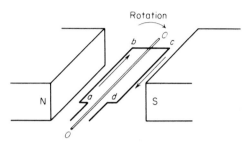

FIGURE 8-8 *Armature coil in the 270° position.*

instantaneous values of generated emf to decrease gradually. When the coil reaches the position shown in Fig. 8-7, the coil is in the neutral plane where the generated emf is again zero.

When the coil is moved in a clockwise direction from the 180° position shown in Fig. 8-7, the coil sides again begin to move across the field. However, the direction of the generated emf is now from *a* to *b* and from *c* to *d*. The values of instantaneous emf gradually increase until a maximum value is generated again at the 270° position shown in Fig. 8-8.

A comparison of Figs. 8-6 and 8-8 shows the directions of the generated emfs in the two coil sides to be reversed. But in both positions the emf in side *ab* adds to that generated in side *cd*.

As the coil is moved from the 270° position, the instantaneous values of generated emf gradually decrease until the coil again reaches the neutral plane, where the emf is zero. The cycle then repeats for each revolution of the coil.

Thus when a coil is rotated in a two-pole magnetic field, an emf is generated in one direction during the first half and in the opposite direction during the second half of each revolution. Such an emf is called an *alternating emf.* If the value of the instantaneous generated emf is plotted for a number of the positions the coil passes through in one revolution, a curve similar to that shown in Fig. 8-9 results.

The single-coil generator may be connected to an external circuit by connecting the coil terminals to two continuous and insulated rings, called

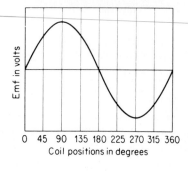

FIGURE 8-9 Curve showing the variation of generated emf in a single coil rotated in a uniform magnetic field.

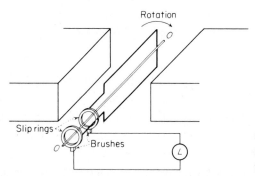

FIGURE 8-10 Slip rings and brushes are used to connect the coil to an external circuit.

slip rings, or *collector rings.* These rings are mounted on the generator shaft and rotate with the coil and the shaft so that two stationary brushes, one bearing on each ring, connect the coil to the external circuit as shown in Fig. 8-10. When the coil is rotated, the generated alternating emf causes a current to flow first in one direction and then the other through the coil and external circuit. Such a current is called an *alternating current.*

8-3 THE SIMPLE DC GENERATOR

It was shown in the preceding section that the emf generated in a rotating armature coil is alternating and that when this coil is connected to an external electric circuit through slip rings, the resulting current is an alternating current.

In a two-pole dc generator, the alternating current that flows in a single armature coil is converted into a direct or unidirectional current for the external circuit by the use of a split ring as shown in Figs. 8-11, 8-12, and 8-13. The two sections or segments that are insulated from each other and from the shaft form a simple *commutator.* Each end of the armature coil is connected to a segment. The action of the commutator is to reverse the armature coil connections to the external circuit at the same instant that the current reverses in the armature coil. This action is described more fully in the following paragraphs.

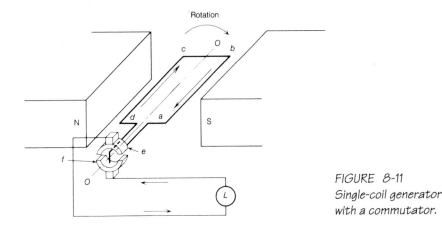

FIGURE 8-11
Single-coil generator
with a commutator.

When the armature coil is being turned in a clockwise direction, as in Fig. 8-11, emf is generated in the coil sides from *b* to *a* and from *d* to *c* as the arrows indicate. The circuit is completed by the lamp connected across the two brushes. The generated emf causes current to flow out through segment *e,* the upper brush, through the lamp, into the lower brush and segment *f,* and around the armature coil as indicated by the arrows. When the coil is turned 90° into the neutral plane as shown in Fig. 8-12, the generated emf and, hence, the current drop to zero. At this point the brushes make contact with both segments. As the coil is moved on past the neutral plane through another 90° to the position shown in Fig. 8-13, emf is generated from *a* to *b* and from *c* to *d.* However, the upper brush now makes contact with segment *f* and the lower brush with segment *e.* A comparison of Figs. 8-11 and 8-13 shows that while the current has reversed in the armature coil it flows out of the upper brush, through the lamp, and into the lower brush in each case. The emf and current drop to zero when the coil is

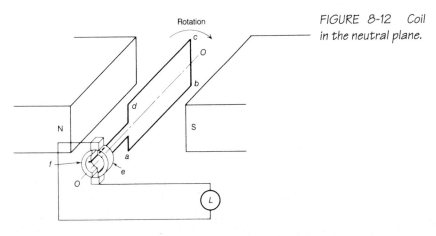

FIGURE 8-12 *Coil
in the neutral plane.*

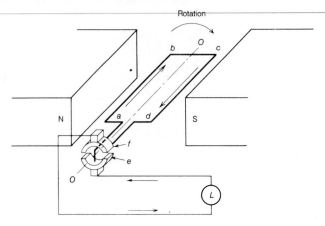

FIGURE 8-13   Coil rotated 180° from the position shown in Fig. 8-11.

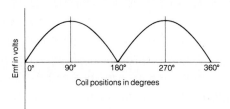

FIGURE 8-14   Variation in brush voltage of a single-coil generator with a commutator.

moved another 90° into the neutral plane, where the connections to the lamp are again reversed.

Although the current through the lamp is always in the same direction, it is not a steady current, since the emf, generated in the armature coil and applied to the brushes, varies from zero to a maximum and back to zero twice each revolution. The variation in brush voltage is shown graphically in Fig. 8-14.

A pulsating direct current such as is produced by a single-coil generator is not suitable for most commercial uses. However, by using a large number of coils and commutator segments, with the coils evenly distributed around the surface of the armature, the brush voltage may be made practically constant.

The voltage generated by a single-turn armature coil is small. For this reason, the coils used in commercial generators consist of several turns in series, thereby increasing the amount of generated emf in direct proportion to the number of turns in the coil.

8-4 MULTICOIL ARMATURES

The method of connecting several armature coils to form a continuous or closed armature winding may be illustrated by the ring winding represented in Fig. 8-15a. A ring winding is formed by winding the armature conduc-

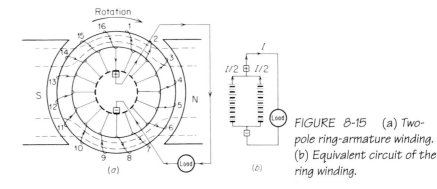

FIGURE 8-15 (a) Two-pole ring-armature winding. (b) Equivalent circuit of the ring winding.

tors around an iron ring or hollow iron cylinder as shown in the diagram. This is an early form of armature winding, which has been replaced by the more efficient, drum type of winding. However, the principle of the two windings is fundamentally the same, and the action of the ring winding is much more clearly represented by a diagram than that of the drum winding. Hence the ring winding is used here as an illustration.

Figure 8-15*a* represents a two-pole ring winding with 16 coils of one turn each. By starting at the bottom brush and tracing through the winding to the upper brush, it may be seen that there are two separate paths through the winding. Coils 8 to 1 form one path, and coils 9 to 16 form the other. When the armature is turned in a clockwise direction, emf is generated away from the reader in the conductors under the face of the north pole and toward the reader in the conductors under the face of the south pole, causing currents to flow in the directions indicated by the arrows. The emf generated in each path is additive; that is, the emf generated in the path 8 to 1 is the sum of the emfs generated in each of the eight coils. Likewise the emf in the path 9 to 16 is the sum of the emfs generated in coils 9 to 16. However, since the two paths are connected in parallel, the voltage between the two brushes is equal to the emf generated in either of the two parallel paths. For example, if the average emf generated in each of the 16 coils is 5V, then the average emf generated in each path is 40 V and the brush volage is 40 V.

Since the two paths through the armature winding are in parallel, any current supplied to the load divides equally between the two paths. For example, if the load current is 30 A, 15 A is supplied by each of the two halves of the armature winding.

The armature winding of Fig. 8-15*a* may be represented by the equivalent circuit of Fig. 8-15*b,* in which each of the 16 coils is represented by a battery cell. The emf generated in each path is the sum of the emfs generated in the eight cells in that path, and the brush voltage is equal to the voltage of either path.

Since current is assumed to flow from the positive terminal of a source through a circuit and back into the negative terminal, the upper brush in Fig. 8-15 is called the positive brush and the lower brush the negative brush.

The drum-type armature winding used in modern dc machines has the advantages of being easier to build and repair and of being more efficient than the ring winding. It is formed by placing the armature conductors, usually in the form of coils, in slots on the surface of a drum-shaped or cylindrical iron core. An armature core designed for drum winding is shown in Fig. 8-2 and a completed drum-wound armature is shown in Fig. 8-3.

In general there are two types of drum-armature windings: the *lap* and the *wave* windings. Each type of winding may be further classified as being a *simplex* (single) or a *multiplex* (double or triple) winding. Both lap and wave windings are closed-circuit windings; that is, they close upon themselves to form a closed electric circuit, just as does the ring winding shown in Fig. 8-15. Both windings are formed by interconnecting a number of separately insulated winding elements or coils, which have been laid in place in the armature core slots.

8-5 METHODS OF FIELD EXCITATION

The general types of dc generators take their names from the type of field excitation used. When a generator is excited from a storage battery or from a separate dc source, it is called a *separately excited* generator. If the field of a generator is connected in parallel with the armature circuit, it is called a *shunt* generator. When the field is in series with the armature, the generator is called a *series* generator. If both shunt and series fields are used, the generator is called a *compound* generator. Compound generators may be connected *short-shunt* with the shunt field in parallel with the armature only or *long-shunt* with the shunt field in parallel with both the armature and series field. Shunt, series, and compound generators are all classified as self-excited generators since they provide their own excitation.

The circuit diagrams of the various types of generators are shown in Fig. 8-16. Field rheostats, as shown in this figure, are adjustable resistances placed in the field circuits to provide a means of varying the field flux and thereby the amount of emf generated by the generator.

8-6 GENERATOR VOLTAGE EQUATIONS

The average generated emf in a generator depends upon the number of magnetic flux lines cut per second. One generator armature conductor making one revolution cuts $\phi \times p$ magnetic flux lines if ϕ is the number of lines per pole and p is the number of poles. When the speed n of the armature is given in revolutions per minute, the number of flux lines cut

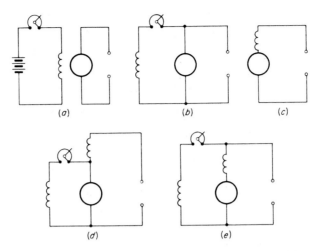

FIGURE 8-16   Circuit diagrams of dc generators: (a)
separately excited, (b) shunt, (c) series, (d) short-shunt
compound, and (e) long-shunt compound generators.

per sec is $\phi \times p \times n/60$. Taking into account the total number of armature conductors Z and the number of parallel paths though the armature b (depending on the type of armature winding), the average emf E may be calculated from the formula

$$E = \frac{p Z \phi n}{60b} \text{ volts} \tag{8-1}$$

For any given generator, all the factors in Eq. (8-1) are fixed values except the flux per pole, ϕ, and the speed n. Therefore Eq. (8-1) may be simplified to the form

$$E = K\phi n \tag{8-2}$$

where K is equal numerically to all the fixed values or constants for a given generator.

Equation (8-2) is merely a restatement of Faraday's law of induction: the value of an induced emf in any circuit is proportional to the rate at which flux is being cut by the circuit. Thus, if the flux per pole of a generator is doubled with the speed remaining constant, the generated emf will be doubled; or if the speed is doubled, the flux remaining constant, the generated emf will be doubled.

In practice, generators are usually operated at nearly constant speed, with the generated emf being adjusted to the desired value by adjusting the field flux. The field flux, being established by the field current, is controlled by the field rheostat.

EXAMPLE 8-1   When a generator is being driven at 1200 rpm, the generated emf is 125 V. What will be the generated emf (*a*) if the field flux is decreased by 10 percent with the speed remaining unchanged, and (*b*) if the speed is reduced to 1100 rpm, the field flux remaining unchanged?

$$(a) \ E_2 = 125 \times 0.90 = 112.5 \text{ V}$$

$$(b) \ \frac{E_1}{E_2} = \frac{n_1}{n_2} \quad \text{or} \quad \frac{125}{E_2} = \frac{1200}{1100}$$

$$E_2 = \frac{1100 \times 125}{1200} = 114.6 \text{ V}$$

The terminal voltage V_t differs from the generated emf of a generator by the voltage drop in the armature series circuit. The armature-circuit resistance consists of the resistance of the armature windings, the series field and commutating pole windings, if used, and the brushes—in other words, the entire resistance between the positive and negative terminals of the generator. At no load, the terminal voltage is equal to the generated emf since there is no IR drop in the armature circuit. However, load current flowing through the armature circuit produces an IR drop, which, of course, increases as the load current increases.

A convenient way of representing the relationship between the generated emf, the IR drop in the armature circuit, and the terminal voltage is shown in the equivalent circuit of Fig. 8-17. The generated emf E is represented as a storage battery with zero internal resistance and the entire armature-circuit resistance by the resistor R_a in series with the battery.

From this circuit, then,

$$V_t = E - I_a R_a \tag{8-3}$$

where I_a is the armature current. Substituting the value of E from Eq. (8-2) in Eq. (8-3) results in the equation

$$V_t = K\phi n - I_a R_a \tag{8-4}$$

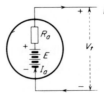

FIGURE 8-17   Equivalent circuit of a generator armature.

This equation is called the *fundamental generator equation* because it contains all the factors that govern the value of the terminal voltage. Note that there are three factors that may affect the generator terminal voltage: (1) ϕ, the flux per pole; (2) n, the speed in rpm; and (3) $I_a R_a$, the voltage drop in the armature circuit.

8-7 THE MAGNETIZATION CURVE

It was shown in Sec. 8-6 that the generated emf is proportional to both the flux per pole and the speed of the generator. If the speed is held constant, the generated emf then depends directly on the flux per pole. Since the flux is produced by the ampere-turns of the field coils, and since the number of turns on the field coils is constant, the flux must depend on the amount of field current flowing. However, the flux is not directly proportional to the field current at all times, owing to the fact that the magnetic circuit of the generator becomes saturated (see Sec. 6-9). The variation of the generated emf with the field current for a given generator driven at a constant speed and at no load may be shown by a curve called the *magnetization curve*. This curve is sometimes called the generator-saturation curve or the no-load characteristic.

The magnetization curve of a shunt generator is shown in Fig. 8-18. With zero field current, that is, with the field circuit open, the generated emf is equal to about 2 V. This is because of the weak flux produced by the residual magnetism in the poles. As the field current is increased, the generated emf increases in a nearly straight line up to about 102 V when the field current is 0.8 A. At this point, the magnetic circuit of the generator approaches saturation. Beyond this point, larger increases in field current are necessary to produce proportionate increases in generated emf. For example, before the field becomes saturated, an increase of field current from 0.4 to 0.6 A increases the generated emf from 60 to 84 V. However, after the field becomes saturated, an increase of field current from 1.2 to 1.4 A raises the generated emf from 125 to 132 V. An increase of 0.2 A in

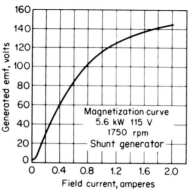

FIGURE 8-18   *Generator magnetization curve.*

field current in the first case causes an increase of 24 V, but the same increase in field current in the second case causes an increase in generated emf of only 7 V.

<table>
<tr><td>

8-8
COMMUTATION

</td><td>

The process of reversing the direction of the current in an armature coil as the commutator segments to which the coil is connected pass under a brush is called *commutation.* During the short time that the commutator segments to which a coil is connected are passing under the brush, the current must be completely reversed so that an arc is not formed as the commutator segments move from under the brush. Arcing at the brushes, if allowed to occur, causes excessive brush and commutator wear. This in turn increases the brush-contact resistance, with a resulting increase in generator losses. Arcing at the brushes, then, must be avoided.

</td></tr>
</table>

Since an armature coil is partially surrounded by iron, it has a considerable amount of inductance. It was shown in Chap. 7 that in an inductive circuit each change in current is opposed by an emf of self-induction. The emf of self-induction in an armature coil, then, opposes the reversal of the coil current while the coil is short-circuited by the brush and maintains a current flow around the short-circuited coil, thereby producing an arc as the commutator segments leave the brush.

Because the emf of self-induction in an armature coil is due to a property of the coil, namely, its inductance, it cannot be prevented. However, it can be neutralized by causing an emf to be induced in the coil *opposite* in polarity to the self-induced emf. This is accomplished by the use of *interpoles.*

Interpoles, or commutating poles, are small poles placed between the main poles of a generator and connected so as to have the same polarity as the following main pole in the direction of rotation. Thus, an armature coil undergoing commutation is caused to move across flux that is in the same direction as that of the following main pole, thereby generating emf in a direction opposite to that of the self-induced emf in the coil.

Since the amount of emf of self-induction present in an armature coil varies with the amount of armature current flowing, the amount of commutating flux necessary also varies. To make the commutating flux produced by the interpoles proportional to the armature current, interpole windings are connected in series with the armature circuit.

<table>
<tr><td>

8-9 VOLTAGE
REGULATION

</td><td>

As load is added to a generator, the terminal voltage will change unless some provision is made to keep it constant. A curve that shows the value of the terminal voltage for different values of load current is called the *load-*

</td></tr>
</table>

voltage characteristic or the *regulation curve*. The values of voltage for various load currents may be calculated or may be obtained from an actual test. Readings of current and voltage obtained from a test are taken at a constant speed and with the field excitation so adjusted that rated voltage is obtained at full load.

Voltage regulation is defined as the difference between the no-load and full-load terminal voltage of a generator and is expressed as a percentage of the full-load value, the full-load voltage being the rated or normal voltage of the generator. Thus

$$\text{Percent regulation} = \frac{\text{no-load voltage} - \text{full-load voltage}}{\text{full-load voltage}} \times 100$$

It is important to know how the terminal voltage of a generator varies with different amounts of load in order to determine the suitability of the generator for a specific use. For example, a generator to be used to supply a lighting circuit should have a very low percentage regulation; that is, its terminal voltage should be very nearly the same at full load as it is at no load. The characteristics of the various types of generators are discussed in the following sections.

8-10 THE SEPARATELY EXCITED GENERATOR

The separately excited generator is supplied field excitation from an independent dc source such as a storage battery or separate dc generator. The connection diagram is shown in Fig. 8-16a. A field rheostat is connected in series with the field to provide a means of varying the field excitation.

Figure 8-19 shows the regulation curve of a typical separately excited generator. It will be noted that as the load current increases the terminal voltage decreases. There are two important reasons for this decrease in terminal voltage:

1. The amount of effective field flux is reduced and the generated emf E thereby decreased by the demagnetizing effect of the armature current.

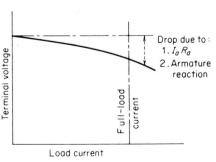

FIGURE 8-19   Regulation curve of a separately excited generator.

Armature currents establish an mmf that distorts and weakens the main field flux, especially in noninterpole machines. This effect of armature current on the field flux of a generator is called armature reaction.

2. There is a voltage drop due to the armature-circuit resistance.

These two factors affecting the terminal voltage are indicated in Fig. 8-19.

The separately excited generator has a decided advantage over the self-excited generator in that it will operate in a stable condition with any field excitation. Thus, a wide range of output voltages may be obtained.

The main disadvantage of a separately excited generator lies in the inconvenience and expense of providing the separate excitation source. For this reason the use of this type of generator is largely confined to use in experimental and testing laboratories and as a component in adjustable-voltage industrial drive systems.

8-11 THE SELF-EXCITED SHUNT GENERATOR

A self-excited generator depends on the residual magnetism of its field poles for its operation (see Sec. 6-10). Ordinarily there is some residual magnetism in the field poles if the poles have at a previous time been magnetized. If, then, a shunt generator is brought up to speed, the armture conductors will cut the small amount of flux present and a small amount of emf will be generated. As the shunt field is connected directly across the brushes (Fig. 8-16b), a current will flow in this winding. If the field resistance is sufficiently low and if the current is in such a direction that it increases the field strength, a higher emf is generated, which in turn causes more current to flow through the field windings. This again increases the field flux and generated emf. At first it would appear that this building-up process would go on indefinitely, but a limit is reached owing to the saturation of the magnetic circuit. As the generator magnetic circuit approaches saturation, smaller and smaller increases in generated emf produce smaller and smaller increases in field current until the building-up process stops. At this point of equilibrium, the field current produces just enough flux to generate the emf that causes the field current to flow. After the voltage has been built up, it may be adjusted to the desired value by an adjustment of the field rheostat.

The regulation curve of a typical shunt generator is shown in Fig. 8-20. This curve shows the performance of a shunt generator to be similar to that of the separately excited generator, except that the voltage of the shunt generator falls off more rapidly with the addition of load. As in the separately excited generator, armature reaction and IR drop in the armature circuit cause the terminal voltage to decrease. Since the

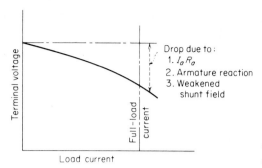

FIGURE 8-20 Regulation curve of a shunt generator.

field excitation is obtained from the generator itself, the decrease in terminal voltage, because of the above-mentioned effects, also decreases the voltage applied to the field circuit. This causes a decrease in field current and field flux, which in turn causes a further decrease in the terminal voltage. These three important components of voltage drop are indicated in Fig. 8-20.

The shunt generator may be used for supplying excitation to ac generators or in other applications where the distance from the generator to its load is short. It is also used for charging storage batteries.

8-12 THE SERIES GENERATOR

The series generator is a self-excited generator with armature, field windings, and load all connected in series (see Fig. 8-16c, p. 103). Thus, the field current and field flux are proportional to the load current. When the generator is running without load, there is a small emf generated owing to the residual magnetism. As an external load is applied and current begins to flow in the field windings, the terminal voltage of the generator is increased. The terminal voltage continues to increase until the magnetic circuit becomes saturated. A series generator regulation curve is shown in Fig. 8-21.

The series generator has very few practical applications. An understanding of the action of the series field is desirable, however, in the study of the compound generator.

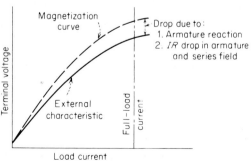

FIGURE 8-21 Regulation curve of a series generator.

8-13 THE COMPOUND GENERATOR

It was pointed out in Sec. 8-11 that the terminal voltage of a shunt generator drops considerably with the addition of load. Many applications, such as lighting circuits, require a practically constant source of voltage, and for such applications the shunt generator is unsuitable, especially if the amount of load is variable.

The decrease in terminal voltage of a shunt generator is due mainly to armature reaction and *IR* drop in the armature circuit. Both of these factors are proportional to the armature current. A generator may be designed to supply a constant terminal voltage automatically by the use of a series-field winding placed on the same poles as the shunt-field winding. The series field, which carries full armature current, supplies an emf to compensate for the decrease in terminal voltage due to armature reaction and armature-circuit *IR* drop. A generator with both a series- and a shunt-field winding is called a *compound generator.* When the series field is so connected that its ampere-turns act in the same direction as those of the shunt field, the generator is said to be a *cumulative-compound generator.*

The cumulative-compound generator combines the characteristics of the shunt and the series generators. At no load the shunt winding provides all the field flux, since there is no current in the series-field winding. As load is increased on the generator, the series field adds an increasing amount of field flux to that of the shunt field. The terminal voltage of the loaded generator will depend on the relative strength of the two fields. If the full-load terminal voltage is equal to the no-load terminal voltage, the generator is said to be *flat-compounded.* If the series ampere-turns at full load are more than enough to compensate for armature reaction and armature-circuit *IR* drop, the terminal voltage is higher than at no load and the generator is said to be *overcompounded.* Similarly, a generator is said to be *undercompounded* when the full-load terminal voltage is less than the no-load terminal voltage. Voltage regulation curves for over-, under-, and flat-compounded generators are shown in Fig. 8-22.

The compound generator is used more extensively than other types of generators because it may be designed to have a wide variety of characteristics. Overcompounded generators are used when they are located some

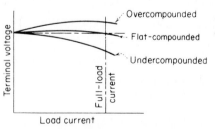

FIGURE 8-22 *Regulation curves of compound generators.*

distance from the loads that they supply. The increase in generator terminal voltage compensates for the voltage drop in the load feeder circuits.

When the shunt field of a compound generator is connected in parallel with the armature only, the connection is called a *short-shunt connection*. If the shunt field is connected in parallel with the circuit containing the armature and the series field, the connection is called a *long-shunt connection*. These connections are shown in Figs. 8-16*d* and 8-16*e*. The operating characteristics are practically the same with either connection.

8-14 EFFICIENCY

The efficiency of any machine is the ratio of the useful power output to the total power input. A part of the energy delivered to a generator or a motor is converted into heat and is therefore wasted. Thus, the useful output is never as large as the total input. There are always unavoidable losses even in the most perfectly constructed machines.

Efficiency is usually expressed as a percentage. Thus the ratio of output to input expressed as a decimal must be multiplied by 100 to change it to a percentage. Of course, output and input must be expressed in the same units. Thus, the definition of efficiency stated as a formula is

$$\text{Percent efficiency} = \frac{\text{output}}{\text{input}} \times 100$$

Suppose it is desired to find the efficiency of a generator whose input is 20 hp and whose terminal voltage is 240 V when it is supplying 50 A to a load. The output of the generator is

$$P = V_t I = 240 \times 50 = 12{,}000 \text{ W}$$

The input to the generator in watts is

$$\text{Input} = 20 \times 746 = 14{,}920 \text{ W}$$

The percent efficiency is

$$\frac{\text{Output}}{\text{Input}} \times 100 = \frac{12{,}000}{14{,}920} \times 100 = 80.4 \text{ percent}$$

8-15 LOSSES AND EFFICIENCY OF A DC MACHINE

The losses of generators and motors consist of the copper losses in the electric circuits of the machine and the mechanical losses due to the rotation of the machine. The important losses may be outlined as follows:

I. Copper losses
   A. Armature I^2R losses
   B. Field losses
      1. Shunt field I^2R
      2. Series field I^2R
      3. Interpole field I^2R

II. Mechanical or rotational losses
   A. Iron losses
      1. Eddy-current loss
      2. Hysteresis loss
   B. Friction losses
      1. Bearing friction
      2. Brush friction
      3. Windage or air friction loss

The copper losses are present because power is used whenever a current is made to flow through a resistance. When an armature current I_a flows through the armature-winding resistance R_a the power used is $I_a^2 R_a$. Likewise the power used in overcoming the resistances of the series field winding and the interpole winding is expressed as $I_a^2 R_s$ and $I_a^2 R_i$, where R_s and R_i are series-field and interpole-field resistances, respectively.

The shunt-field loss is equal to $V_t I_f$ where V_t is the terminal voltage of the machine and I_f is the shunt-field current. This includes the shunt-field rheostat loss.

As the armature rotates in a magnetic field, the iron parts of the armature move across flux lines as do the armature conductors. Since iron is a conductor of electricity, the emf induced in the iron parts causes currents to flow through these iron parts. These circulating currents are called *eddy currents,* and their flow causes the iron through which they flow to become heated. This heat is dissipated to the air and represents energy lost.

To increase the resistance of the paths followed by eddy currents, armature cores are made of laminated steel sheets, stacked and secured in a position perpendicular to the shaft and parallel to the direction of the magnetic field. A thin layer of insulation is placed between the laminations. This insulation may be a type of insulating varnish or merely the oxide that is allowed to form on the surfaces of each lamination. The eddy currents are therefore confined to the individual laminations, and their heating effect is greatly reduced. Laminating the armature core does not increase its reluctance, for the laminations are parallel to the magnetic field.

When a magnetic material is magnetized, first in one direction and then the other, an energy loss takes place owing to the molecular friction in the material. That is, the molecules of the material resist being turned first in one direction and then the other. Energy is thus expended in the material in overcoming this resistance. This loss is in the form of heat and is called *hysteresis loss.* Hysteresis loss is present in a rotating armature core, for the magnetism of the core is continually being reversed as the armature moves through the stationary magnetic field.

Other rotational losses are bearing friction, the friction of the brushes riding on the commutator, and air friction or windage.

Since it is wasteful and sometimes difficult to measure directly the input and output of dc machines, especially in the larger sizes, their efficiencies are often determined by determining their losses and calculating their efficiency.

1 What is *(a)* a generator; *(b)* a motor?

2 Name the parts of the magnetic circuit of a generator.

3 Name the parts that make up the electric circuits of a generator.

4 What are the neutral planes of a generator?

5 Explain why an alternating emf is generated in a coil that is rotated in a magnetic field.

6 If it is desired to bring alternating current out of a generator, what kind of a "take-off" system is used?

7 Describe the action of a commutator.

8 In a single-coil dc generator is the brush voltage constant? Why?

9 Why is more than one coil used on a generator armature?

10 Give three methods of providing field excitation for a dc generator.

11 What is the purpose of the field rheostat?

12 Name the factors that determine the amount of emf generated. Which are variable?

13 When a generator supplies load, the terminal voltage is not the same as the generated emf. Why?

14 Draw an equivalent circuit of a generator armature.

15 What is the fundamental generator equation?

16 The magnetization curve shows the relation between what two things?

17 Explain why the generator magnetization curve is not a straight line.

18 Why is a commutating flux necessary? How can it be obtained?

19 What is an interpole? What is its purpose? How is its winding connected?

20 The regulation curve shows the relation between what two factors?

21 Give the formula for finding the voltage regulation of a generator.

22 Give two reasons for the drop in terminal voltage of a separately excited generator as load is increased.

23 Explain the "building-up" process of a self-excited shunt generator.

24 A certain generator may be connected for either separate or self-excitation. Which of the two connections should be used to obtain the better voltage regulation?

25 Why is a small emf generated in a series generator at no load? What happens to the generated emf as load is added?

26 What is a compound generator? What is a cumulative-compound generator?

27 When is a generator said to be (*a*) flat-compounded; (*b*) overcompounded?

28 When might it be desirable to use an overcompounded generator?

29 Define efficiency. How is it usually expressed?

30 Name three classes of dc machine losses.

31 How are eddy-current losses reduced in dc machines?

PROBLEMS

8-1 How many amperes will a 50-kW 240-V dc generator deliver at full load?

8-2 What is the full-load kilowatt output of a dc generator if the full-load line current is 30 A and the terminal voltage is 115 V?

8-3 A shunt generator generates 100 V when its speed is 900 rpm. What emf does it generate if the speed is increased to 1200 rpm, the field flux remaining unchanged?

8-4 A generator generates an emf of 520 V, has 2000 armature conductors or inductors, a flux per pole of 0.013 weber, a speed of 1200 rpm, and the armature winding has four paths. Find the number of poles.

8-5 If the generated emf of a generator is 125 V and the *IR* drop in the armature circuit is 5 V, what is the terminal voltage?

8-6 A self-excited shunt generator delivers 20 A to a load. Its field current is 1.5 A. What is the armature current?

8-7 A 240-V shunt generator has a field-circuit resistance of 150 Ω. What is the field current when the generator operates at rated voltage?

8-8 A shunt-field winding of a 240-V generator has a resistance of 60 Ω. How much field-rheostat resistance must be added to limit the field current to 3 A when the generator is operating at rated voltage?

8-9 A 120-V shunt generator has a field-circuit resistance of 90 Ω. What is the armature current when the generator supplies 25 A to a load?

8-10 A shunt generator is rated 200 kW at 240 V. (*a*) What is the full-load current? (*b*) If the field-circuit resistance is 100 Ω, what is the field current? (*c*) What is the full-load armature current?

8-11 The terminal voltage of a shunt generator is 115 V when the generated emf is 119 V and the armature current is 20 A. What is the armature-circuit resistance?

8-12 The terminal voltage of a 75-kW shunt generator is 600 V at rated load. The resistance of the shunt-field circuit is 150 Ω, and the armature-circuit resistance is 0.1 Ω. Find the generated emf.

8-13 In a 50-kW 250-V shunt generator, 258 V are generated in the armature when the generator delivers rated load at rated voltage. The shunt-field current is 4 A. Find the resistance of the armature.

8-14 A load that has a resistance of 8 Ω is connected to a separately excited generator that has an armature-circuit resistance of 0.5 Ω. *(a)* What is the armature current when the generated emf is 130 V? *(b)* What is the generator terminal voltage?

8-15 A shunt generator has a full-load terminal voltage of 120 V. When the load is removed, the voltage increases to 150 V. What is the percentage voltage regulation?

8-16 A shunt generator has a field resistance of 50 Ω. When the terminal voltage of the generator is 120 V, the field current is 2 A. How much resistance is cut in on the shunt-field rheostat?

8-17 The no-load terminal voltage of a separately excited generator is 130 V. When the generator is supplying 50 A to a load, the speed and field excitation remaining unchanged, the terminal voltage is 115 V. If the armature-circuit resistance is 0.25 Ω, how much of the 15-V drop is due to the effect of armature reaction?

8-18 A shunt generator has an armature-circuit resistance of 0.4 Ω, a field-circuit resistance of 60 Ω, and a terminal voltage of 120 V when it is supplying a load current of 30 A. *(a)* Find the field current. *(b)* Find the armature current. *(c)* Find the copper losses at the above load. *(d)* If the rotational losses are 350 W, what is the efficiency at the above load?

8-19 Find the efficiency at full load of a 50-kW generator when the input is 80 hp.

8-20 The losses of a 20-kW generator at full load are 4500 W. *(a)* What is the efficiency at full load? *(b)* At no load?

8-21 A shunt generator requires 53-hp input from its prime mover when it delivers 150 A at 240 V. Find the efficiency of the generator.

8-22 The full-load losses of a 20-kW 230-V shunt generator are as follows:

Field I^2R loss	180 W
Armature I^2R loss	1200 W
Windage and friction losses	550 W
Iron loss	500 W

Find the efficiency at full load.

8-23 A shunt generator supplies a load with 106 A at 125 V. The shunt-field resistance is 31.2 Ω, and the armature resistance is 0.11 Ω. Find *(a)* the shunt-field copper loss and *(b)* the armature copper loss.

8-24 From the following data concerning a shunt generator, calculate the efficiency at full load:

Rated kilowatt output.......................... 10 kW

Rated voltage .... 230 V

Armature-circuit resistance .... 0.4 Ω

Field-circuit resistance........................ 192 Ω

Rotational losses at full load................... 750 W

8-25 A short-shunt–connected compound generator delivers 216 A to a load at 250 V. Its shunt-field resistance is 26.8 Ω, its shunt-field rheostat resistance is 6.2 Ω, its series-field resistance is 0.042 Ω, and its armature resistance is 0.096 Ω. Find the copper losses in *(a)* the shunt-field winding, *(b)* the shunt-field rheostat, *(c)* the series field, and *(d)* the armature winding.

8-26 A compound generator, short-shunt–connected, has a terminal voltage of 230 V when the line current is 50 A. The series-field resistance is 0.04 Ω. *(a)* Find the voltage drop across the series field. *(b)* Find the voltage across the armature. *(c)* Find the armature current if the shunt-field current is 2 A. *(d)* If the losses are 1950 W, what is the efficiency?

9
Alternating Current

Over 90 percent of the electric energy used in most countries is generated and distributed as alternating-current energy. Perhaps the greatest single reason for this widespread use of alternating current is the fact that an alternating voltage may be easily raised or lowered in value. Because of this, ac energy may be generated and distributed efficiently at a relatively high voltage and then reduced to a lower usable voltage at the point of use.

Direct-current voltages are not easily changed in value; hence dc energy is usually generated at the voltage at which it is to be used, or more often, ac energy is distributed to the load where it is converted to dc energy by means of rectifiers.

In addition to the fact that alternating voltages are easily changed in value, there are other reasons for the use of alternating currents. Alternating-current generators require no commutators with their accompanying troubles and maintenance. This means that commutation need not limit the size and speed of ac generators as it does dc generators. In general, ac motors and controls are simpler, lighter in weight, and more reliable than dc equipment of similar rating. The more extensive commercial development of ac equipment is still another reason for its use.

Although alternating current is used predominantly, there are some types of work for which direct current must be used or where it has a definite advantage. In mills and factories and in electric-railway service, where efficient speed control of motors is important, direct current has a definite advantage. Direct current is also commonly used for high-intensity light sources in searchlights and projectors, for charging storage batteries, for

117

field excitation of ac generators and synchronous motors, and for many electrochemical processes.

In this chapter, the basic ideas concerning alternating-current systems are presented. Single-phase and three-phase circuit principles are included in Chaps. 10 and 11.

9-1 GENERATION OF AN ALTERNATING EMF

It was shown in the preceding chapter that when an armature coil of a generator is rotated at a constant speed in a uniform magnetic field, an emf is generated in that coil in accordance with Faraday's law. This emf reverses its direction at regular intervals and is continually varying in strength. Such an emf is called an *alternating emf* and the value of this emf at any given time is called the *instantaneous value of emf* (symbol e). The manner in which an alternating emf is generated in a revolving generator armature coil is described further in the following paragraphs.

A simplified cross-sectional view of a two-pole generator is shown in Fig. 9-1a. It is assumed that the armature is being driven in a counterclockwise direction through a uniform magnetic field. For convenience 12 positions 30° apart are marked off on the armature surface. These are positions through which a given armature conductor or coil side will pass successively as the armature makes a complete revolution. Position 0 will be used as a starting point and the point from which all angles are measured. At position 1 the conductor will have moved 30°, at position 2 it will have moved 60°, etc.

Figure 9-1b is a graph showing the variation and direction of the instantaneous values of emf generated in an armature conductor as the armature is rotated. Twelve equally spaced points are marked off on the horizontal line XX' to correspond with the 12 positions that the conductor takes as it makes a complete revolution. The vertical lines at each point (e_1, e_2, e_3, etc.) represent the instantaneous values of the emf generated at each conductor position as the armature completes one revolution.

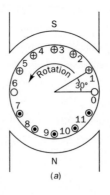

(a)

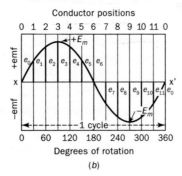

(b)

FIGURE 9-1 (a) Cross-section of a two-pole generator armature. (b) Curve showing variation of generated emf as armature is rotated through a uniform magnetic field.

The emf generated as the conductor is moved through position 0 is zero since a conductor in this position is in the neutral plane of the generator and no magnetic flux lines are being moved across. As the conductor is moved through points 1 and 2, it begins to move across the magnetic field, generating an emf, the direction of which is away from the reader in accordance with the right-hand rule. Vertical lines e_1 and e_2 on the graph represent the values of the instantaneous emf generated at points 1 and 2, respectively. At position 3, or 90°, the conductor is moving at right angles to the field, generating a *maximum value of emf* (symbol E_m), which is represented by the line e_3.

As the conductor is moved from position 3 around to position 6, the direction of the instantaneous emfs remain unchanged, but their values are diminishing as represented by the lines e_4 and e_5 on the graph. At position 6, or 180°, the emf is again zero. As soon as the conductor passes point 6, it begins to move across the field in the opposite direction, reversing the direction of the generated emf. The values of emf generated in this direction are represented below the line XX' on the graph.

At position 9, or 270°, the conductor is again moving across the field at a maximum rate and the emf, e_9 or $-E_m$, is at a maximum. From position 9 to position 0 the emf gradually decreases until it again becomes zero at 360° or the starting point. The cycle of events then repeats as the armature completes each successive revolution.

The curve plotted in Fig. 9-1*b* to represent the values of the emf generated at the various positions represents the variation of the emf generated in a conductor or coil side for one complete revolution in a two-pole generator. It is customary to refer to the values above the horizontal line XX' in such a graph as being in the positive direction and those below the line as being in the negative direction. Most generators generate values of emf whose positive values are exactly like the negative values. When the values of emf generated are plotted as in Fig. 9-1*b* the resulting curve has the same shape as a curve showing the values of the sine of the various angles of rotation made by the armature. For this reason the curve is called a *sine wave.* A brief discussion of trigonometric functions is given in Appendix G.

All ac circuit problems discussed in this text are based on the assumption that the alternating emfs and currents are sine waves.

9-2 ALTERNATING CURRENT

When an alternating emf is applied to a circuit, an alternating current is caused to flow. In general, if a sine-wave emf is applied, the resulting current varies as does a sine wave (transient state ignored). An alternating current may be defined as a current that flows back and forth through a circuit at regular intervals. An alternating current varies in intensity, as does an alternating emf. The value at any given time is called the instantaneous value (symbol i), and the greatest value reached by the current is called its maximum value (symbol I_m).

9-3 CYCLE; FREQUENCY

When the value of an alternating emf or current rises from zero to a maximum, falls to zero, increases to a maximum in the reverse direction, and falls back to zero again, the complete set of values passed through is called a *cycle* (see Fig. 9-1, p. 118). The number of complete cycles passed through in one second is called the *frequency*. The General Conference on Weights and Measures has adopted the name *hertz* (abbreviated Hz) as the unit of frequency. The term hertz supersedes the term *cycles per second* as a unit of frequency, although the latter term is still sometimes used.

The common power frequency used in North America is 60 Hz, but in Europe and in most of Asia and Africa it is 50 Hz. Many early power systems had rated frequencies of 25 Hz, but most of these systems have been converted to 50- or 60-Hz systems. The units kilohertz (kHz) and megahertz (MHz) are used in control, communication, and other high-frequency systems. One kHz and one MHz are equal to 1,000 and 1,000,000 Hz, respectively.

9-4 ELECTRICAL DEGREES: A MEASURE OF TIME

In any generator, a conductor must be moved past a north and a south pole to have one complete cycle generated in it. Therefore, the space through which a conductor must be moved to generate one cycle depends on the number of poles on the generator. In the two-pole generator, a conductor must make one complete revolution, or pass through 360° in space. In a four-pole generator, a conductor needs to be moved through only $1/2$ revolution, or 180°, to generate one cycle. It is evident then that if the number of space degrees passed through by a conductor were used in plotting the emf wave generated, the number of degrees per cycle would vary with the number of poles on the generator.

Since the time required to generate one cycle is constant for a given frequency, regardless of the number of poles on the generator, emf waves are plotted with time as the horizontal axis. Time in seconds may be used, but it is more convenient to divide the time required to generate one cycle into 360 divisions called *electrical degrees*. The relation of time in seconds and time in electrical degrees for a frequency of 60 Hz is shown in Fig. 9-2. The complete cycle is generated in $1/60$ s or in the time required to move an armature conductor past one pair of poles. For this frequency, 360 electrical degrees is equivalent to $1/60$ s. One-half cycle is generated in $1/120$ s or 180

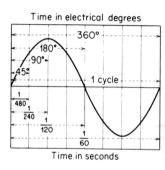

Time in electrical degrees

FIGURE 9-2   Relation between time in seconds and time in electrical degrees for a frequency of 60 Hz.

electrical degrees, $1/4$ cycle in $1/240$ s or 90°, and $1/8$ cycle in $1/480$ s or 45°. It must be kept in mind that electrical degrees used in this sense represent *time*. This concept will be found useful in specifying phase relationships.

9-5 EFFECTIVE VALUES OF ALTERNATING CURRENT AND EMF

As the value of an alternating current is continually fluctuating between a maximum in one direction and a maximum in the other direction, some effective value of an alternating current must be established. A natural question arises at this point. How can an alternating current deliver any power when it is continually changing its direction of flow? The answer to this is that the amount of power used by a dc circuit is $P = I^2R$ regardless of the direction of the current flow through the resistance. This is easily verified experimentally by connecting a resistance to a battery through a reversing switch and measuring the power used with the current first in one direction and then in the other. It will be found that the power is the same in either case. Alternating current is merely a current that reverses its direction of flow periodically. It follows then that power is used during both halves of the cycle when an alternating current flows through a resistance.

The rate at which heat is produced in a resistance forms a convenient basis for establishing an effective value of alternating current. *An alternating current is said to have an effective value of one ampere when it will produce heat in a given resistance at the same rate as does one ampere of direct current.*

The symbol for effective alternating current is I. The rate at which heat is produced in a dc circuit is equal to I^2R. Likewise, the rate at which heat is being produced by an alternating current at any instant is i^2R. Therefore, the heating effect of an alternating current over a period of time depends on the average of the squares of the instantaneous values of current over that period of time. For this reason, the effective value of a sine-wave current is often called the *root-mean-square*, or *rms*, value. It is the square root of the mean or average square of the instantaneous values. By squaring a number of instantaneous values, averaging these squared values, and extracting the square root of this average, the effective value of any sine-wave current may be found. By this method or by mathematical means, it may be shown that the *effective value (I) of any sine-wave current is always 0.707 times the maximum value (I_m)*.

Since alternating currents are caused to flow by alternating emfs, the ratio between effective and maximum values of emfs is the same as for currents. The effective, or rms, value (E) of a sine-wave emf is 0.707 times the maximum value (E_m).

When an alternating current or voltage is specified, it is always the effective value that is meant unless there is a definite statement to the contrary. In practical work, effective values are of more importance than instantaneous or maximum values and the common types of measuring instruments are calibrated to read effective values of current and voltage.

9-6 PHASE RELATIONS

Alternating currents are caused to flow by alternating voltages of the same frequency. When the current and voltage pass through their zero values and increase to their maximum values in the same direction at the same time, the current is said to be *in phase* with the voltage. However, in some types of circuits (to be discussed in later sections) the current and voltage zero and maximum values do *not* occur at the same time. When this happens, the current is said to be *out of phase* with the voltage.

There are three possible phase relations between a current and a voltage in a circuit:

1. The current and voltage may be in phase as shown in Fig. 9-3a.

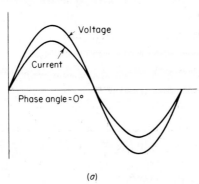

(a)

FIGURE 9-3   Current and voltage are in phase.

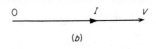

(b)

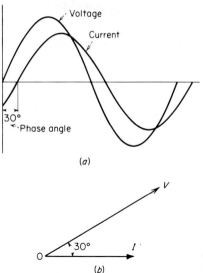

FIGURE 9-4   Current lags the voltage by 30°.

2. The voltage may pass through its zero value and increase to a maximum at some time earlier than the current as in Fig. 9-4*a*. In this case the current is said to *lag* the voltage.

3. The voltage may pass through its zero value and maximum value at some time later than the current as in Fig. 9-5*a*. In this case the current is said to *lead* the voltage.

The amount of time that a current leads or lags the voltage varies in different circuits from the *inphase* condition to a lead or lag of ¼ cycle, or 90°. Since time may be measured in electrical degrees (see Sec. 9-4), this

FIGURE 9-5   Current leads the voltage by 30°.

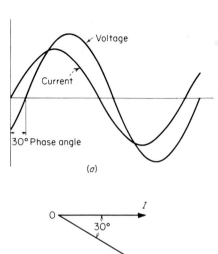

difference in time, or *phase difference,* of a current and voltage is commonly expressed in electrical degrees and is called the *phase angle.* The usual symbol for the phase angle is the Greek letter theta (θ).

Just as the current and voltage of a circuit may have a phase difference, two or more voltages or currents in the same circuit may be out of phase. This difference in phase must be taken into account when such voltages or currents are added or subtracted. Methods of combining currents and voltages that are out of phase will be discussed in Secs. 9-8 and 9-9.

9-7 PHASOR DIAGRAMS

It has been shown in previous sections how alternating currents and voltages and their phase relations may be represented by means of sine waves. This method of representation is quite cumbersome, and a simplified method is universally employed in ac circuit analysis in which currents and voltages are represented by straight lines having definite direction and length. Such lines are called *phasors* and diagrams in which these lines are used to represent sine-wave currents, voltages, and their phase relations are called *phasor diagrams.* Phasor diagrams may be drawn to represent either maximum or effective values of sine-wave quantities. Because effective values are of much more importance, however, phasor diagrams are nearly always drawn to represent effective values.

When drawing phasor diagrams, certain conventions need to be adopted to ensure consistent and accurate results. Here are some of the more common conventions or usual practices in regard to phasor diagrams:

1. It is common practice to consider the counterclockwise direction as being the positive direction of rotation of phasors. That is, phasors rotated in a counterclockwise direction from a given phasor are said to lead the given phasor, while phasors rotated in a clockwise direction are said to lag the given phasor.
2. For series circuits in which the current is common to all parts of the circuit, the current phasor is usually drawn on a horizontal line and used as a reference line for other phasors in the same diagram.
3. Since, in a parallel circuit, a common voltage is applied to all branches of the circuit, the voltage phasor is drawn on the horizontal reference line and other phasors are referred to the common voltage phasor.
4. It is not necessary to use the same scale for current and voltage phasors; in fact, it is often desirable to draw the current phasor to a larger scale than the voltage phasor when the values of current being represented are small. However, if several current phasors are to be used in the same diagram, they should all be drawn to the same scale.

Likewise, all voltage phasors in the same diagram should be drawn to the same scale.

5. To distinguish between current and voltage phasors, the current phasors may be drawn with closed arrowheads and the voltage phasors with open arrowheads.

To illustrate the methods used in drawing phasor diagrams, the sine-wave currents and voltages of Figs. 9-3a, 9-4a, and 9-5a are represented by phasor diagrams in Figs. 9-3b, 9-4b, and 9-5b, respectively.

Figure 9-3 (see p. 122) represents a circuit in which the current and voltages are in phase; that is, the phase angle is 0°. The phasor diagram showing this relation is drawn in Fig. 9-3b. The phasor representing the effective current is drawn to some convenient scale on a horizontal line, starting at the point O. Likewise, the phasor representing the effective voltage is drawn starting at the same point O and extending in the same direction as the current phasor. The fact that the two phasors lie along the same line indicates that the current and voltage being represented are in phase.

In Fig. 9-4 (see p. 123) the current is shown lagging the voltage by 30°. The current phasor is drawn as before on the horizontal reference line. However, to represent the fact that the current lags the voltage by 30°, the voltage phasor is drawn 30° ahead of (or in the counterclockwise direction from) the current phasor as in Fig. 9-4b.

Figure 9-5 (see p. 123) represents a circuit in which the current leads the voltage by 30°. The current phasor is drawn on the horizontal reference line as before but with the voltage phasor 30° behind (or in the clockwise direction from the current phasor) as in Fig. 9-5b.

Thus, phasor diagrams, when drawn to scale, provide a simple means of representing both the phase relations and the magnitudes of the currents and voltages involved in a given circuit.

9-8 ADDITION OF PHASORS

A phasor has both direction and magnitude. Therefore, the addition of quantities that are represented as phasors, such as alternating currents and voltages, must take into account both the direction and magnitude of those quantities.

The phasor sum of two phasors may be found by constructing a triangle, the two phasors forming two sides of the triangle and their sum forming the third side. To construct such a triangle, the first phasor is drawn to scale starting at some point O. For convenience, this phasor is usually drawn as a horizontal line to the right of the origin O. The second phasor is then drawn to scale from the arrow end of the first phasor. If the second phasor is defined as leading the first phasor by θ degrees, the second phasor is drawn θ degrees in a counterclockwise direction from the first phasor. If it lags the first phasor, it is drawn at the designated angle in a clockwise direction from the horizontal. The phasor sum is then the line drawn from the origin to the arrow end of the second phasor.

Following are several examples that illustrate the addition of two phasors. Several special conditions are illustrated that are often encountered. However, it should be noted that the procedure in constructing each of the diagrams is the same. The two original phasors are placed without changing their direction so as to form two sides of a triangle. The third side of the triangle is then the phasor sum of the two original phasors.

Addition of Inphase Phasors. The phasor sum of two inphase phasors is their direct or arithmetic sum.

EXAMPLE 9-1 Find the phasor sum of a current of 3 A and a current of 4 A, the two currents being in phase.

Phasor sum = $3 + 4 = 7$ A

Addition of Phasors Out of Phase by 180°. The phasor sum of two phasors 180° out of phase is their numerical difference, the direction of the result being the same as the larger phasor.

EXAMPLE 9-2 A voltage V_1 of 125 V leads a voltage V_2 of 100 V by 180°. Find their phasor sum.

Phasor sum = $125 - 100 = 25$ V

The direction of the resultant phasor is the same as the direction of V_1.

Addition of Phasors Out of Phase by 90°. The phasor sum C of two phasors A and B that are out of phase 90° is

$$C = \sqrt{A^2 + B^2}$$

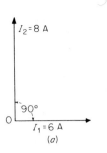

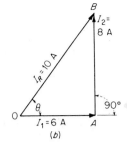

EXAMPLE 9-3

A current I_2 of 8 A leads a current I_1 of 6 A by 90° as shown in Fig. 9-6a. Find their phasor sum.

GRAPHICAL SOLUTION: Lay off I_1 to scale along the horizontal to the right of the origin as in Fig. 9-6b. Starting at the arrow end of I_1, lay off to scale I_2, 90° ahead of or in a counterclockwise direction from I_1. The phasor sum is the length of the line OB. The resultant phasor I_R leads I_1 by the angle θ.

MATHEMATICAL SOLUTION: Since the triangle OAB is a right triangle,

$$I_R = \sqrt{(I_1)^2 + (I_2)^2} = \sqrt{6^2 + 8^2} = 10 \text{ A}$$

and $\tan \theta = \dfrac{8}{6} = 1.33$

From tables or by calculator $\theta = 53.13°$. (For trigonometric functions and a table of values, see Appendixes G and H.)

Addition of Phasors Out of Phase by Any Angle.

EXAMPLE 9-4

Add two voltages of 100 V each that are out of phase by 30°.

GRAPHICAL SOLUTION: The two phasors drawn from the same origin are shown in Fig. 9-7a. Lay off to scale one phasor on the horizontal to the right of the origin as in Fig. 9-7b. Starting at the arrow end of the first phasor, scale off the second phasor 30° in a counterclockwise direction from the horizontal. Draw a line from the origin O to the end of the second phasor. The length of this line is the phasor sum, to scale, of the two phasors. Measure angle θ.

**EXAMPLE 9-4
(CONTINUED)**

MATHEMATICAL SOLUTION: In Fig. 9-7b, drop a perpendicular from C to the horizontal. The phasor sum OC is then the hypotenuse of the right triangle OBC.

$$\text{Side } CB = 100 \sin 30° = 100 \times 0.5 = 50$$

$$\text{Line } AB = 100 \cos 30° = 100 \times 0.866 = 86.6$$

$$\text{Side } OB = OA + AB = 100 + 86.6 = 186.6$$

$$\text{Side } OC = \sqrt{(OB)^2 + (CB)^2}$$

$$= \sqrt{(186.6)^2 + (50)^2} = 193.2 \text{ V}$$

$$\tan \theta = \frac{50}{186.6} = 0.268$$

$$\theta = 15°$$

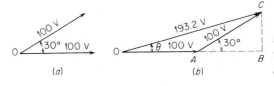

FIGURE 9-7 Addition of two phasors out of phase by 30°. See Example 9-4.

(a)   (b)

EXAMPLE 9-5

A voltage of 120 V leads a voltage of 240 V by 120° (Fig. 9-8a). What is their phasor sum?

GRAPHICAL SOLUTION: The graphical construction is shown in Fig. 9-8b. The phasor sum is line OB at the angle θ from the horizontal.

MATHEMATICAL SOLUTION: In Fig. 9-8b, draw the perpendicular line BA. The phasor sum OB is then the hypotenuse of right triangle OAB.

$$\text{Side } AB = 120 \sin 60° = 120 \times 0.866 = 103.9$$

$$\text{Line } AC = 120 \cos 60° = 120 \times 0.5 = 60$$

$$\text{Side } OA = OC - AC = 240 - 60 = 180$$

$$\text{Side } OB = \sqrt{(OA)^2 + (AB)^2}$$

$$= \sqrt{(180)^2 + (103.9)^2} = 207.5 \text{ V}$$

$$\tan \theta = \frac{103.9}{180} = 0.577$$

$$\theta = 30°$$

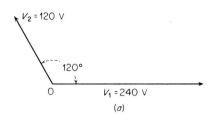

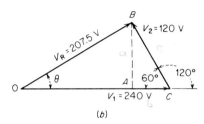

FIGURE 9-8 Addition of two phasors out of phase by 120°. See Example 9-5.

If three phasors are to be added, the sum of two of the phasors may be found by the methods outlined above, and their sum may then be added to the third phasor. This method may be extended to any number of phasors.

9-9 SUBTRACTION OF PHASORS

A phasor B may be subtracted from a phasor A by reversing the phasor B (rotating it by 180°) and adding it to the phasor A. This can best be explained by an example.

EXAMPLE 9-6

Subtract a current of 10 A from a current of 17.3 A when the 17.3-A current leads the 10-A current by 30° (Fig. 9-9a).

GRAPHICAL SOLUTION: Draw I_1, I_2, and $-I_2$ as shown in Fig. 9-9b. The phasor-difference current is the line OB, which leads I_2 by the angle θ.

MATHEMATICAL SOLUTION: In Fig. 9-9b, drop a perpendicular from B to the horizontal. The line OB is then the hypotenuse of the right triangle OBC.

Side $BC = 17.3 \sin 30° = 17.3 \times 0.5 = 8.66$

Side $AC = 17.3 \cos 30° = 17.3 \times 0.866 = 15$

Side $OC = 15 - 10 = 5$

Line $OB = \sqrt{(8.66)^2 + (5)^2} = 10$ A

$\tan θ = \dfrac{8.66}{5} = 1.73$

$θ = 60°$

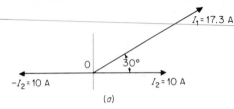

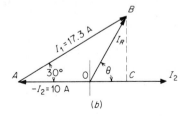

FIGURE 9-9 Subtraction of two phasors out of phase by 30°. See Example 9-6.

For those who may wish to use the complex quantity notation in the study of ac circuits in this book and for those who wish to extend their study beyond the scope of this book, a brief discussion of the complex quantity notation used to represent phasors is presented in Appendix I.

9-10 PROPERTIES OF AN ELECTRIC CIRCUIT The value of a direct current flowing in a circuit for a given constant value of impressed voltage is determined wholly by the *resistance* of the circuit. The value of an alternating current flowing in a circuit depends not only on the resistance of the circuit, but on the *inductance* and *capacitance* of the circuit as well. Resistance offers the same kind of opposition to the flow of an alternating current as it does to direct current, and for most practical purposes, the ac resistance and dc resistance of a circuit are considered to be the same. Inductance and capacitance affect only a *change* in current or voltage. Since alternating currents and voltages are continually changing in magnitude and direction, both inductance and capacitance affect the flow of an alternating current. The nature of the circuit properties and the effects of each in ac circuits are discussed in the following sections.

9-11 RESISTANCE IN AN AC CIRCUIT When a dc potential is applied to a resistance, the value of the current that flows, according to Ohm's law, is proportional to the voltage: the greater the voltage, the greater the current. The same is true when an alternating voltage is applied to a resistance. As the voltage is increased, the current is increased; as the voltage is decreased, the current is decreased. The maximum value of current and voltage occur at the same time, and their zero values occur at the same time. Thus, the current and the voltage are in phase and the *phase angle is zero in a purely resistive circuit*.

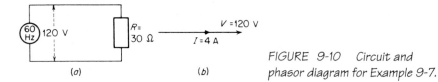

FIGURE 9-10   Circuit and phasor diagram for Example 9-7.

When a circuit contains resistance only, the effective current flowing in that circuit is equal to the effective applied voltage divided by the resistance of the circuit, or

$$I = \frac{V}{R} \tag{9-1}$$

What is the effective current flowing through a resistance of 30 Ω when a 120-V 60-Hz voltage is applied (Fig. 9-10a)?   **EXAMPLE 9-7**

$$I = \frac{V}{R} = \frac{120}{30} = 4\,\text{A}$$

The phasor diagram is shown in Fig. 9-10b.

9-12
INDUCTANCE
Inductance is a circuit property just as is resistance. Any circuit that is capable of producing flux has inductance, as was pointed out in Chap 7. Any change in the value of current flowing in such a circuit also causes a change in the flux linking the circuit, thereby causing an emf to be induced. Since the emf is induced in the circuit owing to a change of current in the circuit itself, and since any induced emf opposes the change that produces it, it is called the *counter emf of self-induction.* Thus, an opposition to any *change* in the current flowing is present in a circuit containing inductance.

The unit of inductance is the *henry,* and the symbol is *L.* As was stated in Sec. 7-7, a circuit has an inductance of one henry if, when the current is changing at the rate of one ampere per second, the average counter emf induced is one volt.

An inductive circuit is usually in the form of a coil, very often wound around a magnetic core. Motor, generator, and transformer windings are examples of circuits that have inductance. However, a long, straight wire also has a certain amount of inductance, and calculations of transmission-line performance must take into account the inductance of the line.

When an electric circuit is energized with a dc potential, the current increase from zero to its steady-state value is opposed by the circuit emf of self-induction. Similarly, when the circuit is deenergized, the current decrease from its steady-state value to zero is opposed by the circuit emf of self-induction. However, the flow of a steady direct current in the circuit is opposed only by the circuit resistance.

When an inductive circuit is energized by an alternating emf, the resulting current is changing in value continuously and therefore is continuously inducing an emf of self-induction. Because an induced emf opposes the continuous change in the current flowing, its effect is measured in ohms. This opposition of an inductive circuit to the flow of alternating current is called *inductive reactance* and is represented by the symbol X_L. The current through a circuit that contains only inductive reactance is

$$I = \frac{V}{X_L} \qquad (9\text{-}2)$$

where I = effective current, amperes
X_L = inductive reactance, ohms
V = effective voltage across the reactance, volts

The value of inductive reactance in any circuit depends on the inductance of the circuit and on the rate at which the current through the circuit is changing. The rate of change of current depends on the frequency of the applied voltage. Inductive reactance in ohms may be calculated from the formula

$$X_L = 2\,\pi f L \qquad\qquad X_L \text{ measured in} \qquad (9\text{-}3)$$
$$\text{ohms.}$$

where f = frequency, hertz
L = inductance, henries
$\pi = 3.14$

(Note: The decimal equivalent of π [pi] correct to five decimal places is 3.14159. The simplified value used in this text is 3.14.)

The amount of any induced emf in a circuit depends on how fast the flux that links the circuit is changing. In the case of self-induction, a counter emf is induced in a circuit because of a change of current and flux in the circuit itself. This counter emf, according to Lenz's law, opposes any change in current, and its value at any time depends on the rate at which the current and flux are changing at the time. In an inductive circuit in which the resistance is negligible in comparison to the inductive reactance, the voltage applied to the circuit must at all times be equal and opposite to the emf of self-induction.

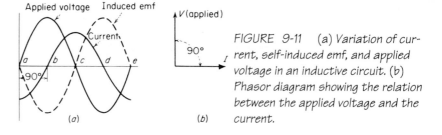

Applied voltage　Induced emf

Current

(a)

V (applied)

90°

I

(b)

FIGURE 9-11　(a) Variation of current, self-induced emf, and applied voltage in an inductive circuit. (b) Phasor diagram showing the relation between the applied voltage and the current.

Figure 9-11 shows the variation of the induced counter emf with the variation of a sine-wave current in an inductive circuit. At points a, c, and e the current is not changing in value, and the self-induced emf is zero. Between the points a and c the current is increasing in value from its negative maximum to its positive maximum. Therefore, during this time, the induced emf is negative in value, opposing the increase in current. Between the points c and e, the current is decreasing from positive maximum to negative maximum so that the induced emf is positive and is attempting to oppose the decrease in current. The current changes at its greatest rate at the points b and d, causing the induced emf to reach its maximum values at these points. The value of the self-induced emf then varies as a sine curve and lags the current by 90° as shown in Fig. 9-11. Since the applied voltage must be equal and opposite to the self-induced emf at all times, it is evident from Fig. 9-11 *that the current lags the applied voltage by 90° in a purely inductive circuit.* If, as is usually the case, resistance is present in the circuit, the angle of lag will be less than 90°, depending on the relative values of resistance and inductive reactance in the circuit.

A 0.2-H choke coil with negligible resistance is connected to a 120-V 60-Hz supply (Fig. 9-12a). Find (a) the inductive reactance of the coil, and (b) the current flowing; (c) draw the phasor diagram showing the phase relations between the current and the applied voltage.

EXAMPLE 9-8

(a) $X_L = 2\pi fL = 2 \times 3.14 \times 60 \times 0.2 = 75.4 \ \Omega$

(b) $I = \dfrac{V}{X_L} = \dfrac{120}{75.4} = 1.59 \ A$

(c) The phasor diagram showing the current lagging the voltage by 90° is drawn in Fig. 9-12b.

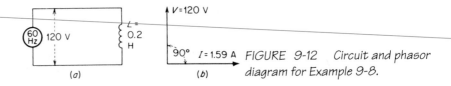

FIGURE 9-12 *Circuit and phasor diagram for Example 9-8.*

9-14
CAPACITORS

Any two conductors that are separated by an insulating material form a simple *capacitor*. The conductors are called the *plates* of the capacitor, and the insulating material is called the *dielectric*. A type of capacitor often used in power and communication systems has plates of aluminum foil interleaved with sheets of dielectric material such as paper or a plastic material. Alternate plates are connected to each of two terminals so that each terminal and plate to which it is attached is separated electrically from the other plate. Thus each group of plates forms a large conductor surface that is separated from another large conductor surface by the dielectric. The interleaved plates and dielectric material are formed into cylindrical rolls and sealed into metal or paper containers. Capacitors suitable for high-voltage service are placed in metal containers filled with an insulating fluid before they are sealed.

Capacitors are used extensively in telephone and other communication circuits. Electric utility companies are using them in increasing numbers on transmission and distribution circuits to correct for some of the effects of inductive loads. Capacitors play an important part in the operation of various other kinds of electric apparatus, such as automotive ignition systems, motors, and control apparatus.

Figure 9-13 shows a capacitor *AB* connected in series with two galvanometers (sensitive current-detecting instruments) *C* and *D*, with the switch *S* arranged either to connect the capacitor to the battery or to short-circuit the terminals of the capacitor. Immediately after the switch is closed to the left, connecting the capacitor to the battery, galvanometer *C* indicates a momentary current flow toward plate *A* and galvanometer *D* indicates a current flow away from plate *B*. A movement of electric charge or electrons takes place in the circuit until the potential difference between plates *A* and *B* becomes the same as the battery potential difference. This causes plate *A* to become positively charged and plate *B* to become negatively charged. Hence, a capacitor in this condition is said to be charged. If the switch is opened, the capacitor retains its charge indefinitely (if there is no leakage between the plates), since plate *B* is left with an excess of electrons and plate *A* has a deficiency of electrons.

If the switch *S* is closed to the right while the capacitor is still charged, the galvanometers indicate a flow of current out of plate *A* and

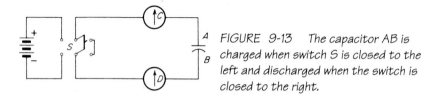

FIGURE 9-13 *The capacitor AB is charged when switch S is closed to the left and discharged when the switch is closed to the right.*

into plate *B*. This flow of current is momentary, being only long enough to neutralize the charged atoms on each plate. The capacitor is then said to be *discharged*.

The amount of charge that a capacitor receives for each volt of applied potential is called the *capacitance* of the capacitor. The unit of capacitance is the *farad* (F), and the symbol is *C*. A capacitor has a capacitance of one *farad* when an applied potential of one *volt* causes the capacitor to take a charge of one *coulomb*. The farad is much too large a unit for practical use. Therefore the smaller units of *microfarad* (μF) and *picofarad* (pF) are usually used. One microfarad is 0.000 001 farad and one picofarad is 0.000 000 000 001 farad.

The capacitance of a capacitor is directly proportional to the area of the plates and inversely proportional to their separation. It has been found by experiment that the material used for the dielectric also affects the capacitance. For example, when glass is substituted for air as a dielectric, the capacitance increases approximately eight times. Likewise capacitors using mica, paraffin, transformer oil, paper, and various other insulating materials for dielectrics have a higher capacitance than do similar air capacitors. These materials are said to have a higher *dielectric constant* than air.

One ampere is defined as the rate of flow of an electric charge of one coulomb per second. If the voltage applied to a one-farad capacitor is increased at the rate of one volt every second, the increase in charge on the capacitor will be one coulomb every second. Since one coulomb of electricity or electric charge is moved every second, the average current flowing into the capacitor must be one ampere. Likewise, if the potential being applied to a one-farad capacitor is being decreased by one volt every second, the average current flowing out of the capacitor will be one ampere. Thus a current flows in a circuit containing a capacitor while the applied voltage of that circuit is changing, even though there is a discontinuity in the circuit. *The amount of current flowing depends on the capacitance of the circuit and the rate at which the applied voltage is changing.* If either the capacitance of a circuit or the rate at which the applied voltage is changing (the frequency) is increased, the amount of current flowing is also increased.

9-15 CAPACITORS IN SERIES

It may be shown mathematically that the combined capacitance C of several capacitors C_1, C_2, C_3, . . . in series may be found by the relationship

$$\frac{1}{C} = \frac{1}{C_1} + \frac{1}{C_2} + \frac{1}{C_3} + \cdots \qquad (9\text{-}4)$$

The reciprocal of the combined capacitance of several capacitors in series is equal to the sum of the reciprocals of the capacitances of the separate capacitors.

9-16 CAPACITORS IN PARALLEL

When capacitors are connected in parallel, the effect is the same as increasing the number of plates, and the capacitance of the circuit is increased. In a parallel circuit the combined capacitance of several capacitors C_1, C_2, C_3, . . . is

$$C = C_1 + C_2 + C_3 + \cdots \qquad (9\text{-}5)$$

When capacitors are connected in parallel, their total capacitance equals the sum of the capacitances of the separate capacitors.

9-17 ACTION OF A CAPACITOR IN AN AC CIRCUIT

In a dc circuit containing a capacitor, there is only a momentary flow of current immediately after the voltage is applied. As soon as the voltage across the capacitor reaches its steady-state value, the current flow ceases. However, when a capacitor is connected to an ac supply in series with an ac ammeter, the ammeter indicates a flow of current as long as the supply is connected. The circuit appears to be complete and not one that has a discontinuity in it. The reason for this is that the current alternately flows into and out of the capacitor as the capacitor is charged, discharged, charged in the opposite direction, and discharged again.

The variation of an alternating voltage applied to a capacitor, the charge on the capacitor, and the current flowing are represented in Fig. 9-14. Since the current flowing in a circuit containing capacitance depends on the rate at which the voltage changes, the current flow is greatest at points a, c, and e; at these points the voltage is changing at its maximum rate. Between the points a and b, while the voltage and charge on the capacitor are both increasing, the current flow is into the capacitor but decreasing in value. At point b the capacitor is fully charged, and the current is zero. From b to c, the voltage and charge both decrease as the capacitor discharges, its current flowing in a direction opposite to the voltage. From c to d, the capacitor begins to charge in the opposite direction and the voltage and current are again in the same direction. The capacitor is fully charged at d, where the current flow is again zero. From d to e the capacitor discharges, the flow of current being opposite to the voltage. The

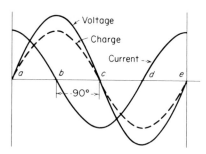

FIGURE 9-14 *Variation of voltage, charge, and current in a capacitor.*

cycle then repeats. Fig. 9-14 shows the current to be *leading* the applied voltage by 90°. This is true in any purely capacitive circuit and is in contrast to a purely inductive circuit in which the current *lags* the voltage by 90°. Therefore, the effect of capacitance in a circuit is exactly opposite to the effect of inductance.

9-18 CAPACITIVE REACTANCE

Capacitive reactance is the opposition offered by a capacitor or by any capacitive circuit to the flow of current. It was shown in Sec. 9-14 that the current flowing in a capacitive circuit is directly proportional to the capacitance and to the rate at which the applied voltage is changing. The rate at which the voltage changes is determined by the frequency of the supply. Therefore, if either the frequency or the capacitance of a given circuit is increased, the current flow increases. This is equivalent to saying that if either the frequency or capacitance is increased, the opposition to the current flow is *decreased*. Therefore capacitive reactance, which is the opposition to current flow, is *inversely* proportional to frequency and capacitance. Capacitive reactance, X_c ,is measured in ohms, as are resistance and inductive reactance, and may be calculated by the formula

$$X_C = \frac{1}{2\pi fC} \tag{9-6}$$

where f = frequency, hertz
$\pi = 3.14$
C = capacitance, farads

When the capacitance is expressed in microfarads $C_{\mu F}$, then

$$X_C = \frac{1,000,000}{2\pi fC_{\mu F}} \text{ or } \frac{10^6}{2\pi fC_{\mu F}} \tag{9-7}$$

Likewise, when the capacitance is expressed in picofarads C_{pF}, then

$$X_C = \frac{1,000,000,000,000}{2\pi fC_{pF}} \text{ or } \frac{10^{12}}{2\pi fC_{pF}} \tag{9-8}$$

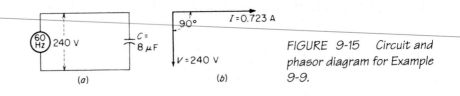

FIGURE 9-15 Circuit and phasor diagram for Example 9-9.

The current flowing in a circuit containing only capacitive reactance is

$$I = \frac{V}{X_C} \qquad (9\text{-}9)$$

where I = effective current, amperes
V = effective voltage across the capacitive reactance, volts
X_C = capacitive reactance, ohms

EXAMPLE 9-9 An 8-μF capacitor is connected to a 240-V 60-Hz circuit (Fig. 9-15). (a) Find the capacitive reactance, (b) find the current flowing, and (c) draw the phasor diagram.

(a) $X_C = \dfrac{10^6}{2\pi f C_{\mu F}} = \dfrac{10^6}{2 \times 3.14 \times 60 \times 8} = 331.7 \ \Omega$

(b) $I = \dfrac{V}{X_C} = \dfrac{240}{331.7} = 0.723 \ \text{A}$

(c) The phasor diagram showing the current leading the voltage by 90° is drawn in Fig. 9-15b.

9-19 REACTANCE Inductive reactance causes the current to *lag* behind the applied voltage, while capacitive reactance causes the current to *lead* the voltage. Therefore, when inductive reactance and capacitive reactance are connected in series, their effects neutralize each other and their combined effect is then their difference. The combined effect of inductive reactance and capacitive reactance is called *reactance* and is found by subtracting the capacitive reactance from the inductive reactance, or as an equation

$$X = X_L - X_C$$

where X = reactance, ohms
X_L = inductive reactance, ohms
X_C = capacitive reactance, ohms

9-20 IMPEDANCE *Impedance* is the total opposition offered by a circuit to the flow of current. It is the combined effect of the *resistance* and the *reactance* of a circuit. Impedance, symbol Z, is measured in ohms. The impedance of an ac circuit is equal to the effective applied voltage divided by the effective current that flows, or

$$Z = \frac{V}{I} \tag{9-11}$$

also
$$V = IZ \quad \text{and} \quad I = \frac{V}{Z}$$

In a circuit containing resistance only, the IR drop is in phase with the current. In a circuit containing inductive reactance only, the IX_L drop leads the current by 90°, which is, of course, equivalent to saying that the current lags the IX_L drop by 90°. Likewise, in a capacitive circuit, the IX_C drop lags the current by 90°. In a circuit containing both resistance and reactance, the total voltage drop, or the IZ drop, is equal to the sum of IR and IX drops. Since the IR and IX drops are out of phase by 90°, this phase difference must be taken into account when they are added. As was shown in Sec. 9-8, the sum of two phasor quantities such as IR and IX that are out of phase by 90° is $\sqrt{(IR)^2 + (IX)^2}$. Thus,

$$IZ = \sqrt{(IR)^2 + (IX)^2}$$

and canceling I from each term results in

$$Z = \sqrt{R^2 + X^2} \tag{9-12}$$

Because of their different effects, then, resistance and reactance cannot be added arithmetically but must be combined according to the relationship given in Eq. (9-11).

Since reactance has been defined as $X_L - X_C$, the complete expression for the impedance of a series circuit is

$$Z = \sqrt{R^2 + (X_L - X_C)^2} \tag{9-13}$$

When a circuit contains a negligible amount of X_C compared with its R and X_L, the above expression reduces to

$$Z = \sqrt{R^2 + (X_L - 0)^2} = \sqrt{R^2 + X_L^2}$$

Likewise, when a circuit contains a negligible amount of X_L, the impedance becomes

$$Z = \sqrt{R^2 + (0 - X_C)^2} = \sqrt{R^2 + (-X_C)^2}$$

The minus sign has no effect on the magnitude of Z, since a minus number squared is equal to a positive number.

EXAMPLE 9-10

A resistance of 30 Ω is connected in series with an inductive reactance of 60 Ω and a capacitive reactance of 20 Ω. What is the impedance of the circuit?

$$Z = \sqrt{R^2 + (X_L - X_C)^2} = \sqrt{(30)^2 + (60 - 20)^2}$$

$$= \sqrt{900 + 1600} = \sqrt{2500} = 50 \ \Omega$$

9-21 GRAPHICAL REPRESENTATION OF RESISTANCE, REACTANCE, AND IMPEDANCE

The relations expressed by the formula

$$Z = \sqrt{R^2 + X^2}$$

may be represented by a right triangle, where Z is the hypotenuse and R and X form the other two sides of the triangle, since the hypotenuse of a right triangle is equal to the square root of the sum of the squares of the other two sides (see Appendix G). The value of the resistance is drawn to a convenient scale on the horizontal to form the base line. In a circuit containing only resistance and inductive reactance, the value of the reactance is laid off on a vertical line above the resistance line as in Fig. 9-16. If the circuit contains capacitive reactance, the reactance line is laid off below the horizontal line as in Fig. 9-17. If the circuit contains both X_L and X_C, the reactance is laid off on the vertical above or below the horizontal, depending on whether X_L or X_C is the larger. Such a triangle is called an

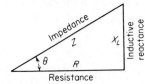

FIGURE 9-16 Impedance triangle: resistance and inductive reactance in series.

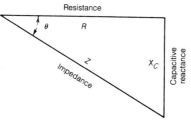

FIGURE 9-17   Impedance triangle:
resistance and capacitive reactance in
series.

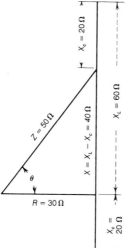

FIGURE 9-18   Construction of the impedance
triangle for Example 9-10.

impedance triangle, and when it is carefully drawn to scale it affords an
excellent check on the mathematical calculations of impedance. Figure
9-18 shows the impedance triangle for Example 9-10.

**REVIEW
QUESTIONS**

1 Give four advantages of alternating current over direct current.
2 List some applications in which direct current is preferred to alternating current.
3 Describe briefly the generation of an alternating emf. In general, what two factors will determine the amount of emf generated in a conductor?
4 What is meant by an instantaneous value of emf? What is the symbol? What is the unit?
5 What is meant by a maximum value of emf? What is the symbol? What is the unit?
6 When a conductor is revolved in a uniform magnetic field and the instantaneous values of emf are plotted against degrees of rotation, what is the resulting curve called?

7 Define an alternating current.

8 What is the symbol for (*a*) an instantaneous value of current; (*b*) a maximum value?

9 What is a cycle? What is frequency? What is the unit of frequency?

10 How many electrical degrees correspond (*a*) to one cycle; (*b*) to one-half cycle; (*c*) to two cycles? In a 60-Hz generator, what fraction of a second does it take to generate one cycle?

11 How many cycles are generated in a coil that makes one revolution in a 24-pole generator?

12 Define an effective ac ampere.

13 What is the ratio of the effective to the maximum values of a sine-wave current?

14 What is the effective value of current sometimes called?

15 What is the symbol for effective current?

16 Do common ac ammeters and voltmeters read instantaneous, maximum, or effective values?

17 What is meant when it is said that a current and voltage are in phase?

18 What are three possible phase relations between a current and a voltage?

19 What is the phase angle of a circuit?

20 What is a phasor? What is a phasor diagram?

21 Two phasors are drawn starting at a common point. Phasor 1 extends to the right from the point along the horizontal, and phasor 2 is drawn upward from the point at right angles to phasor 1. Does phasor 1 lead or lag phasor 2 by 90°?

22 Draw a phasor diagram of a current of 10 A lagging a voltage of 125 V by 45°.

23 What are the three circuit properties?

24 What is the phase relation of the current and voltage in a circuit containing resistance only?

25 Why does inductance affect the flow of an alternating current and not the flow of a steady direct current?

26 What is the unit of inductance? What is the symbol?

27 What is inductive reactance? What is the unit? What is the symbol?

28 What is the phase relation between (*a*) the applied voltage and the emf of self-induction; (*b*) the applied voltage and the current in a circuit containing inductance only?

29 Describe a parallel-plate capacitor.

30 What is capacitance? What is the unit? What is the symbol?

31 Upon what three things does the capacitance of a capacitor depend?

32 Why does a circuit containing a capacitor appear to be complete when an alternating emf is applied?

33 What is the phase relation between the applied voltage and the current flowing in a circuit containing capacitance only?

34 What is the unit of capacitive reactance? What is the symbol?

35 What is the effect on the capacitive reactance if the capacitance of a circuit is doubled, the frequency remaining constant?

36 To what is the reactance of a circuit equal?

37 Impedance is the combined effect of what two things in a circuit?

38 Show how an inductive reactance of 16 Ω and a resistance of 12 Ω may be combined graphically to form an impedance of 20 Ω.

39 Draw to scale an impedance triangle of a circuit in which the resistance is 3 Ω, the inductive reactance is 5 Ω, and the capacitive reactance is 2 Ω.

PROBLEMS

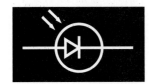

9-1 An alternating current has a maximum value of 50 A. Find the effective value of the current.

9-2 Find the maximum value of an alternating current whose effective value is 100 A.

9-3 Find the maximum value of an alternating voltage whose effective value is 240 V.

9-4 Standard ac ammeters and voltmeters read effective values of current and voltage. If the maximum value of current in a circuit is 7 A, find the reading of an ammeter that is placed in the circuit.

9-5 A voltmeter reads 480 V. Find the maximum value of the voltage wave. What is the effective value of the wave?

9-6 A current of 20 A lags a current of 40 A by 90°. Find the phasor sum of the two currents.

9-7 What is the phasor sum of two currents of 10 A each that are out of phase by 60°?

9-8 Two voltages of 120 V each are in phase. What is their phasor sum?

9-9 Two voltages of 120 V each are out of phase by 120°. What is their phasor sum?

9-10 Three voltages of 240 V are displaced symmetrically in time phase by 120°. What is their phasor sum?

9-11 Subtract a current of 10 A from a current of 14.14 A when the 14.14-A current leads the 10-A current by 45°.

9-12 A coil has an inductance of 0.2 H. Find its inductive reactance on a line frequency of 60, 120, and 600 Hz.

9-13 A coil has an inductive reactance of 40 Ω at a frequency of 60 Hz. Find the inductance of the coil.

9-14 A coil has an inductance of 0.05 H. The inductive reactance is 100 Ω. Find the line frequency.

9-15 A certain coil has an inductance of 0.30 H. What is its reactance at 60 and at 25 Hz?

9-16 The coil in the preceding problem is connected across 115 V. What current will flow when the frequency is (a) 60 Hz; (b) 25 Hz? Neglect the resistance of the coil.

9-17 A capacitor has a capacitance of 40 μF. What is its capacitive reactance at 60 and at 120 Hz?

9-18 Determine the capacity of a capacitor if its capacitive reactance on a 60 Hz system is 80 Ω.

9-19 What is the capacitive reactance of a 0.5-μF capacitor at 60 Hz and at 600 kHz (600,000 Hz)?

9-20 What is the capacitive reactance of a 100-pF capacitor at a frequency of 500 MHz?

9-21 A series ac circuit has a capacitive reactance of 40 Ω and a resistance of 20 Ω. Compute the impedance.

9-22 A series circuit has negligible resistance. Its X_C is 100 Ω, and its X_L is 40 Ω. Compute the impedance.

9-23 Find the impedance of a series circuit if it has 30 Ω resistance and 20 Ω capacitive reactance.

9-24 What is the impedance of a coil that has a resistance of 8 Ω and an inductive reactance of 20 Ω?

9-25 Find the impedance of a circuit that consists of the coil of Prob. 9-24 in series with a capacitor that has a capacitive reactance of 32 Ω.

9-26 A coil has a resistance of 3 Ω and an X_L of 30 Ω. Find the impedance of the circuit if a capacitor having an X_C of 30 Ω is connected in series with the coil.

9-27 A coil having a resistance of 2 Ω and an inductance of 0.25 H is connected to a 12-V storage battery. What steady-state current will flow through the coil?

9-28 Three capacitors of 4, 8, and 16 μF each are connected so that each capacitor forms the side of a triangle. Point A is the connection between the 4- and 8-μF sides; point B between the 8- and 16-μF sides; point C between the 4- and 16-μF sides. Find the capacitance between points A and B, B and C, and C and A.

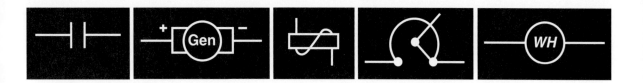

10

Single-Phase Circuits

Power in a dc circuit is equal to the product of the current and voltage. The power in an ac circuit at any instant is equal to the product of the values of current and voltage at that instant. When an alternating current and voltage are in phase, the average power over a complete cycle is equal to the product of the effective current and voltage. However, when reactance is present in the circuit, the current and voltage are out of phase so that at times during each cycle the current is negative while the voltage is positive. This results in a value of power in the circuit that is less than the product of I and V.

10-1 POWER AND POWER FACTOR

The product of the effective values of current and voltage in an ac circuit is expressed in *voltamperes* (VA) or in *kilovoltamperes* (kVA), 1 kVA being equal to 1000 VA. The useful or *actual power* is measured in watts and is the value obtained when the voltamperes of the circuit are multiplied by a factor called the *power factor*. Thus, power in a single-phase ac circuit is

$$P \text{ (in watts)} = VI \times \text{power factor} \qquad (10\text{-}1)$$

$$P \text{ (in kilowatts)} = \frac{VI}{1000} \times \text{power factor} \qquad (10\text{-}2)$$

By transposing Eq. (10-1)

$$\text{Power factor} = \frac{P}{VI} \qquad (10\text{-}3)$$

Thus, power factor may be defined as the ratio of the actual power in watts to the voltamperes of an ac circuit.

The value of the power factor depends on how much the current and voltage are out of phase. When the current and voltage are in phase, power is equal to $I \times V$, or in other words the power factor is unity. When current and voltage are out of phase by 90°, as in a purely capacitive or inductive circuit, the power factor is zero, resulting in a zero value of actual power. In circuits containing both resistance and reactance, the value of the power factor is some value between 1 and zero, its value depending on the relative values of resistance and reactance in the circuit.

Power factor may be expressed either as a decimal or as a percentage. For example, a power factor of 0.8 may be expressed as 80 percent and a power factor of 1.0 as 100 percent. Some typical average power factors are incandescent lights, 95 to 100 percent; large induction motors carrying rated load, 85 to 90 percent; fractional-horsepower induction motors, 60 to 75 percent, all with the current lagging the voltage.

The current flowing in an ac circuit may be considered as being made up of two components: a component in phase with the voltage and a component out of phase with the voltage by 90° as shown in Fig. 10-1. The inphase component is called the *active* component since this value when multiplied by the voltage gives the useful or actual power of the circuit. The out-of-phase component is called the *reactive* component, or the wattless component, since it contributes nothing to the actual power of the circuit. The product of the reactive component of the current and the voltage is called *reactive power* or, more properly, *reactive voltamperes,* and is measured in *vars,* a word coined from the first letters of the words *volt amperes reactive.* A larger unit of reactive power is the *kilovar,* one kilovar being equal to 1000 vars.

The angle θ (theta) in Fig. 10-1 is called the phase angle. It is evident from Fig. 10-1 that the larger the phase angle θ the greater the reactive component and the smaller the active component for a given value of total current.

In Fig. 10-1, the cosine of the phase angle θ is the ratio of the active current to the total current, or

$$\cos \theta = \frac{\text{active } I}{I}$$

or
$$\text{active } I = I \cos \theta$$

FIGURE 10-1   *Two components of current in an ac circuit.*

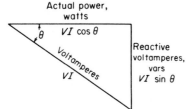

FIGURE 10-2 *Relation among power, voltamperes, and reactive voltamperes.*

Then, since the actual power is the voltage multiplied by the active component of current,

$$P = V \times \text{active } I$$

or $$P = VI \cos \theta \qquad (10\text{-}4)$$

Since power has already been shown to be EI multiplied by a power factor, it follows that the *power factor of an ac circuit is equal to the cosine of the phase angle.*

The values of the cosines of angles from $0°$ to $90°$ are given in Appendix H. Note that the values range from 1.000 for an angle of $0°$ to 0.000 for an angle of $90°$.

The relation among the power in watts, the voltamperes, and the reactive voltamperes, or vars, may be represented by a triangle as shown in Fig. 10-2, where the angle θ is the phase angle of the circuit being represented. The base of the triangle represents the actual power, the altitude the reactive voltamperes, and the hypotenuse the voltamperes. Since the voltamperes are equal to VI, the actual power is $VI \cos \theta$, and the reactive voltamperes are $VI \sin \theta$. Also the following relation is true:

$$\text{Voltamperes} = \sqrt{(\text{power})^2 + (\text{reactive VA})^2}$$

From the definition of the effective ac ampere (see Chap. 9), power is also equal to the effective current squared multiplied by the resistance of the circuit, or

$$P = I^2 R \qquad (10\text{-}5)$$

This is true in any single-phase ac circuit regardless of the phase relation between the current and the voltage.

10-2 CURRENT AND VOLTAGE RELATIONS IN SERIES CIRCUITS

Current and voltage relations in ac circuits containing resistance only, inductance only, and capacitance only were considered in the preceding chapter. In practice most circuits are combinations of two, or often all three, of these basic types of circuits. For example, transformer, generator, and motor windings all contain resistance and inductance. Even though the resistance and inductance are distributed throughout the windings, it is convenient when studying such circuits to consider the circuit as being made up of a pure resistance in series with a pure inductance.

Two rules similar to those of dc series circuits apply to the current and voltage relations of an ac series circuit:

1. In an ac series circuit, the current has the same value in all parts of the circuit.
2. The voltage applied to an ac series circuit is equal to the *phasor* sum of the voltages across the several parts of the circuit.

10-3 RESISTANCE AND INDUCTANCE IN SERIES

Figure 10-3a represents a circuit with resistance and inductance in series. The phasor diagram is drawn in Fig. 10-3b. The current phasor is used as a reference, since the current is the same in all parts of a series circuit. Since the voltage across a resistance is in phase with the current through the resistance, the phasor $V_R = IR$ is drawn along the current phasor. However, the voltage across an inductance leads the current through the circuit by 90°. Therefore, $V_L = IX_L$ is drawn 90° ahead of the current phasor I. The applied voltage V is equal to the phasor sum of V_R and V_L. The phasor addition may be performed by moving the phasor V_L to the arrow end of V_R and completing the triangle, the hypotenuse of the triangle representing V or IZ. The angle θ, or the angle between the applied voltage V and the current I, is the phase angle of the circuit.

Power in the circuit of Fig. 10-3a may be calculated from the formula

$$P = VI \times \text{power factor}$$

Since the power factor is the cosine of the phase angle, the power factor from Fig. 10-3b is

$$\text{Power factor} = \cos \theta = \frac{V_R}{V} = \frac{IR}{IZ} = \frac{R}{Z} \qquad (10\text{-}6)$$

Hence, the power factor of a series circuit is equal to the ratio of the resistance to the impedance of the circuit.

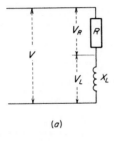

(a)

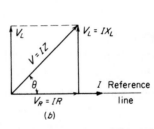

(b)

FIGURE 10-3 (a) Circuit with resistance and inductance in series. (b) Phasor diagram.

The following example will illustrate the method of calculating the current, power factor, and power of a simple *RL* (resistance-inductance) series circuit.

The circuit represented in Fig. 10-4 has a resistance of 6 Ω in series with an inductance of 0.0212 H. The circuit is connected to a 120-V 60-Hz supply. Find (*a*) the circuit impedance, (*b*) the current flowing, (*c*) the voltage across the resistance, (*d*) the voltage across the inductance, (*e*) the power factor, and (*f*) the power.

EXAMPLE 10-1

$$(a)\, X_L = 2\pi f L$$
$$= 2 \times 3.14 \times 60 \times 0.0212 = 8\,\Omega$$

$$Z = \sqrt{R^2 + X^2} = \sqrt{6^2 + 8^2} = 10\,A$$

$$(b)\quad I = \frac{V}{Z} = \frac{120}{10} = 12\,A$$

$$(c)\; V_R = IR = 12 \times 6 = 72\,V$$
$$(d)\; V_L = IX_L = 12 \times 8 = 96\,V$$

$$V = \sqrt{V_R^2 + V_L^2} = \sqrt{72^2 + 96^2} = 120\,V \;\text{(check)}$$

Note that *V* is not equal to the arithmetic sum of V_R and V_L (72 + 96 = 168).

The phasor diagram is drawn in Fig. 10-4*b*.

$$(e)\quad \text{Power factor} = \cos\theta = \frac{R}{Z} = \frac{6}{10} = 0.6 \text{ or } 60 \text{ percent}$$

$$(f)\; P = VI\cos\theta = 120 \times 12 \times 0.6 = 864\,W$$
$$\text{also } P = I^2R = (12)^2 6 = 864\,W \;\text{(check)}$$

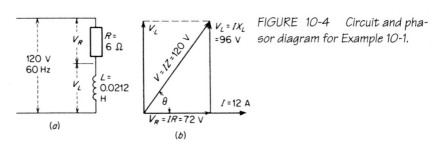

FIGURE 10-4 Circuit and phasor diagram for Example 10-1.

10-4 RESISTANCE AND CAPACITANCE IN SERIES

A capacitance and resistance in series are shown in Fig. 10-5. The current, being common to both branches, is used as a reference in the phasor diagram. The phasor representing the voltage across the resistance V_R is drawn in phase with the current. However, the current through a capacitance leads the voltage across the capacitance by 90°. Thus, the phasor V_C is drawn 90° behind the current. Again the applied voltage V is equal to the phasor sum of the voltages across each part of the circuit. The phasor addition by the triangle method is shown in Fig. 10-5b.

The power factor is equal to the cosine of the phase angle, or R/Z. The power taken by the circuit is equal to $VI \cos \theta$, or I^2R. Example 10-2 illustrates the method of calculating RC (resistance-capacitance) series circuits.

EXAMPLE 10-2

A resistance of 80 Ω in series with a 50-μF capacitor is connected to a 120-V 60-Hz source. Find (a) the circuit impedance, (b) the current flowing, (c) the voltage across the resistance, (d) the voltage across the capacitor, (e) the power factor, and (f) the power. The circuit diagram and phasor diagram are shown in Fig. 10-6a and b.

$$(a) \quad X = \frac{1}{2\pi f C} = \frac{1,000,000}{2\pi f C_{\mu F}}$$

$$= \frac{1,000,000}{2 \times 3.14 \times 60 \times 50} = 53.1 \ \Omega$$

$$Z = \sqrt{R^2 + X^2} = \sqrt{80^2 + 53.1^2} = 96 \ \Omega$$

$$(b) \quad I = \frac{V}{Z} = \frac{120}{96} = 1.25 \ A$$

$$(c) \quad V_R = IR = 1.25 \times 80 = 100 \ V$$

$$(d) \quad V_C = IX_C = 1.25 \times 53.1 = 66.4 \ V$$

$$V = \sqrt{V_R^2 + V_C^2} = \sqrt{100^2 + 66.4^2} = 120 \ V \ (\text{check})$$

$$(e) \quad \text{Power factor} = \cos \theta = \frac{R}{Z} = \frac{80}{96} = 0.833$$

$$(f) \quad P = VI \cos \theta = 120 \times 1.25 \times 0.833 = 125 \ W$$

$$P = I^2R = (1.25)^2 \times 80 = 125 \ W \ (\text{check})$$

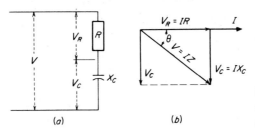

FIGURE 10-5 (a) Circuit with resistance and capacitance in series. (b) Phasor diagram.

10-5
RESISTANCE,
INDUCTANCE,
AND
CAPACITANCE IN
SERIES

This type of circuit may be solved by the same methods as those used in Secs. 10-3 and 10-4. The current phasor is used as a reference in series circuits in general because it is common to all elements of the circuit. The voltage drop across the resistance of the circuit ($V_R = IR$) is in phase with the current, and its phasor is drawn along the current phasor. The phasors representing the voltage drop across the inductance ($V_L = IX_L$) and the voltage drop across the capacitance ($V_C = IX_C$) are drawn 90° ahead of and 90° behind the current phasor, respectively. The phasor sum of the voltages V_R, V_L, and V_C is equal to the impressed voltage V. The power factor and power are also found by the same methods used in the simpler series circuits. Example 10-3 illustrates.

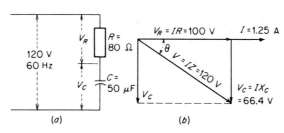

FIGURE 10-6 Circuit and phasor diagram for Example 10-2.

Figure 10-7*a* shows a circuit diagram of a circuit containing a resistance of 20 Ω, an inductance of 0.10 H, and a 100-μF capacitor all connected in series. The applied voltage is 240 V at 60 Hz. Find (*a*) the circuit impedance, (*b*) the current flow, (*c*) the voltage across each part, (*d*) the power factor, (*e*) the voltamperes, and (*f*) the power.

EXAMPLE 10-3

(*a*) $X_L = 2\pi fL = 2 \times 3.14 \times 60 \times 0.10 = 37.7\,\Omega$

$X_C = \dfrac{1,000,000}{2\pi fC_{\mu F}} = \dfrac{1,000,000}{2 \times 3.14 \times 60 \times 100} = 26.5\,\Omega$

$X = X_L - X_C = 37.7 - 26.5 = 11.2\,\Omega$

**EXAMPLE 10-3
(CONTINUED)**

$$Z = \sqrt{R^2 + X^2} = \sqrt{20^2 + 11.2^2} = 22.9\,\Omega$$

(b) $I = \dfrac{V}{Z} = \dfrac{240}{22.9} = 10.48\text{ A}$

(c) $V_R = IR = 10.48 \times 20 = 209.6\text{ V}$

$V_L = IX_L = 10.48 \times 37.7 = 395\text{ V}$

$V_C = IX_C = 10.48 \times 26.5 = 278\text{ V}$

The phasor diagram is drawn in Fig. 10-7b showing the position of the three voltages relative to the current. As a check on the work, the three voltages may be added as phasors by first adding V_L and V_C and then adding the result to V_R. Since V_L and V_C are exactly opposed, their resultant is $V_L - V_C$ as shown. This resultant is added to V_R by the triangle method as shown.

Mathematically

$$V = \sqrt{V_R^{\,2} + (V_L - V_C)^2}$$

$$= \sqrt{(209.6)^2 + (395 - 278)^2} = 240\text{ V (check)}$$

(d) Power factor $= \cos\theta = \dfrac{R}{Z} = \dfrac{20}{22.9} = 0.873$

The current lags the voltage since X_L is greater than X_C.

(e) Voltamperes $= VI = 240 \times 10.48 = 2515\text{ VA}$

(f) $P = VI \cos\theta = 2515 \times 0.873 = 2195\text{ W}$

or $P = I^2R = (10.48)^2 \times 20 = 2195\text{ W (check)}$

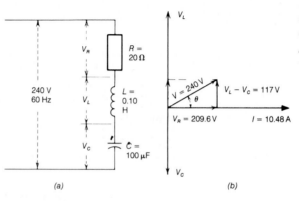

FIGURE 10-7
Circuit and phasor diagram for Example 10-3.

10-6 IMPEDANCES IN SERIES

When several electric devices are connected in series, the total impedance of the circuit is the phasor sum of the impedances of the separate parts. This total impedance may be found by adding arithmetically all the resistances, all the inductive reactances, and all the capacitive reactances and combining these totals, using the relationship

$$Z = \sqrt{R^2 + (X_L - X_C)^2}$$

EXAMPLE 10-4

Find the total impedance of two coils and a capacitor in series. Coil 1 has a resistance of 15 Ω and an inductive reactance of 35 Ω. Coil 2 has a resistance of 20 Ω and an inductive reactance of 45 Ω. The capacitor has a capacitive reactance of 60 Ω and has negligible resistance.

The total resistance of the circuit is the sum of the resistances of each part, or

$$R = 15 + 20 = 35 \; \Omega$$

The total inductive reactance is

$$X_L = 35 + 45 = 80 \; \Omega$$

The net reactance of the circuit is

$$X_L - X_C = 80 - 60 = 20 \; \Omega \text{ (inductive)}$$

The combined impedance is

$$Z = \sqrt{R^2 + X^2} = \sqrt{35^2 + 20^2} = 40.3 \; \Omega$$

10-7 SERIES RESONANCE

A series circuit in which the inductive and capacitive reactances are equal is said to be *resonant*. In such a circuit, the net reactance is zero and the impedance is equal to the resistance.

As an example, consider a circuit consisting of a 10-Ω resistance, a 30-Ω inductive reactance, and a 30-Ω capacitive reactance all connected in series to a 120-V 60-Hz supply. Since $X_L = X_C$, the reactance of the circuit is zero and the impedance is equal to the resistance, i.e., 10 Ω. The power factor of the circuit is unity since power factor is equal to R/Z. The current in the circuit is V/Z or 12 A. The voltage drop V across the resistance is IR or 120 V. Likewise, the voltage drop V across each reactance is IX or 360 V. However, the voltage across the inductive reactance is 180° out of phase with the voltage across the capacitive reactance, and while the effective value of each of these voltages may be relatively large, their phasor sum is zero.

If the frequency of the supply voltage for the circuit just described were lower than 60 Hz, the X_L of the circuit would be less than 30 Ω, the X_C would be greater than 30 Ω, and the total impedance would be greater than 10 Ω. Likewise, the impedance of the circuit would be greater than 10 Ω at frequencies greater than 60 Hz. In other words, a given series circuit is resonant at only one frequency and, at the resonant frequency, the impedance is at a minimum. Since at resonance $X_L = X_C$ then

$$2\pi fL = \frac{1}{2\pi fC}$$

Solving this equation for f results in the following value of frequency at resonance:

$$f = \frac{1}{2\pi\sqrt{LC}} \tag{10-7}$$

Resonance is usually to be avoided in power circuits because of the possibility of excessive voltages existing across parts of the circuit. However, communications circuits take advantage of the phenomenon in the tuning of circuits to obtain selectivity or the ability to respond strongly to a given frequency and discriminate against others.

10-8 PARALLEL CIRCUITS

Practically all lighting and power circuits are constant-voltage circuits with the loads connected in parallel. As in parallel dc circuits, the voltage is the same across each branch of a parallel ac circuit. The total line current supplied to the circuit is equal to the *phasor* sum of the branch currents. The total current supplied to a parallel circuit may be found by finding the current taken by each branch and adding these currents, taking into account their phase relations, or by finding an equivalent impedance and dividing the applied voltage by the equivalent impedance.

The aid offered by a phasor diagram in the solution of parallel-circuit problems is indispensable. Examples 10-5, 10-6, and 10-7 will illustrate the principles involved in parallel-circuit calculations.

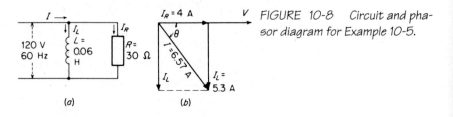

FIGURE 10-8 Circuit and phasor diagram for Example 10-5.

(Resistance and inductance in parallel.) A 0.06-H inductance coil **EXAMPLE 10-5**
with negligible resistance is connected in parallel with a 30-Ω resistance as in Fig. 10-8. When a voltage of 120 V at 60 Hz is applied, find (a) the current, (b) the power factor, and (c) the power taken by the circuit.

$$(a) \quad X_L = 2\pi fL = 2 \times 3.14 \times 60 \times 0.06 = 22.62 \; \Omega$$

$$I_L = \frac{V}{X_L} = \frac{120}{22.62} = 53 \; A$$

$$I_R = \frac{V}{R} = \frac{120}{30} = 4 \; A$$

The phasor diagram is drawn in Fig. 10-8b. In a parallel circuit the voltage phasor is used as a reference, since the voltage is the same for all branches of the circuit. Hence, the voltage phasor is laid off along the horizontal line. The current I_R is in phase with the voltage, and the current I_L lags the voltage by 90° as shown in the phasor diagram. The total current I is the phasor sum of I_R and I_L, the phasor addition being made as shown.

Mathematically

$$I = \sqrt{I_R^2 + I_L^2} = \sqrt{4^2 + 5.3^2} = 6.57 \; A$$

(b) The total current I lags behind the applied voltage by the angle θ. The power factor of the circuit is

$$\text{Power factor} = \cos\theta = \frac{I_R}{I} = \frac{4}{6.57} = 0.609$$

(c) $P = VI \cos\theta = 120 \times 6.57 \times 0.609 = 480 \; W$

(Resistance, inductance, and capacitance in parallel.) A resistance of **EXAMPLE 10-6**
80 Ω, an inductance of 0.15 H, and a capacitance of 25 μF are connected in parallel, and a voltage of 240 V at 60 Hz is applied. Find (a) the current, (b) the power factor, (c) the power, and (d) the equivalent or combined impedance of the combination.

The circuit and phasor diagrams are shown in Fig. 10-9. Note that the voltage phasor is used as a reference and that the current taken by the inductance lags the voltage by 90° and that taken by the capacitance leads the voltage by 90°.

EXAMPLE 10-6
(CONTINUED)

$$(a)\, I_R = \frac{V}{R} = \frac{240}{80} = 3 \text{ A}$$

$$X_L = 2\pi fL = 2 \times 3.14 \times 60 \times 0.15 = 56.55\,\Omega$$

$$I_L = \frac{V}{X_L} = \frac{240}{56.55} = 4.24 \text{ A}$$

$$X_C = \frac{1{,}000{,}000}{2\pi fC_{\mu F}} = \frac{1{,}000{,}000}{2 \times 3.14 \times 60 \times 25} = 106.1\,\Omega$$

$$I_C = \frac{V}{X_C} = \frac{240}{106.1} = 2.26 \text{ A}$$

The total current is the phasor sum of the branch currents. Combining I_L and I_C, which are in opposition, we have

$$I_L - I_C = 4.24 - 2.26 = 1.98 \text{ A}$$

and the total current is

$$I = \sqrt{I_R^{\,2} + (I_L - I_C)^2} = \sqrt{3^2 + 1.98^2} = 3.6 \text{ A}$$

$$(b)\quad \text{Power factor} = \cos\theta = \frac{I_R}{I} = \frac{3}{3.6} = 0.833$$

As I_L is larger than I_C, the total current lags voltage by the angle θ.

$$(c)\quad P = VL \cos\theta = 240 \times 3.6 \times 0.833 = 720 \text{ W}$$

also

$$P = (I_R)^2 R = 3^2 \times 80 = 720 \text{ W (check)}$$

$$(d)\quad Z = \frac{V}{I} = \frac{240}{3.6} = 66.7\,\Omega$$

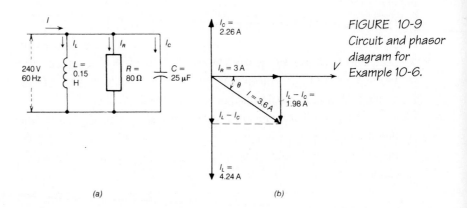

(a)

(b)

FIGURE 10-9
Circuit and phasor diagram for Example 10-6.

A resistance of 30 Ω is connected in parallel with a coil that has a resistance of 12 Ω and an inductive reactance of 16 Ω. Find (a) the line current and (b) the power taken by the circuit when it is connected to a 240-V supply as in Fig. 10-10.

EXAMPLE 10-7

$$(a)\ Z_2 = \sqrt{R^2 + X^2} = \sqrt{12^2 + 16^2} = 20\ \Omega$$

$$I_2 = \frac{V}{Z_2} = \frac{240}{20} = 12\ \text{A}$$

$$I_1 = \frac{V}{Z_1} = \frac{240}{30} = 8\ \text{A}$$

The phasor diagram is drawn in Fig. 10-10b. The current I_1 is in phase with the voltage, while I_2 lags the voltage by an angle ϕ whose cosine is R_2/Z_2, or

$$\cos \phi = \frac{R_2}{Z_2} = \frac{12}{20} = 0.6$$

$$\phi = 53.13° \text{ (from tables or by calculator)}$$

The phasor addition of I_1 and I_2 is performed by completing the triangle OBC as shown in Fig. 10-10b.

$$OA = 8$$
$$AC = 12$$
$$AB = 12 \cos 53.13° = 12 \times 0.6 = 7.2$$
$$OB = OA + AB = 8 + 7.2 = 15.2$$
$$BC = 12 \sin 53.13° = 12 \times 0.8 = 9.6$$

$$I = OC = \sqrt{(OB)^2 + (BC)^2}$$

$$= \sqrt{15.2^2 + 9.6^2} = 18\ \text{A}$$

$$(b)\ \text{Power factor} = \cos \theta = \frac{OB}{OC} = \frac{15.2}{18} = 0.8444$$

$$P = VI \cos \theta = 240 \times 18 \times 0.8444 = 3648\ \text{W}$$

also
$$P_1 = I_1^2 R_1 = (8)^2 \times 20 = 1920\ \text{W}$$
$$P_2 = I_2^2 R_2 = (12)^2 \times 12 = 1728\ \text{W}$$
$$P = P_1 + P_2 = 1920 + 1728 = 3648\ \text{W (check)}$$

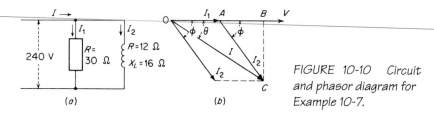

240 V

$R = 30\ \Omega$

$R = 12\ \Omega$
$X_L = 16\ \Omega$

(a)

(b)

FIGURE 10-10   Circuit and phasor diagram for Example 10-7.

10-9 PARALLEL RESONANCE

Resonance occurs in a parallel circuit when the reactive current in the inductive branches is equal to the reactive current in the capacitive branches. Since inductive and capacitive reactive currents are equal and opposite in phase, they cancel each other at parallel resonance.

If a coil and a capacitor, each with negligible resistance, were connected in parallel to an ac source and the reactances were adjusted so that at the frequency of the supply they were exactly equal, current would flow in the coil and in the capacitor, but the total current drawn from the supply would be negligible. Stating this another way, the parallel circuit would present an almost infinite impedance to the source. For this condition, the circuit is said to be oscillating; that is, the capacitor is alternately charging and discharging through the coil. This condition may be approached but not actually achieved in practical circuits because all coils and capacitors have some resistance and some current will always be drawn from the supply as determined by the amount of this resistance, even though the net reactance of the circuit is zero.

The expression for the resonant frequency in a parallel circuit in which the branch resistances are zero or very small is

$$f = \frac{1}{2\pi\sqrt{LC}}$$

which is identical to Eq. (10-7) for series resonant circuits.

Parallel resonant circuits are used extensively in communication equipment circuits.

10-10 POWER-FACTOR CORRECTION

More current is required to supply a given kilowatt load at a low power factor than at unity power factor. Since both voltage drop and line losses in any power line supplying a load are proportional to the line current, it follows that the lower the power factor, the higher are the line drop and losses to transmit a given kilowatt load. Because of the large number of induction motors and other inductive devices used on modern power systems, the power factors of many such systems are quite low, resulting in substantial voltage drops and line losses.

Many power suppliers correct for low power factors by the use of power-factor-corrective capacitors connected to their systems at various points. Since the current drawn by a capacitor leads the applied voltage by 90° (neglecting the small amount of resistance in the capacitor), the current drawn by such a device effectively cancels the effect of the inductive reactive component of the load devices on the same circuit. An installation of power-factor-corrective capacitors on a distribution system is shown in Fig. 10-11.

To illustrate the effect of a power-factor-corrective capacitor, consider the circuit of Example 10-7. The total current drawn by the circuit is 18 A, and the current lags the applied voltage by an angle whose cosine is 0.8444. From the phasor diagram in Fig. 10-10*b*, this current has an active component *OB* of 15.2 A in phase with the voltage and a reactive component *BC* of 9.6 A lagging the voltage by 90°. If a capacitor with a current rating of 9.6 A is connected in parallel with the resistor and coil, the resulting circuit power factor is unity. This is because the current drawn by the capacitor leads the voltage by 90° and is therefore 180° out of phase with the reactive component of the load current. Since there is no net reactive component in the circuit, the total line current is thus reduced from 18 to 15.2 A. Even though the total line current has been reduced, the resistor and coil are still supplied the same currents as before and the kilowatt load on the circuit is unchanged. The following example further illustrates the application of power-factor-corrective capacitors.

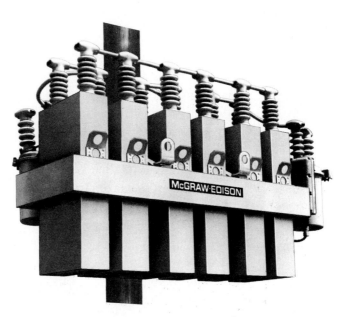

FIGURE 10-11   Assembly of power-factor corrective capacitors. (McGraw-Edison Company.)

EXAMPLE 10-8   A single-phase circuit has a load of 600 kVA at a power factor of 75 percent. If capacitors are to be added to the circuit to improve the power factor to 100 percent, what will be the rating of the capacitors in kilovars?

$$\text{Circuit kW} = \text{kVA} \times \text{power factor}$$
$$= 600 \times 0.75 = 450 \text{ kW}$$
$$\text{kVA}^2 = \text{kW}^2 + \text{kvar}^2$$

$$\text{kvar} = \sqrt{\text{kVA}^2 - \text{kW}^2}$$

$$= \sqrt{600^2 - 450^2} = 397 \text{ kvar}$$

Since the load reactive is 397 kvar, the capacitor bank rating must be 397 kvar to correct the circuit power factor to 100 percent.

Additional information on power-factor correction, including the use of synchronous motors for this purpose, is included in Sec. 17-8.

REVIEW QUESTIONS

1 In an ac circuit, how can the power at any instant be calculated? Is the instantaneous power a constant value over an entire cycle?
2 Draw a sine-wave voltage in phase with a sine-wave current. Is the instantaneous power zero at any time during a complete cycle? When is the instantaneous power at a maximum?
3 Under what condition is the power in an ac circuit equal to VI?
4 In general, what unit is used to express the product of volts and amperes in an ac circuit?
5 Define power factor. How is it used in finding the power of an ac circuit? What are its maximum and minimum values?
6 What is meant when it is said that an alternating current has an active and a reactive component?
7 In a circuit in which the current and voltage are in phase, what is the value of the reactive component of the current? What is the value of the reactive voltamperes, or vars, of the circuit?
8 What is the phase angle in degrees of a circuit in which the vars are equal to the watts? What is the phase angle of a circuit in which the voltamperes are equal to the watts?

9 When the resistance and current of an ac circuit are known, how may power be calculated regardless of power factor?

10 Give the current and voltage rules for an ac series circuit.

11 When a series circuit contains inductance and resistance, does the current lead or lag the applied voltage? What is the phase relation between the voltage across the inductance and the current through the inductance?

12 How can the power factor of a series ac circuit be calculated from the circuit constants?

13 What is the phase relation between current and voltage in an *RC* series circuit?

14 Describe the procedure for finding the current, power factor, and power of a resistance-inductance-capacitance series circuit when the circuit constants, the frequency, and the voltage are given. Under what conditions in this circuit will the current lead the voltage?

15 When several impedances are connected in series, how is the combined impedance found?

16 Give the current and voltage rules for a parallel ac circuit.

17 Is the line current necessarily the arithmetic sum of the branch currents in a parallel ac circuit? When is this the case?

18 If the applied voltage and the branch currents of a parallel circuit are known, explain how the equivalent impedance of the circuit might be found.

19 Explain how the power factor of an inductive circuit may be improved.

PROBLEMS

10-1 When a voltage of 416 V is applied to a certain load, the current that flows is 20 A. What are (*a*) the voltamperes of the circuit; (*b*) the kilovoltamperes?

10-2 The power taken by a motor is 800 W when the current is 9 A. If the line voltage is 120 V, at what power factor is the motor operating?

10-3 A certain load draws 85 kW and 105 kVA from an ac supply. At what power factor is it operating?

10-4 What full-load power can a 5-kVA single-phase generator supply at a power factor of 90 percent? What is the full-load line current if the terminal voltage is 120 V?

10-5 A single-phase motor operates at a power factor of 70 percent lagging. The line voltage is 240 V, and the line current is 20 A. Find the voltamperes and the actual power in watts taken by the motor.

10-6 A single-phase circuit has a unity power factor. The applied voltage is 120 V. The line current is 30 A. Find the actual power in watts.

10-7   The power consumed in a single-phase circuit is 4 kW. The line current is 40 A and the voltage is 120 V. Find the voltamperes and the power factor of the circuit.

10-8   A single-phase motor is operating at a power factor of 0.8 lagging on a 240-V line. The motor consumes 8 kW. Find the motor line current.

10-9   A motor draws 1000 W and 1500 VA from a single-phase line. What is the value of the reactive voltamperes, or vars, drawn by the motor? What is the power factor?

10-10 Find the kilovoltamperes of a circuit when the actual power is 1200 W and the reactive voltamperes are 600 vars. What is the power factor?

10-11 An inductive reactance of 12 Ω is connected in series with a resistance of 16 Ω. A 115-V 60-Hz voltage is applied. Find (*a*) the impedance, (*b*) the current flowing, (*c*) the power factor, (*d*) the voltamperes, and (*e*) the power.

10-12 Find the voltage applied to a coil that has a reactance of 8 Ω and a resistance of 3 Ω when a current of 11 A is flowing through the coil. What is the power used by the coil?

10-13 A resistance of 40 Ω and a capacitance of 50 μF are connected in series across a 120-V 60-Hz line. Find (*a*) the impedance, (*b*) the current, (*c*) the power factor, (*d*) the voltamperes, and (*e*) the power.

10-14 A circuit consisting of a capacitor and a resistance in series takes 60 W at a power factor of 28 percent (leading) from a 120-V 60-Hz line. Find (*a*) the current, (*b*) the impedance, (*c*) the resistance, and (*d*) the capacitance of the circuit.

10-15 A 100-μF capacitor is connected to a 120-V 60-Hz line. Find the current that will flow. If a resistance of 10 Ω is connected in series with the capacitor, what will be the new value of current?

10-16 A series circuit consists of a resistance of 10 Ω, an 80-μF capacitor, and an inductance of 0.1 H. What current will flow when the circuit is connected to a 240-V 60-Hz supply?

10-17 A capacitor, a resistor, and a coil with negligible resistance are connected in series. The voltage drop across the capacitor is 50 V, across the resistor 80 V, and across the coil 20 V. Find the voltage drop across the entire circuit.

10-18 A resistance of 40 Ω and a capacitive reactance of 60 Ω are connected in series across 120-V circuit. Find (*a*) the current, (*b*) the power factor, and (*c*) the power.

10-19 When a resistance of 4 Ω and an inductance of unknown value are connected in series across 120-V 60-Hz supply lines, the current is 20 A. Find the value of the inductance.

10-20 The following meter readings were taken in an inductive single-phase circuit: wattmeter, 2400 W; voltmeter, 240 V; ammeter, 15 A; frequency meter, 60 Hz. Find (*a*) the voltamperes, (*b*) the power factor, (*c*) the impedance, (*d*) the resistance, (*e*) the inductive reactance, (*f*) the inductance, and (*g*) the value of a series capacitor that will make the power factor 100 percent.

10-21 A 120-V ac motor delivers 5 hp at an efficiency of 85 percent. The motor operates at a power factor of 80 percent lagging. Find (*a*) the power input and (*b*) the current drawn by the motor.

10-22 Find the total current required by a circuit having a resistance of 20 Ω, an inductive reactance of 10 Ω, and a capacitive reactance of 15 Ω, all connected in parallel to a 120-V line.

10-23 A resistance of 16 Ω is connected in parallel with a coil that has a resistance of 4 Ω and a reactance of 12 Ω. The supply voltage is 120 V. Find (*a*) the total current, (*b*) the power factor, and (*c*) the power taken by the circuit.

10-24 A resistance of 20 Ω is connected in parallel with an inductive reactance of 10 Ω to a 100-V 60-Hz line. Find (*a*) the current through the resistance, (*b*) the current through the reactance, and (*c*) the reactance of a capacitor that, when connected in parallel with the circuit, will improve the power factor of the circuit to 100 percent.

10-25 A coil with a resistance of 5 Ω and an inductive reactance of 12 Ω is connected in parallel with a capacitor that has a capacitive reactance of 10 Ω and negligible resistance. When the parallel combination of coil and capacitor is connected to a 120-V 60-Hz supply, find (*a*) the current in each branch, (*b*) the total line current, (*c*) the total power in watts, and (*d*) the circuit power factor.

10-26 A coil having an inductive reactance of 8 Ω and a resistance of 0.5 Ω is connected in series with a 5.5-Ω resistor. The coil and the 5.5-Ω series resistor are connected in parallel with a 20-Ω resistor to a 120-V 60-Hz supply. Determine (*a*) the current flowing through the coil, (*b*) the current flowing through the 20-Ω resistor, (*c*) the total line current, (*d*) the power taken by the circuit, and (*e*) the power factor of the circuit.

10-27 The input to a motor is 5.5 kVA at 80 percent power factor. What must be the reactive voltampere rating of a capacitor connected across the motor terminals to correct the power factor to 100 percent?

10-28 A 2-hp 60-Hz motor has a full-load input of 2200 W and 12 A at 230 V. (*a*) Find the reactive voltampere rating of a capacitor to be connected across the terminals of the motor to correct the full-load power factor to 100 percent. (*b*) What is the size of the capacitor in microfarads? (*c*) What is the total current drawn by the motor and the capacitor?

10-29 A 5-kvar (5000-var) capacitor added in parallel to a given 440-V load corrects the circuit power factor to unity, and the resulting line current is 5 A. (*a*) What was the line current drawn by the load before the addition of the capacitor? (*b*) What was the power factor of the original 440-V load?

10-30 A single-phase circuit has a variable load. At light-load periods the load is 200 kVA at 75 percent power factor, and at peak-load periods it is 500 kVA at 85 percent power factor. What is the rating in kilovars of a capacitor bank that will raise the light-load power factor to unity? What will be the peak-load power factor with this capacitor bank added to the circuit?

11

Three-Phase Circuits

A three-phase circuit is merely a combination of three single-phase circuits. Because of this fact, current, voltage, and power relations of balanced three-phase circuits may be studied by the application of single-phase rules to the component parts of the three-phase circuit. Viewed in this light, the analysis of three-phase circuits is a little more difficult than that of single-phase circuits.

11-1 REASONS FOR USE OF THREE-PHASE CIRCUITS

In a single-phase circuit, the power is of a pulsating nature. At unity power factor the power in a single-phase circuit is zero twice each cycle. When the power factor is less than unity, the power is negative during parts of each cycle. Although the power supplied to each of the three phases of a three-phase circuit is pulsating, it may be proved that the total three-phase power supplied a balanced three-phase circuit is constant. Because of this, the operating characteristics of three-phase apparatus, in general, are superior to those of similar single-phase apparatus.

Three-phase machinery and control equipment are smaller, lighter in weight, and more efficient than single-phase equipment of the same rated capacity. In addition to the above-mentioned advantages offered by a three-phase system, the distribution of three-phase power requires only three-fourths as much line conductor material as does the single-phase distribution of the same amount of power.

#82

11-2 GENERATION OF THREE-PHASE EMFS

A three-phase electric circuit is energized by three alternating emfs of the same frequency and differing in time phase by 120 electrical degrees. Three such sine-wave emfs are shown in Fig. 11-1. These emfs are generated in three

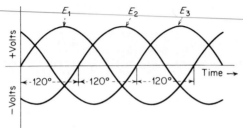

FIGURE 11-1   Three sine-wave emfs differing in phase by 120 electrical degrees such as are used for energizing a three-phase circuit.

separate sets of armature coils in an ac generator. These three sets of coils are mounted 120 electrical degrees apart on the generator armature. The coil ends may all be brought out of the generator to form three separate single-phase circuits. However, the coils are ordinarily interconnected either internally or externally to form a three-wire or four-wire three-phase system.

There are two ways of connecting the coils of three-phase generators, and in general, there are two ways of connecting devices of any sort to a three-phase circuit. These are the *wye connection* and the *delta connection*. Most generators are wye-connected, but loads may be either wye-connected or delta-connected.

11-3 VOLTAGE RELATIONS IN A WYE-CONNECTED GENERATOR

Figure 11-2a represents the three coils or phase windings of a generator. These windings are so spaced on the armature surface that the emfs generated in them are 120° apart in time phase. Each coil is lettered S and F (start and finish). In Fig. 11-2a all the coil ends marked S are connected to a common point N, called the neutral, and the three coil ends marked F are brought out to the line terminals A, B, and C to form a three-wire three-phase supply. This type of connection is called the *wye connection* (sometimes called the *star connection*). Often the neutral connection is brought out to the terminal board, as shown by the dotted line in Fig. 11-2a, to form a four-wire three-phase system.

The voltages generated in each phase of an ac generator are called the *phase voltages* (symbol E_P or V_P). If the neutral connection is brought out of the generator, the voltage from any one of the line terminals A, B, or C to the neutral connection N is a phase voltage. The voltage between any two of the three line terminals A, B, or C is called a line-to-line voltage or, simply, a *line voltage* (symbol E_L or V_L).

The order in which the three voltages of a three-phase system succeed one another is called the *phase sequence* or the phase rotation of the voltages. This is determined by the direction of rotation of the generator but may be reversed outside the generator by interchanging any two of the three line wires (not a line wire and a neutral wire).

It is helpful when drawing circuit diagrams of a wye connection to arrange the three phases in the shape of a Y as shown in Fig. 11-2b. Note

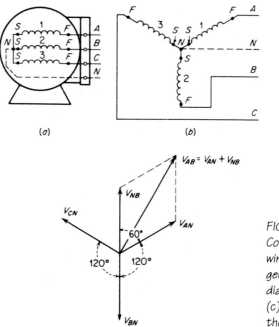

FIGURE 11-2 (a) Connection of the phase windings in a wye-connected generator. (b) Conventional diagram of a wye connection. (c) Phasor diagram showing the relation of phase and line voltages.

that the circuit of Fig. 11-2*b* is exactly the same as that of Fig. 11-2*a*, with the *S* end of each coil connected to the neutral point and the *F* end brought out to the terminal in each case. After a circuit diagram has been drawn with all intersections lettered, a phasor diagram may be drawn as in Fig. 11-2*c*. The phasor diagram shows the three phase voltages V_{AN}, V_{BN}, and V_{CN}, which are 120° apart.

It should be noted in Fig. 11-2 that each phasor is lettered with two subscripts. The two letters indicate the two points between which the voltage exists, and the order of the letters indicates the relative polarity of the voltage during its positive half-cycle. For example, the symbol V_{AN} indicates a voltage V between the points A and N with the point A being positive with respect to point N during its positive half-cycle. In the phasor diagram shown, it has been assumed that the generator terminals were positive with respect to the neutral during the positive half-cycle. Since the voltage reverses every half-cycle, either polarity may be assumed if this polarity is assumed consistently for all three phases. It should be noted that if the polarity of point A with respect to N (V_{AN}) is assumed for the positive half-cycle, then V_{NA} when used in the same phasor diagram should be drawn opposite to, or 180° out of phase with, V_{AN}.

The voltage between any two line terminals of a wye-connected generator is the difference between the potentials of these two terminals with respect to the neutral. For example, the line voltage V_{AB} is equal to the voltage A with respect to neutral (V_{AN}) minus the voltage B with respect to neutral (V_{BN}). To subtract V_{BN} from V_{AN}, it is necessary to reverse V_{BN} and add this phasor to V_{AN}. The two phasors V_{AN} and V_{NB} are equal in length and are 60° apart, as shown in Fig. 11-2c. It may be shown graphically or proved by geometry that V_{AB} is equal to $\sqrt{3}$, or 1.73, multiplied by the value of either V_{AN} or V_{NB}. The graphical construction is shown in the phasor diagram. Therefore, in a balanced wye connection

$$V_L = \sqrt{3}\, V_P = 1.73 V_P \qquad (11\text{-}1)$$

(Note: The value of $\sqrt{3}$ correct to five decimal places is 1.73205. The approximate value used in this text is 1.73.)

11-4 CURRENT RELATIONS IN A WYE-CONNECTED GENERATOR

The current flowing out to the line wires from the generator terminals A, B, and C (Fig. 11-2) must flow from the neutral point N out through the generator coils. Thus, the current in each line wire (I_L) must equal the current in the phase (I_P) to which it is connected. In a wye connection

$$I_L = I_P \qquad (11\text{-}2)$$

11-5 VOLTAGE RELATIONS IN A DELTA-CONNECTED GENERATOR

A *delta-connected* generator is shown in Fig. 11-3a. This connection is formed by connecting the S terminal of one phase to the F terminal of the adjacent phase. The line connections are then made at the common points between phases as shown. The conventional circuit diagram in which the three coils are arranged in the shape of the Greek letter delta (Δ) is shown in Fig. 11-3b. An inspection of the diagram shows that the voltage generat-

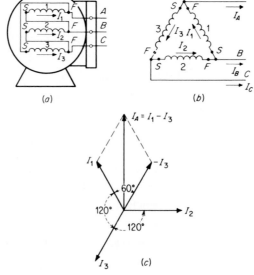

FIGURE 11-3 (a) Connection of the phase windings in a delta-connected generator. (b) Conventional diagram of a delta connection. (c) Phasor diagram showing the relation of phase and line currents.

ed in each phase is also the voltage between two line wires. For example, the voltage generated in phase 1 is also the voltage between lines A and B. Therefore in a delta connection

$$V_L = V_P \qquad (11\text{-}3)$$

11-6 CURRENT RELATIONS IN A DELTA-CONNECTED GENERATOR

The phase currents in the delta connection of Fig. 11-3b are I_1, I_2, and I_3. The phasor diagram representing these currents is shown in Fig. 11-3c. To find the current in any of the three line wires it is necessary to add as phasors the currents flowing in the two phases to which that line is connected. For example, the current flowing out toward the load through line A must be

$$I_A = I_1 + (-I_3)$$

Since I_1 and $-I_3$ are two equal phasors 60° apart, their phasor sum is $\sqrt{3}$, or 1.73, times the value of either I_1 or $-I_3$ (Fig. 11-3c). Therefore, in the delta connection,

$$I_L = \sqrt{3}\, I_P = 1.73 I_P \qquad (11\text{-}4)$$

11-7 POWER IN THREE-PHASE CIRCUITS

From the formula for power in a single-phase circuit, the power developed in each phase (P_P) of either a delta or wye connection is

$$P_P = V_P I_P \cos \theta$$

where θ is the angle between the phase current and the phase voltage. The power developed in all three phases of a balanced three-phase connection is then

$$P = 3P_P = 3\, V_P I_P \cos \theta \qquad (11\text{-}5)$$

But in a wye connection

$$I_P = I_L \quad \text{and} \quad V_P = \frac{V_L}{\sqrt{3}}$$

Thus, the three-phase power in a wye-connected system in terms of line voltage and current is

$$P = 3\frac{V_L}{\sqrt{3}} I_L \cos \theta$$

$$= \sqrt{3}\, V_L I_L \cos \theta$$

$$= 1.73 V_L I_L \cos \theta$$

In a delta connection

$$V_P = V_L \quad \text{and} \quad I_P = \frac{I_L}{\sqrt{3}}$$

Again the three-phase power in terms of line current and voltage is

$$P = 3V_L \frac{I_L}{\sqrt{3}} \cos \theta$$

$$= \sqrt{3} V_L I_L \cos \theta$$

$$= 1.73 V_L I_L \cos \theta$$

Thus, the expression for three-phase power in a balanced system, either wye- or delta-connected, is equal to

$$P = 1.73 \ V_L I_L \cos \theta \qquad (11\text{-}6)$$

If $V_P I_P \cos \theta$ is the power per phase in watts, then $V_P I_P$ represents the voltamperes per phase. It follows, then, that the three-phase voltamperes are equal to $3V_P I_P$, or $1.73 \ V_L I_L$. The total three-phase kilovoltamperes in either a wye or delta system may be found by

$$kVA = \frac{1.73 \ V_L I_L}{1000} \qquad (11\text{-}7)$$

The power factor of a balance three-phase system either wye- or delta-connected is defined as the cosine of the angle between the phase voltage and the phase current. From Eq. (11-6) also, the power factor is seen to be

$$\text{Power factor} = \cos \theta = \frac{P}{1.73 \ V_L I_L} \qquad (11\text{-}8)$$

or is equal to the ratio of the three-phase power to the three-phase voltamperes.

EXAMPLE 11-1 A three-phase wye-connected generator has a terminal voltage of 480 V and delivers a full-load current of 300 A per terminal at a lagging power factor of 75 percent. Find (a) the voltage per phase, (b) the full-load current per phase, (c) the kilovoltampere rating, and (d) the full-load power in kilowatts.

$$(a) \quad V_P = \frac{V_L}{1.73} = \frac{480}{1.73} = 277 \text{ V}$$

EXAMPLE 11-1
(CONTINUED)

(*b*) As the generator is wye-connected, the phase current is equal to the line current.

$$I_P = I_L = 300 \text{ A}$$

(*c*) $\quad \text{kVA} = \dfrac{1.73 \, V_L I_L}{1000} = \dfrac{1.73 \times 480 \times 300}{1000} = 249.1 \text{ kVA}$

(*d*) $\quad P(\text{kW}) = \dfrac{1.73 \, V_L I_L \times \text{power factor}}{1000}$

$$= \text{kVA} \times \text{power factor} = 249.1 \times 0.75$$

$$= 186.8 \text{ kW}$$

EXAMPLE 11-2

Each phase of a three-phase delta-connected generator supplies a full-load current of 100 A at a voltage of 240 V and at a power factor of 0.6, lagging. Find (*a*) the line voltage, (*b*) the line current, (*c*) the three-phase kilovoltamperes, and (*d*) the three-phase power in kilowatts.

(*a*) The line voltage is equal to the phase voltage in a delta system.

$$V_L = V_p = 240 \text{ V}$$

(*b*) $\quad I_L = 1.73 \, I_p = 1.73 \times 100 = 173 \text{ A}$

(*c*) $\quad \text{kVA} = \dfrac{1.73 \, V_L I_L}{1000} = \dfrac{1.73 \times 240 \times 173}{1000} = 72 \text{ kVA}$

(*d*) $\quad P = \text{kVA} \times \text{power factor} = 72 \times 0.6 = 43.2 \text{ kW}$

11-8 WYE- AND DELTA-CONNECTED LOADS

Not only may generators be connected in either wye or delta, but various types of loads such as motor windings, lamps, or transformers may also be wye- or delta-connected. The same current, voltage, and power relations that are used for three-phase generators are used for three-phase load connections. The following examples will illustrate these relations for load connections. (See also Sec. 12-17 for additional discussion of three-phase load connections.)

EXAMPLE 11-3 Three resistances of 15 Ω each (power factor = 100 percent) are wye-connected to a 240-V three-phase line. Find (*a*) the current through each resistance, (*b*) the line current, and (*c*) the power taken by the three resistances.

The circuit diagram is shown in Fig. 11-4. The phase voltage or the voltage applied to each resistance is

$$V_P = \frac{V_L}{1.73} = \frac{240}{1.73} = 138.7 \text{ V}$$

(*a*) The current in each resistance is (no reactance present)

$$I_P = \frac{V_P}{Z_P} = \frac{V_P}{R_P} = \frac{138.7}{15} = 9.25 \text{ A}$$

(*b*) As the line and phase currents are equal,

$$I_L = I_P = 9.25 \text{ A}$$

(*c*) The power taken by one resistance is

$$P_P = V_P I_P \times \text{power factor} = 138.7 \times 9.25 \times 1 = 1280 \text{ W}$$

The three-phase power is

$$P = 3P_P = 3 \times 1280 = 3840 \text{ W}$$

or $$P = 1.73 V_L I_L \times \text{power factor}$$

$$= 1.73 \times 240 \times 9.25 \times 1 = 3840 \text{ W (check)}$$

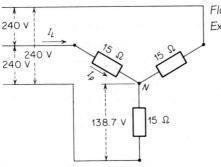

FIGURE 11-4   *Circuit diagram for Example 11-3.*

Repeat the preceding problem if the three resistances are reconnected in delta.

EXAMPLE 11-4

(*a*) The circuit diagram is shown in Fig. 11-5. Now the voltage applied to each phase is equal to the line voltage, or 240 V.

$$I_P = \frac{V_P}{Z_P} = \frac{V_P}{R_P} = \frac{240}{15} = 16 \text{ A}$$

(*b*) The line current is

$$I_L = 1.73I_P = 1.73 \times 16 = 27.7 \text{ A}$$

(*c*) The three-phase power is

$$P = 1.73 \, V_L I_L \times \text{power factor}$$

or

$$= 1.73 \times 240 \times 27.7 \times 1 = 11,520 \text{ W}$$

$$P = 3V_P I_P \times \text{power factor}$$

$$= 3 \times 240 \times 16 \times 1 = 11,520 \text{ W (check)}$$

A 15-hp 230-V three-phase induction motor has a full-load current of 42 A per terminal. (*a*) What is the full-load kilovoltampere input? (*b*) If the full-load power factor is 85 percent, what is the kilowatt input?

EXAMPLE 11-5

(*a*) $\text{kVA} = \dfrac{1.73 V_L I_L}{1000}$

$$= \frac{1.73 \times 230 \times 42}{1000} = 16.71 \text{ kVA}$$

(*b*) P (in kW) = kVA × power factor

$$= 16.71 \times 0.85 = 14.21 \text{ kW}$$

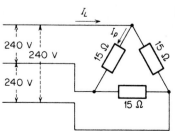

FIGURE 11-5   *Circuit diagram for Example 11-4.*

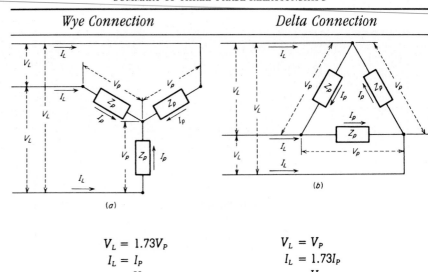

Wye Connection · *Delta Connection*

$$V_L = 1.73V_P \qquad V_L = V_P$$
$$I_L = I_P \qquad I_L = 1.73I_P$$
$$I_P = \frac{V_P}{Z_P} \qquad I_P = \frac{V_P}{Z_P}$$

For Either Connection

$$\text{Voltamperes per phase} = V_P I_P$$

$$\text{Kilovoltamperes per phase} = \frac{V_P I_P}{1000}$$

$$\text{Three-phase voltamperes} = 3V_P I_P$$
$$= 1.73\,V_L I_L$$

$$\text{Three-phase voltamperes} = \frac{3V_P I_P}{1000}$$
$$= \frac{1.73\,V_L I_L}{1000}$$

$$\text{Power factor} = \text{cosine of angle between } V_P \text{ and } I_P$$
$$= \frac{P}{1.73\,V_L I_L}$$

$$\text{Power per phase} = V_P I_P \cos\theta \text{ (in watts)}$$
$$= \frac{V_P I_P \cos\theta}{1000} \text{ (in kilowatts)}$$

$$\text{Three-phase power} = 3V_P I_P \cos\theta \text{ (in watts)}$$
$$= 1.73\,V_L I_L \cos\theta \text{ (in watts)}$$

Also
$$= 3V_P I_P \times \text{power factor (in watts)}$$
$$= 1.73\,V_L I_L \times \text{power factor (in watts)}$$

1 What is a three-phase circuit?

2 What are some advantages of three-phase systems over single-phase systems?

3 In the three-phase generator, the three emfs are generated how many electrical degrees out of phase?

4 What are two commonly used connections in a three-phase circuit?

5 Explain briefly how generator phase windings are connected in wye.

6 How is a wye-connected generator connected to supply (*a*) a three-wire three-phase system; (*b*) a four-wire three-phase system?

7 In a wye connection, what is the line-to-neutral voltage called? What is meant by a line voltage?

8 What is meant by phase sequence? How may it be reversed?

9 If V_{AB} represents a line voltage in a phasor diagram, what would be the symbol for the voltage that is equal and opposite to V_{AB} in the same diagram?

10 Give the relationship between phase and line voltages in a wye connection. How are the phase and line circuits related?

11 Show how the phase windings of an ac generator may be connected in delta.

12 What is the relation between phase and line voltages in a delta connection? Between phase and line currents?

13 Give two three-phase power formulas. Since power in a single-phase circuit is equal to I^2R, how could this relationship be used in finding three-phase power?

14 Give the formula for finding three-phase voltamperes.

15 If the three-phase watts and voltamperes of a circuit are known, how could the three-phase vars be found?

PROBLEMS

Note: Balanced three-phase connections are assumed in all the following problems.

11-1 The phase voltage of a wye-connected generator is 120 V. What is the line voltage?

11-2 If the terminal voltage of a wye-connected generator is 13,200 V, what is the voltage generated in each of the phases?

11-3 What is the *maximum* value of the line voltage of a wye connection if the effective value of the phase voltage is 127 V?

11-4 For a given load, the phase current of a wye-connected generator is 50 A. What is the current flowing in the line wires?

11-5 Each phase of a wye-connected generator delivers a current of 30 A at a phase voltage of 254 V and a power factor of 80 percent, lagging. What is the generator terminal voltage? What is the power developed in each phase? What is the three-phase power developed?

11-6 At full load, each of the three phases of a wye-connected generator delivers 150 A at 1329 V at a power factor of 75 percent, the current lagging. Find (*a*) the terminal voltage rating, (*b*) the kilovoltampere rating, and (*c*) the kilowatt rating.

11-7 It is desired that a 10,000-kVA, three-phase, 60-Hz generator have a rated terminal voltage of 13,800 V when it is wye-connected. What must be the voltage rating of each phase? Find the kilovoltampere rating per phase and the rated current per terminal.

11-8 The line current of a delta connection is 40 A. What is the phase current?

11-9 The phase current in a delta connection is 50 A. Find the line current.

11-10 The voltage per phase of a delta-connected generator is 450 V. What is the terminal or line voltage?

11-11 If the generator of Prob. 11-6 is reconnected for delta operation, find (*a*) the new voltage rating, (*b*) the new kilovoltampere rating, and (*c*) the new kW rating.

11-12 A three-phase motor takes a current of 20 A per terminal. The line voltage is 480 V, and the power factor is 80 percent, lagging. What is the power used by the motor?

11-13 In a three-phase delta-connected load, the phase current is 20 A and the phase voltage is 240 V. The power factor is 90 percent, lagging. Find the total three-phase power in kilowatts.

11-14 A three-phase load consumes 60 kW when connected to a 480-V line. The line current is 100 A. Find the kilovoltamperes and the power factor of the load.

11-15 A wye-connected load draws 20 kW from a 416-V line. The line current is 30 A. Find the phase voltage and the power factor.

11-16 A three-phase generator delivers 150 A at 7200 V. What is the kilovoltampere output. If the power output is 1500 kW, what is the power factor?

11-17 Find the current rating of a three-phase 460-V, 60-Hz, 50-hp induction motor, if it operates at an 85 percent lagging power factor. The full-load input is 38.8 kW.

11-18 If a generator delivers 26,000 kW at a power factor of 80 percent, lagging, what is its kilovoltampere output?

11-19 A wye-connected load of three 10-Ω resistors (power factor = 100 percent) is connected to a 208-V three-phase supply. Find (*a*) the voltage applied to each resistor, (*b*) the line current, and (*c*) the total power used.

11-20 Three loads, each having a resistance of 16 Ω and an inductive reactance of 12 Ω, are wye-connected to a 240-V three-phase supply. Find (*a*) the impedance per phase, (*b*) the current per phase, (*c*) the three-phase kilovoltamperes, (*d*) the power factor, and (*e*) the three-phase power in kilowatts.

11-21 Work Prob. 11-19 if the resistors are delta-connected.

11-22 Work Prob. 11-20 if the loads are delta-connected.

11-23 A delta-connected load draws 28.8 kW from a three-phase supply. The line current is 69.2 A and the power factor is 80 percent, lagging. Find the resistance and the reactance of each phase.

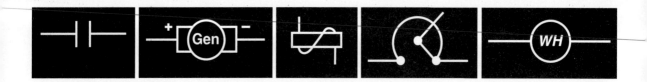

12

Transformers and Regulators

Transformers provide a simple means of changing an alternating voltage from one value to another. If a transformer receives energy at a low voltage and delivers it at a higher voltage, it is called a *step-up* transformer. When a transformer is supplied energy at a given voltage and delivers it at some lower voltage, it is called a *step-down* transformer. Any transformer may be operated as either a step-up or step-down transformer; however, the voltage for which the transformer is designed must be applied in either case.

To transmit a given amount of energy, less current is required at a high voltage than at a low voltage. This means that energy may be transmitted with less I^2R or line loss when high transmission voltages are used. In order to secure high transmission voltages, such as 345,000 or 765,000 V, step-up transformers are used at the generating station, since it is not feasible to generate voltages this high. At the points where energy is to be used, step-down transformers are used to reduce the high transmission voltage to safe and usable values. Thus, transformers make possible the economical transmission of electric energy over long distances.

Since a basic transformer has no moving parts, it requires little attention and the maintenance expense is low. Transformer efficiencies are high, running as high as 98 or 99 percent at full load in the larger sizes.

12-1
TRANSFORMER
TERMINOLOGY

A simple transformer consists of two coils wound on a closed iron core as represented in Fig. 12-1. The coils are insulated from each other and from the core. Energy is supplied to one winding, called the *primary winding,*

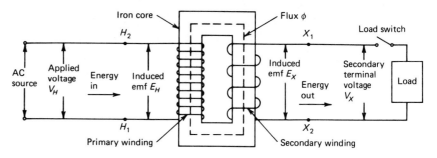

FIGURE 12-1   *Elementary diagram of a step-down transformer.*

and is delivered to the load from the other winding, called the *secondary winding*. When the transformer is used as a step-up transformer, the low-voltage winding is the primary. In a step-down transformer, the high-voltage winding is the primary.

Standard markings have been adopted for transformer terminals. The terminals of the high-voltage winding are marked H_1, H_2 . . . ; the terminals of the low-voltage winding are marked X_1, X_2 . . . . . It is convenient, therefore, to call the high-voltage winding the H winding and the low-voltage winding the X winding and to designate the number of turns of each winding as T_H and T_X.

12-2 THEORY OF OPERATION: NO LOAD

When an alternating voltage V_H is applied to the primary (H) winding of the step-down transformer represented in Fig. 12-1, with the load switch open, a small current called the *exciting current* flows. As in any inductive circuit, the current is limited by the counter emf of self-induction that is induced in the winding. Transformer windings are designed to have an inductance high enough to make the counter emf practically equal to the applied voltage at no load. This limits the no-load or exciting current to a very low value.

The exciting current causes an alternating flux to be set up in the core. This alternating flux cuts across the turns of both the primary and secondary windings as it increases and decreases in alternate directions, thereby inducing an emf in both windings. As noted previously, the emf induced in the primary winding opposes the applied voltage V_H. Since the turns of both windings are cut by the same flux, the emf induced in each turn of both windings is the same. If E_H is the emf induced in the primary winding and E_X is the emf induced in the secondary winding, then the voltage per turn in the two windings is E_H/T_H and E_X/T_X, respectively, and $E_H/T_H = E_X/T_X$.

If the resistance of the primary winding is small, as it usually is, E_H will be very nearly equal to the applied voltage V_H. Neglecting this small difference and noting that the secondary terminal voltage V_X will be equal to E_X since there is no current flowing, then $V_H/T_H = V_X/T_X$. Cross-multiplying and dividing by V_X/T_X results in

$$\frac{V_H}{V_X} = \frac{T_H}{T_X} \tag{12-1}$$

This equation shows that the voltages of each of the windings of a transformer are directly proportional to the number of turns in each winding.

EXAMPLE 12-1   A transformer with 200 turns on the primary winding is to be wound to step the voltage down from 240 to 120 V. Find the number of turns required on the secondary winding.

$$\frac{V_H}{V_X} = \frac{T_H}{T_X}$$

$$\frac{240}{120} = \frac{200}{T_X}$$

$$240\,T_X = 24,000$$

$$T_X = 100$$

12-3 OPERATION UNDER LOAD   When the load switch in the secondary circuit of the transformer in Fig. 12-1 is closed, a current I_X, equal to V_X divided by the load impedance, will flow. By Lenz's law, any current caused to flow by an induced emf flows in such direction as to oppose the action that causes the emf to be induced. In the case of a transformer, this means that I_X will always flow in a direction such that its magnetizing action will oppose the magnetizing action of the primary winding. The current I_X, then, tends to reduce the flux in the transformer core. However, if the flux is reduced, the counter emf E_H is reduced, thereby permitting more primary current I_H, to flow, which restores the flux to its original value.

If more load is added, causing the transformer secondary current I_X to increase, its increased demagnetizing action lowers the flux still more, which permits more primary current to flow. Conversely, when the secondary load is decreased, the demagnetizing action of I_X decreases, causing a decrease in primary current. Thus the magnetizing action of the primary winding adjusts itself with each change in secondary winding current.

From the above discussion the following relations are evident:

Primary ampere-turns = secondary ampere-turns

or
$$I_H T_H = I_X T_X \tag{12-2}$$

If both sides of Eq. (12-2) are divided by $I_H T_X$, then

$$\frac{I_X}{I_H} = \frac{T_H}{T_X} \tag{12-3}$$

that is, the ratio of the currents in a transformer is inversely proportional to the ratio of the turns.

When a load current flows from the secondary winding of a transformer, there is a small potential drop in the transformer as a result of its impedance. Thus, the terminal voltage is slightly lower than the induced emf. However, this difference is often neglected, and V_X is assumed to be equal to E_X. Thus Eq. (12-1) still applies. Combining Eqs. (12-1) and (12-3) results in

$$\frac{V_H}{V_X} = \frac{I_X}{I_H}$$

and
$$V_H I_H = V_X I_X \tag{12-4}$$

Equation (12-4) shows that the voltampere input of a transformer is equal to the voltampere output.

It should be noted in connection with Eqs. (12-1) to (12-4) that they are approximate equations only. They are true only for an ideal transformer, that is, a transformer with no losses. However, they are sufficiently accurate for most practical purposes since the losses in most transformers are very small.

A transformer supplies a load with 30 A at 240 V. If the primary voltage is 2400 V, find (*a*) the secondary voltamperes, (*b*) the primary voltamperes, and (*c*) the primary current. **EXAMPLE 12-2**

(*a*) $V_X I_X = 30 \times 240 = 7200$ VA

(*b*) $V_H I_H = V_X I_X = 7200$ VA

(*c*) $2400 I_H = 7200$ VA

$$I_H = \frac{7200}{2400} = 3 \text{ A}$$

Transformers normally replicate the primary voltage and current waveforms in the secondary winding. That is, when a sine wave is applied to the primary terminals, a sine-wave voltage or current is produced in the secondary. The only difference is the peak magnitude due to the effect of the transformer turns ratio. Figure 12-2*a* illustrates the current waveform for a single-phase transformer with a resistive load under normal conditions.

However, transformer cores are magnetic and can become saturated. (see Chap. 6). Saturation can take place when the transformer is first energized, when a higher than rated voltage is applied to the primary, or when the kVA rating of the transformer is exceeded. The result is that the transformer current waveform becomes distorted as shown in Fig. 12-2*b*. Note that the magnetizing component of the saturated transformer current is reactive and causes the waveform in Fig. 12-2*a* to not only be distorted, but also causes the wave to have the more lagging power factor shown in Fig. 12-2*b*.

Since a transformer has a magnetic core, it has a magnetization curve similar to that of a generator. Figure 12-3 shows a typical trans-

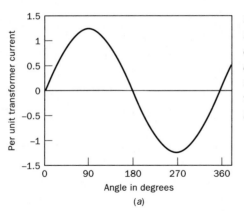

(*a*)

FIGURE 12-2   Transformer current waveforms illustrating effects of saturation: (a) transformer current for a resistive load for an unsaturated transformer; (b) transformer current for a resistive load for a saturated transformer.

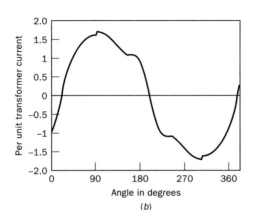

(*b*)

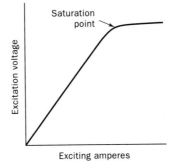

FIGURE 12-3   *Typical single-phase transformer magnetization curve.*

former magnetization curve, which has the characteristic shape of a magnetic material and has a saturation point. As long as the transformer operates along the linear portion of the magnetization curve, the transformer reproduces the primary waveform since, in this part of the curve, there is a constant relationship between the mmf and the flux.

When the primary saturates the iron core, the transformer operates "above the knee" of the magnetization curve and the current waveform no longer has the perfect sine-wave form that is required, as shown in Fig. 12-2*b*. Since it is not a pure sine wave, the waveform is made up of many harmonics or frequencies having varying magnitudes and phases. The result can be that secondary equipment operation may not be as desired. Harmonics are discussed in Chap. 13.

12-5 TRANSFORMER CONSTRUCTION

There are two different transformer core shapes in common use, namely, the *core type* and the *shell type* as shown in Fig. 12-4. The cores of both types are fabricated of special low-loss steel and are laminated to minimize core losses.

In the core-type construction shown in Fig. 12-4*a,* the windings surround the laminated iron core. For the sake of simplicity, the primary winding of the core-type transformer represented in Fig. 12-1 is shown on one leg of the core and the secondary on the other. Commercial transformers are not constructed in this manner because a large amount of the flux pro-

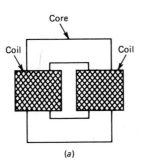
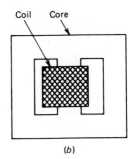

FIGURE 12-4   *Single-phase transformer core configurations: (a) core type; (b) shell type.*

FIGURE 12-5   Transformer core and coil assembly, core type. (General Electric Company.)

duced by the primary winding does not cut the secondary winding, or it is said that the transformer has a large *leakage flux*. To keep the leakage flux to a minimum, the windings are divided with half of each winding being placed on each leg of the core. The completed core and coil assembly of a core-type transformer is shown in Fig. 12-5.

Transformer shell-type construction is represented in Fig. 12-4*b*. In this construction the iron core surrounds the windings. A completed core and coil assembly of a shell-form distribution transformer is shown in Fig. 12-6.

The core and coil assemblies of the transformers shown in Figs. 12-5 and 12-6 are designed to be immersed in insulating oil in a steel tank. In addition to having insulating properties, the oil conducts heat from the core and coils to the surface of the tank where it is given off to the sur-

FIGURE 12-6   Transformer core and coil assembly, shell type. (General Electric Company.)

rounding air. Connections from the transformer coils to the external circuits are made through insulating bushings usually made of porcelain.

Coils of transformers are wound with copper or aluminum wire or strap. For heavy-current windings, several strands of conductor are paralleled to reduce eddy-current loss in the conductors. Coil insulation materials used are cotton, cellulose, special papers, polyester glass tapes, or other similar materials. Completed coils are dried in an oxygen-free atmosphere to remove all moisture. Coils of oil-immersed transformers are then impregnated with dry insulating oil while they are in a vacuum.

Two basic types of transformer windings are commonly used. These are the *concentric* and the *pancake* types. Concentric windings are cylindrical in form with one winding placed inside the other with the necessary insulation between them. The low-voltage winding is normally placed on the inside next to but insulated from the core. Pancake windings are built up with primary and secondary sections interleaved. In both types, spacers are provided between adjacent coils to permit ventilation or the circulation of the cooling liquid.

12-6 RATINGS In addition to voltage ratings for both primary and secondary windings, transformers are rated as to the kilovoltamperes they can deliver at rated secondary voltage and rated frequency without exceeding a given temperature rise. The capacity rating is therefore independent of the secondary-load power factor, as illustrated by Example 12-3.

What is the full-load kilowatt output of a 5-kVA 2400/120-V transformer at (*a*) 100 percent, (*b*) 80 percent, and (*c*) 30 percent power factor? (*d*) What is the full-load current output? **EXAMPLE 12-3**

(*a*) $P = \text{kVA} \times \text{power factor} = 5 \times 1.0 = 5 \text{ kW}$

(*b*) $P = 5 \times 0.8 = 4 \text{ kW}$

(*c*) $P = 5 \times 0.3 = 1.5 \text{ kW}$

(*d*) $I = \dfrac{VA}{V} = \dfrac{5000}{120} = 41.7 \text{ A}$

Full-rated current of 41.7 A is supplied by the transformer at the three different power factors even though the kilowatt output is different in each case.

Included on a transformer nameplate in addition to kVA capacity and voltage rating data are the connection diagrams, percent impedance, polarity data, cooling class, data on winding taps, and basic insulation level. Basic insulation level is an indication of the ability of the insulation system to withstand transient overvoltages.

Transformers may be classified as *distribution transformers* or as *power transformers*. The term *distribution transformer*, as generally accepted, is applied to transformers rated 500 kVA and smaller, with high-voltage ratings of 67,000 V and below and with low-voltage ratings of 15,000 V and below. Transformers with higher kilovoltamperes and voltage ratings are referred to as power transformers.

12-7 TRANSFORMER LOSSES AND EFFICIENCY

The efficiency of a transformer is the ratio of the useful power output to the total power input. Since the input to a transformer is equal to its useful output plus its losses, the efficiency equation may be written in either of the following forms:

$$\text{Percent efficiency} = \frac{\text{power output}}{\text{power input}} \times 100 \text{ percent} \qquad (12\text{-}5)$$

$$= \frac{\text{power output}}{\text{power output} + \text{losses}} \times 100 \text{ percent} \qquad (12\text{-}6)$$

From the above, it is evident that the efficiency of a transformer can be determined for any given load by making a direct measurement of its power input and its power output and computing its efficiency using Eq. 12-5. Because of limitations of available testing facilities, it is sometimes difficult to perform direct input-output load measurements, especially for very large transformers. When input-output measurements are not feasible, the losses of a transformer may be measured or calculated and its efficiency determined by using Eq. 12-6.

Although transformers are highly efficient, some losses are present in all transformers. There are two classes of losses, namely, *load losses,* which are principally I^2R losses in the transformer windings, and *no-load losses,* which are primarily core losses.

Load or I^2R losses are present because power is used whenever a current is made to flow through a resistance. Load currents flowing through transformer windings produce a power or I^2R loss that varies with the load being supplied by the transformer. Load losses can be calculated for any given load if the resistances of both windings are known or can be measured. If R_H and R_X are the high- and low-voltage winding resistances, then the load loss is

$$\text{Load loss} = I_H^2 R_H + I_X^2 R_X \qquad (12\text{-}7)$$

No-load or core losses are due to the effects of hysteresis and eddy currents in the iron core of the transformer. These effects are similar to those occurring in generators and motors as described in Sec. 8-15. Core losses in a transformer can be determined by energizing one winding of the transformer with the other winding open. The power input to the open-circuited transformer as measured with a wattmeter supplemented with a frequency meter, voltmeter, and ammeter is then the no-load or core loss. The core loss of a transformer is essentially constant for all loads when rated frequency and voltage are applied to the transformer.

12-8 TRANSFORMER COOLING

When a transformer is operating under load, heat is generated in both the windings and in the core owing to the losses mentioned in Sec. 12-7. Means must be provided to extract this heat from the transformer.

Transformers may be grouped into two general classifications depending upon the method of cooling used. Transformers may be the *liquid-immersed type* or the *dry type*. The liquid normally used in the liquid-immersed transformers is oil, except that when the use of oil may present a fire hazard the liquid used is a nonflammable type. Dry-type transformers use air or an inert gas as a cooling medium. Both liquid-immersed and dry-type transformers are available with several different variations of the basic cooling method. Service conditions and economics dictate the selection of the specific cooling method for each application.

The simplest type of oil-immersed transformer is the self-cooled type. Heat generated in the core and coil assembly is conducted by convection through the surrounding oil to the tank, where it is given off to the surrounding atmosphere. In the larger sizes, fins or radiators are provided to present a larger tank surface to the air. Pole-type distribution transformers as well as smaller station-type power transformers are usually self-cooled.

In order to obtain more kVA capacity from a given core and coil assembly, supplemental cooling is used on larger power transformers. The *self-cooled, forced-air-cooled transformer* is provided with electrically operated, thermostatically controlled cooling fans to supplement the natural cooling system. The fans move air around fins or radiators through which the cooling oil circulates. The fans operate only when the oil temperature exceeds a certain value such as during high-transformer-load periods or during periods of high ambient-air temperature.

Still greater cooling may be obtained on large power transformers by the use of pumps to circulate the oil through radiators. The oil pumps may be the only type of supplemental cooling, or they may be arranged to supplement forced-air cooling.

FIGURE 12-7 *Three-phase oil-immersed trans-*
former with supplemental cooling. (Westinghouse
Electric Corporation.)

A completed three-phase oil-immersed power transformer with supplemental cooling is shown in Fig. 12-7. Cooling fans can be seen mounted on
the radiator at the rear of the transformer.

Smaller dry transformers are the ventilated self-cooled type, which are
cooled by the natural circulation of air around the core and coils. On larger
dry-type transformers, fans thermostatically controlled by winding temperature are sometimes used to supplement the natural cooling. In extremely
dusty, dirty, or hazardous locations, sealed dry-type transformers are used.
The insulating gas in these self-cooled transformers may be air, nitrogen,
or other inert gas.

12-9
TRANSFORMER
IMPEDANCE

A characteristic of transformers on which data is needed for the calculation
of transformer losses, voltage regulation, and system short-circuit currents
is called the *percent impedance* of the transformer.

Transformer percent impedance is determined by a test normally
made on a transformer immediately after it is manufactured. The secondary winding is short-circuited, and a reduced voltage is applied to the
primary winding. The primary voltage is then increased until rated secondary winding current flows. The voltage required to cause rated secondary current to flow is called the impedance voltage. This voltage, when
expressed as a percentage of the rated primary voltage, is called the percent

impedance of the transformer. Percent impedance of power transformers typically is 5 to 10 percent depending on the transformer voltage ratings.

The subject of system short-circuit currents and the relationship of transformer impedance to these currents is discussed in Chap. 19.

12-10 POLARITY AND PARALLEL OPERATION

It is sometimes desirable to operate two transformers in parallel. One of the requirements for successful parallel operation is that the polarities of the two transformers must be the same. In addition to having identical polarities, the two transformers must have the same or very similar voltage ratings, transformation ratios, and impedances.

As indicated in Sec. 12-1, standard terminal designations have been adopted for transformers. The high-voltage terminals of a two-winding single-phase transformer are marked H_1 and H_2 and the low-voltage terminals X_1 and X_2. Furthermore, the lead *polarity* or the relative instantaneous direction of currents in the leads is standardized. In a standard single-phase transformer, at any given instant, if current enters the H_1 lead, current leaves the X_1 lead in the same direction as though the two leads formed a continuous circuit.

Thus to parallel two correctly marked transformers, both H_1 terminals should be connected to one high-voltage line and both H_2 terminals to the other high-voltage line. Then the X_1 terminals should be connected to the same low-voltage line and the X_2 terminals to the remaining low-voltage line. Two transformers connected for parallel operation in this manner are shown in Fig. 12-8. It should be noted that *if terminals of unlike polarity should be connected to the same line, the two secondary windings would be short-circuited on each other with a resulting excessive current flow.*

The physical location of single-phase transformer terminals is shown in Fig. 12-9, which represents plan views of two transformers. If terminals are arranged as shown in Fig. 12-9a, the transformer is said to have *additive polarity,* and if arranged as shown in Fig. 12-9b, the transformer is

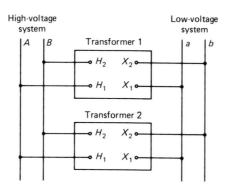

FIGURE 12-8 *Two transformers connected for parallel operation.*

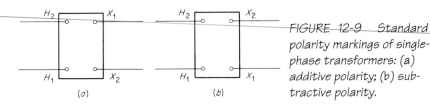

FIGURE 12-9 Standard polarity markings of single-phase transformers: (a) additive polarity; (b) subtractive polarity.

said to have *subtractive polarity*. Note that for either polarity, the H_1 terminal is on the right-hand side when the transformer is viewed from the high-voltage side of the tank.

The following description of a test to determine transformer polarity provides a good explanation of the terms *additive* and *subtractive polarity*. The connections shown in Fig. 12-10 may be used to make a simple test to determine transformer polarity if the terminals are not marked or if it is wished to verify markings. If the terminals are not marked, the right-hand and left-hand high-voltage terminals, as viewed from the high-voltage side of the transformer, may be marked H_1 and H_2, respectively. For convenience, the low-voltage terminals may be marked X_A and X_B temporarily as shown. Adjacent terminals such as H_2 and X_A are then connected, and a voltmeter is connected across the other two terminals H_1 and X_B . An alternating voltage is then applied to the high-voltage winding of the transformer. If the voltmeter reads less than the value, of the applied voltage, the polarity is subtractive and the terminals X_A and X_B may be marked as the X_2 and X_1 terminals, respectively. If the voltmeter reads more than the applied voltage, the polarity is additive and the terminals X_A and X_B are marked X_1 and X_2, respectively.

12-11 INSTRUMENT TRANSFORMERS

It is not a safe practice to connect instruments, meters, or control apparatus directly to high-voltage circuits. Instrument transformers are universally used to reduce high voltages and currents to safe and usable values for the operation of such apparatus.

Instrument transformers perform two functions: (1) They act as *ratio devices*, making possible the use of standardized low-voltage and low-current meters and instruments. (2) They act as *insulating devices*, to protect the apparatus and the operating personnel from high voltages. There are two kinds of instrument transformers: *voltage transformers* and *current transformers*.

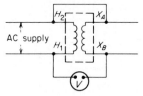

FIGURE 12-10 Connection for checking the polarity of a transformer.

FIGURE 12-11 Indoor-type voltage transformer. (General Electric Company.)

12-12 VOLTAGE TRANSFORMERS

Voltage transformers supply a voltage to meters, instruments, or control devices that has a definite ratio to the line voltage. These transformers formerly were called *potential transformers*. The voltage transformer operates on the same principle as a power transformer. When the primary winding is connected across the line, a current flows, which sets up a flux in the core. The flux linking the secondary winding induces an emf that is proportional to the ratio of primary to secondary turns. Voltage transformers are commonly designed to have a rated voltage of 115 or 120 V at the secondary terminals when rated voltage is applied to the primary winding. Since the amount of load supplied by voltage transformers is small, voltampere ratings are small, common ratings being 200, 600, and 1000 VA. A 7200- to 120-V 1000-VA voltage transformer is shown in Fig. 12-11.

Voltage transformers suitable for metering and control on modern high-voltage transmission systems having rated voltages of 350 kV and higher become very bulky and costly. For this reason, the less costly coupling-capacitor voltage transformer (CCVT) has come into use on high-voltage systems as a replacement for the conventional voltage transformer. The CCVT consists of a capacitor divider and an electromagnetic unit connected from line to ground on the system as shown schematically in Fig. 12-12. The unit is designed and interconnected to provide a secondary voltage proportional to and in phase with the applied primary voltage.

12-13 CURRENT TRANSFORMERS

The purpose of a current transformer is to provide a means of reducing line current to values that may be used to operate standard low-current measuring and control devices, with these devices being completely insulated from the main circuits. As the current transformer is used in conjunction with current-measuring devices, its primary winding is designed to be connected

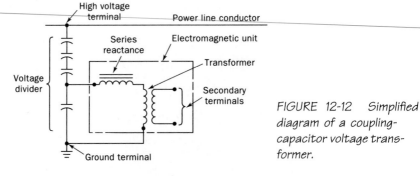

High voltage
terminal

Power line conductor

Series
reactance

Electromagnetic unit

Transformer

Voltage
divider

Secondary
terminals

Ground terminal

FIGURE 12-12 Simplified
diagram of a coupling-
capacitor voltage trans-
former.

in series with the line. It is therefore necessary that the impedance of the
primary winding be made as low as possible. This is done by using a few
turns of low-resistance wire capable of carrying rated line current. Current
transformers designed for use in very high current circuits may have a primary winding consisting of only one turn.

Since a current transformer is used ordinarily to decrease current, the
secondary contains more turns than the primary and the ratio of primary to
secondary current is inversely proportional to the ratio of primary to secondary turns. Current transformers are usually designed so that when rated
current flows in the primary, 5 A will flow in the secondary. A current transformer, having a ratio of 30 to 5 A, is shown in Fig. 12-13.

Since the impedance of the primary winding of a current transformer
is low, the current through the primary depends primarily on the load connected to the primary circuit rather than on the secondary load of the
transformer itself. When the secondary circuit is closed, the secondary current establishes an mmf that opposes the primary mmf, thereby limiting the
flux density in the core. When the secondary circuit is opened while the primary is energized, the secondary demagnetizing mmf is no longer present.
This means that the flux density may become very high, causing a dangerously high voltage to be induced in the open secondary winding. To prevent
injury to the apparatus and operating personnel, it is necessary that the
secondary terminals of a current transformer be short-circuited before
removing or inserting an instrument in the secondary circuit while the pri-

FIGURE 12-13 Indoor-type current
transformer. (General Electric
Company.)

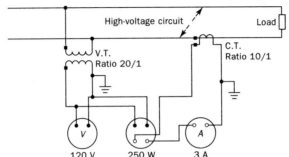

FIGURE 12-14
Instrument-
transformer connec-
tions.

mary winding is energized. Modern current transformers are provided with a device for short-circuiting the secondary winding when changes must be made in the secondary circuit. However, anyone working with a secondary circuit of a current transformer should always make certain that the short-circuiting device is closed before the secondary wiring is disturbed.

Figure 12-14 illustrates the method of connecting instrument transformers into a high-voltage circuit. As shown, the secondary circuits of both the voltage and current transformers are connected to ground. This is a general safety practice followed when instrument transformers are connected into high-voltage circuits.

Instrument transformer voltage, current, or voltampere ratings must not be exceeded or they will become saturated and the secondary waveform will become distorted. This will cause the instruments and protective equipment connected to the transformers to receive false or inaccurate indications.

Find the load current, voltage, power, and power factor in the circuit **EXAMPLE 12-4** of Fig. 12-14, as indicated by the instrument readings.

$$\text{Current} = 3 \times 10 = 30 \text{ A}$$

$$\text{Voltage} = 120 \times 20 = 2400 \text{ V}$$

$$\text{Power} = \frac{250 \times 10 \times 20}{1000} = 50 \text{ kW}$$

$$\text{Kilovoltamperes} = \frac{30 \times 2400}{1000} = 72 \text{ kVA}$$

$$\text{Power factor} = \frac{50}{72} \times 100 = 69.4 \text{ percent}$$

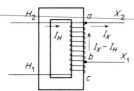

FIGURE 12-15   Step-down autotransformer.

12-14 AUTO-TRANSFORMERS

A transformer in which the primary and secondary windings are connected *electrically* as well as *magnetically* is called an autotransformer. Figure 12-15 shows a connection diagram of an autotransformer. If this transformer is to be used as a step-down transformer, the entire winding *ac* forms the primary winding and the section *ab* forms the secondary winding. In other words, the section *ab* is common to both primary and secondary. As in the standard two-winding transformer, the ratio of voltage transformation is equal to the ratio of primary to secondary turns if the losses and exciting currents are neglected and

$$\frac{V_H}{V_X} = \frac{T_H}{T_X} = \frac{I_X}{I_H}$$

Figure 12-16 represents an autotransformer winding with a total of 220 turns, with the sections *ab* and *bc* having 150 and 70 turns, respectively. If a voltage of 440 V is applied to the winding *ac*, the voltage across each turn will be 2 V. The voltage from *a* to *b* will then be 150 × 2, or 300 V.

When a noninductive load of 30 Ω is connected to winding *ab*, a current, I_X, of 300/30 or 10 A flows and the power output of the transformer is 300 × 10 or 3000 W. Neglecting the transformer losses, the power input must be 3000 W and the primary current 3000/440 or 6.82 A.

An application of Kirchhoff's current law to point *a* shows that when I_X is 10 A and I_H is 6.82 A, then the current from *b* and *a* must be 3.18 A. Similarly, the current from *b* to *c* must be 6.82 A.

Thus the section of the winding that is common to both primary and secondary circuits carries only the difference in primary and secondary currents. In effect, the transformer in the example transforms only 3.18 × 300 = 954 W rather than the full circuit power of 3000 W. The percentage

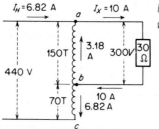

FIGURE 12-16   Step-down autotransformer winding supplying load.

of power transformed is $100 \times 954/3000$ or 31.8 percent. This is the same as the percent voltage difference between the primary and secondary voltages or $100 \times (440 - 300)/440$ or 31.8 percent. Since only a part of the circuit power or kVA is transformed by an autotransformer, it is smaller and more efficient than a two-winding transformer having a similar rating.

For some applications that require a multivoltage supply, an autotransformer in which the winding is tapped at several points is used. The connections from the various taps are brought out of the tank to terminals or to a suitable switching device so that any one of several voltages may be selected.

Autotransformers are used when voltage transformations of near unity are required. Such an application of an autotransformer is in "boosting" a distribution voltage by a small percentage to compensate for line drop. Another common application is in the starting of ac motors, in which case the voltage applied to the motor is reduced during the starting period.

Autotransformers are not safe, however, for supplying a low voltage from a high-voltage source; for, if the winding that is common to both primary and secondary should accidentally become open, the full primary voltage will appear across the secondary terminals. The requirements of safety codes should always be followed whenever autotransformers are applied.

12-15 NO-LOAD TAP CHANGING

For small adjustments of transformer ratio, to compensate for line drop, high-voltage windings of transformers are often equipped with tap-changing devices. The winding is tapped at several points, and connections from these points are made either to a tap-changing switch or to a terminal block inside the transformer tank. The transformers shown in Figs. 12-5 and 12-6 are both equipped with tap-changing switches, which are mounted on an insulating block directly above the windings. By operating the switch, changes are made in the number of active turns in the high-voltage winding. These changes must be made with the transformer deenergized except in certain types of transformers in which special provision is made for tap changing under load.

12-16 SINGLE-PHASE CONNECTIONS

Distribution transformers are normally wound with the secondary or low-voltage winding in two sections, as shown schematically in Fig. 12-17. When the two low-voltage sections are connected in parallel, the transformer may be used to supply a two-wire 120-V load with the two load wires connected to terminals X_1X_3 and X_2X_4 as in Fig. 12-17a.

However, the more commonly used transformer connection, shown in Fig. 12-17b, is the series connection, which is used to supply a three-wire

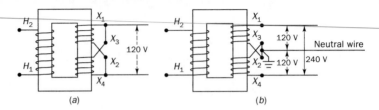

FIGURE 12-17 *The divided low-voltage winding connected (a) in parallel to supply a 120-V two-wire system or (b) in series to supply a 120/240-V three-wire grounded neutral system.*

120/240-V distribution system. In addition to the two load wires connected to terminals X_1 and X_4, a third wire called the neutral wire is connected from the common X_2X_3 transformer terminal to the load. As protection against dangerously high abnormal voltages, the neutral wire is connected to ground at both the transformer and at the load in accordance with electrical safety codes. Either 120- or 240-V loads may be supplied from the three-wire system as described in Sec. 2-12.

Earlier transformers were made with four insulated leads brought out of the transformer tank, and the desired series or parallel connection of the windings was made outside the tank. In more modern transformers, the connections are made inside the tank and only three secondary leads are brought out of the transformer.

12-17
THREE-PHASE
CONNECTIONS

Single-phase transformers can be connected to form three-phase transformer banks for raising or lowering the voltages of three-phase systems. Four common methods of connecting three transformers for three-phase transformations are the delta-delta, wye-wye, delta-wye, and wye-delta connections. The first three of these are shown in Fig. 12-18; the wye-delta is not shown since it is simply the reverse of the delta-wye connection.

To match transformer polarities correctly in a three-phase transformer bank, the terminals must be connected symmetrically as shown in Fig. 12-18. If, in a wye-connected bank, an H_1 or X_1 terminal of one transformer is connected to the system neutral, then the corresponding H_1 or X_1 terminal of all the transformers should be connected to the same neutral point and the remaining H_2 or X_2 terminals brought out as line connections. In a delta connection, H_1 should always be connected to H_2 and X_1 to X_2 and the line connections made at these junctions.

An advantage of the delta-delta connection shown in Fig. 12-18*a* is that if one transformer in the bank becomes damaged or is removed from service, the remaining two can be operated in what is known as the open-

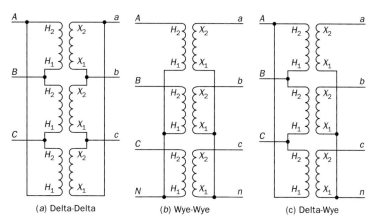

FIGURE 12-18 *Three-phase transformer connections: (a) delta-delta; (b) wye-wye; (c) delta-wye.*

delta or V connection. By being operated in this manner, the bank still delivers three-phase currents and voltages in their correct phase relationships, but with the capacity of the bank reduced to 57.7 percent of what it was with all three transformers in service.

In the wye-wye connection shown in Fig. 12-18*b*, only 57.7 percent (or 1/1.73) of the line-to-line voltage is impressed on each winding, but full-line current flows in each transformer winding. Power circuits supplied from a wye-wye bank often create serious disturbances in nearby communication circuits. Because of this and other disadvantages, the wye-wye connection is used infrequently.

The delta-wye connection shown in Fig. 12-18*c* is well adapted for stepping up voltages at generating stations since the voltage is increased by the transformer ratio multiplied by the factor of 1.73. Similarly, the wye-delta connection is often used for stepping down voltages.

The low-voltage connections of transformer banks that supply power and lighting systems may be either wye- or delta-connected depending on the size and voltage requirements of the loads. Small and medium-sized three-phase motor loads are sometimes supplied from 240-volt delta-connected transformer secondaries. To provide a 120-volt supply for lighting circuits, the midpoint of one of the transformers is brought out and grounded to form the neutral of what is called a *four-wire delta system*. The connection of this system is shown in Fig. 12-19*a*. As shown, a three-wire three-phase power circuit is formed by lines *A, B,* and *C* and a three-wire single-phase lighting circuit is formed by lines *B, N,* and *C*. The center-tapped transformer should have a larger kilovoltampere rating than the other two transformers.

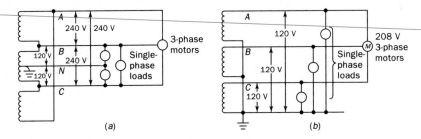

FIGURE 12-19 (a) Four-wire delta-connected transformer secondaries.
(b) Four-wire wye-connected transformer secondaries.

When a large number of single-phase loads are to be served from a three-phase transformer bank, the wye-connected low-voltage winding is desirable, since the single-phase loads can be balanced evenly on all three phases. Figure 12-19*b* shows the method of connecting both three-phase and single-phase loads to four-wire wye-connected secondaries. The single-phase loads are connected from the three line wires *A, B,* and *C* to the neutral *N* or grounded wire, while the three-phase loads are connected to lines *A, B,* and *C.* This connection, with phase and line voltages of 120 and 208 V, is used in underground networks in metropolitan areas to supply both single- and three-phase loads. Similarly, this connection is used in commercial and industrial systems with phase and line voltages of 277 and 480 V or 240 and 416 V, respectively.

In addition to the more widely used delta and wye three-phase connections, the T connection is sometimes used in distribution systems. The connection is made by using two single-phase transformers, often mounted in the same tank, connected as shown in Fig. 12-20. As shown, the end of one winding connects to the midpoint of the other to form a configuration similar to the letter T. Although not a symmetrical connection, the T connection approximates the performance of the delta and wye connections with a less costly transformer.

**12-18
THREE-PHASE
TRANSFORMERS**

Three-phase voltages may be transformed by means of three-phase transformers. The core of a three-phase transformer is normally designed with three legs, a primary and a secondary winding of one phase being placed on

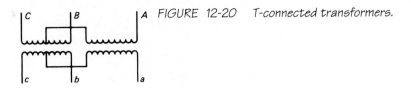

FIGURE 12-20 T-connected transformers.

each leg. The three-legged-core transformer is adequate for all the more commonly used three-phase connections since the fluxes established by the three windings are 120° apart in time phase. At any time, two core legs act as the return path for the flux in the third leg and the net sum of the flux is zero. A core and coil assembly of a three-phase dry-type distribution transformer with a three-legged core is shown in Fig. 12-21.

During system-fault conditions or periods of highly unbalanced loads, the sum of the individual fluxes may not total zero in a transformer with a three-legged core. During such unbalanced conditions, some of the magnetic flux is forced to return through the transformer tank, resulting in severe overheating of the tank. This condition occurs most often in wye-wye–connected three-phase transformers operating with both neutrals connected to earth. To rectify this problem, transformers used for this service are designed with four- or five-legged cores that provide a return path for the unbalanced magnetic fluxes.

Because the three windings can be placed compactly on one core, the three-phase transformer occupies less space than do three single-phase transformers with the same total kilovoltampere rating. The three-phase transformer is slightly more efficient than the single-phase transformer and in the larger ratings has a lower first cost. A disadvantage of the three-phase transformer lies in the fact that when one phase becomes defective, the entire three-phase unit must be removed from service. When one transformer in a bank of three single-phase transformers becomes defective, it may be removed from service and the other two transformers may often be reconnected to supply service on an emergency basis until repairs can be made.

FIGURE 12-21   Three-phase dry-type transformer core and coil assembly. (Westinghouse Electric Corporation.)

Three-phase transformers may also be connected according to the four combinations mentioned for three single-phase transformers. These connections are made inside the tank, and for delta-delta connections only three high-voltage and three low-voltage leads need to be brought outside the tank. Four connections are brought out from wye-connected windings.

FEEDER-VOLTAGE REGULATORS

Much of the load on an electrical system is at some distance from the generating station. Because of this distance, there is a drop in voltage between the generating station and the point of utilization. Furthermore, this voltage drop is not constant but is proportional to the amount of load on the circuit at any one time. As the load varies, the voltage drop varies also.

Most electric equipment is designed to operate at a given voltage, and the efficiency of operation of this equipment is affected appreciably if the applied voltage deviates from the rated value. This is especially true with domestic loads such as incandescent lighting, electric ranges, and electric heaters. With 90 percent of rated voltage applied to an incandescent lamp, for example, the illumination output is only about 70 percent of rated value. Heating time of electric heaters and ranges is about 120 percent of normal when 90 percent of rated voltage is applied. While induction-motor operation is not affected as greatly by voltage variation as lamps and heaters, there are changes in efficiency, heating, and speed occasioned by variations in voltage.

Feeder-voltage regulators are used extensively as an aid to maintaining a reasonably constant voltage at the point of use on an electrical system. A voltage regulator is in effect an autotransformer with a variable ratio.

12-19 VOLTAGE-REGULATOR PRINCIPLE
The basic principle of the voltage regulator is illustrated by the diagram shown in Fig. 12-22. The transformer shown has a voltage ratio of 2400 to 240. By means of the reversing switch, the transformer may be connected as an autotransformer, and, depending on the switch position, the voltage induced in the low-voltage winding may be added to or subtracted from the

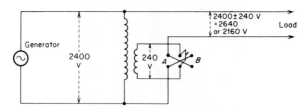

FIGURE 12-22   Circuit for illustrating the principle of the feeder voltage regulator.

voltage applied to the high-voltage winding. Thus, when the generator voltage is 2400 V, as shown in Fig. 12-22, the output voltage of the transformer is either 2640 or 2160 V. When the generator voltage is 10 percent low, the transformer may be used to raise the voltage 10 percent. When the generator voltage is 10 percent high, the transformer may be used to lower the voltage by 10 percent.

Feeder-voltage regulators are constructed with the two separate windings as shown in Fig. 12-22 and are provided with means for varying the amount of voltage induced in the low-voltage or series winding in small increments. These small increments of voltage may be added to or subtracted from the regulator input voltage, so that it is possible to maintain a nearly constant output voltage from the regulator.

As may be seen from Fig. 12-22, the autotransformer used as a voltage regulator actually transforms only 10 percent of the circuit kilovoltamperes, so the kilovoltampere rating of the regulator need be only 10 percent of the circuit kilovoltamperes. If, for example, the circuit full-load current in Fig. 12-22 were 100 A, the regulator kilovoltampere rating would be 240×100 or 24,000 VA (24 kVA), while the circuit rating would be 2400×100 or 240,000 VA (240 kVA). In general, the kilovoltampere rating of a single-phase regulator is the product of the rated load amperes and the percent regulation expressed in kilovolts. For a three-phase regulator this product is multiplied by 1.73.

12-20 STEP-VOLTAGE REGULATORS

Step-type voltage regulators are tapped autotransformers with a tap-changing-under-load switching mechanism to change tap settings and thereby change the ratio between the primary and secondary windings. Step regulators consist of three major parts: the transformer, the tap-changing mechanism, and the control mechanism. A typical single-phase, distribution-type, step-voltage regulator is shown in Fig. 12-23.

The transformer section of a regulator is a conventional transformer with the primary (shunt) and secondary (series) windings wound on an iron core. The series winding is tapped at regular intervals, the number of taps depending on the type of regulator. Pole-mounted regulators for regulating small loads may have taps arranged for 16 or 32 steps with a total range of regulation of 10 or 20 percent. Station-type regulators used for regulating entire feeder circuits or substations usually have 32 steps with a 20 percent range in regulation. Voltage is adjusted in steps of 5/8 of 1 percent in these regulators.

A schematic diagram of connections of a single-phase 32-step regulator is shown in Fig. 12-24. The taps of the series winding are connected to wide stationary contacts arranged physically in a circle on an insulating panel in the tap-changing mechanism. Two contacts, *A* and *B*, which are mounted a fixed distance apart on a movable arm, are used to make contact with the stationary contacts. The two movable contacts are spaced so that for any operating position, they both make contact with the same stationary contact, or one of them contacts each of two adjacent stationary contacts.

The operation of the tap-changing mechanism of a 32-step 20 percent regulator may be explained by referring to Fig. 12-24. The two movable contacts *A* and *B,* as shown in the diagram, are in contact with stationary contact number 1, the maximum LOWER position. The load current divides equally through the two halves of the reactor, which is connected to the two movable contacts. The movable arm of the switch assembly is operated by a motor that drives a spring-loaded quick-operating mechanism for changing taps. The switch arm and movable contacts are latched in place until the

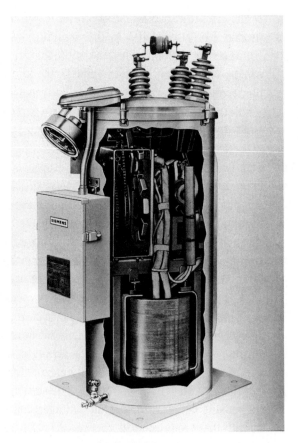

FIGURE 12-23 *Single-phase step-type voltage regulator for pole mounting. (Siemens Energy & Automation, Inc.)*

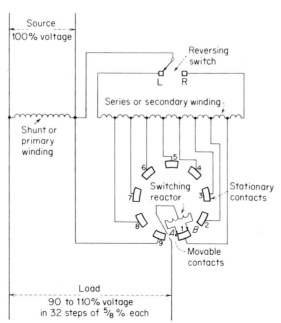

Source
100% voltage

Reversing
switch

L   R

Series or secondary winding

Shunt or
primary
winding

5
6   4

Switching   Stationary
7   reactor   3   contacts

8   2

9   A B

Movable
contacts

Load
90 to 110% voltage
in 32 steps of $5/8$ % each

FIGURE 12-24
Connections of a single-
phase 32-step voltage
regulator.

driving springs are fully loaded, at which time the latch releases and the switch arm moves with a quick snap to the next position, where it is latched in place again.

A change of the switch arm from tap 1 to tap 2 to raise voltage is made in two operations. First, the movable contact arm snaps to a point where contact B is on contact 2 but A is still on contact 1. In this position, the reactor acts as an autotransformer, and the resulting voltage is halfway between the voltages of taps 1 and 2. It should be noted that the load current is not interrupted during the changing of the switch position, since the movable contact A is in contact with stationary contact 1 during the instant that the movable contact B is moving from 1 to 2. During this instant, all the load current is carried by one-half of the reactor. The next movement of the tap changer moves both A and B to contact 2, and the resulting voltage is that of tap 2. As the tap-changing switch arm is moved on around toward contact 9, the voltage is raised in $5/8$ percent steps until the number 9 contact is reached. When the two movable contacts are on stationary contact 9, the regulator is in the NEUTRAL position, and the series winding is completely out of the circuit, so that voltage is neither raised nor lowered by the regulator. As the mechanism is moved through the neutral position to the RAISE positions, the reversing switch is operated automatically, reversing the polarity of the series winding. Thus the first RAISE position is contact 1, and the voltage is raised in $5/8$ percent steps as the contact arm moves around to contact 8, which is the maximum RAISE position. The full range of plus and minus 10 percent voltage regulation in thirty-two ⁵/₈ percent steps with one neutral position is obtained with only eight taps in the series winding.

Step regulators are manufactured in both single- and three-phase units. The transformer section of the three-phase unit is assembled on a three-phase core, and the tap-changer switch is of a three-pole design, so that taps on all three phases are changed simultaneously. Step regulators are available in voltage ratings as high as 69,000 V but are normally used on lower voltage distribution circuits. Kilovoltampere ratings of single-phase regulators range from approximately 25 to 833 kVA. Three-phase regulators are available in even larger sizes.

12-21 VOLTAGE-REGULATOR CONTROL AND ACCESSORY EQUIPMENT

Voltage regulators ordinarily are equipped with controls and accessories that operate the regulator automatically to maintain a predetermined output voltage. Modern regulators have self-contained instrument and auxiliary transformers to supply the voltage and current indications for the control system and the necessary power for the tap-changing operating motor.

A solid-state voltage sensing device in the regulator-control system initiates changes in the regulator position. This device is connected to the internally mounted voltage transformer to sense the regulator output voltage. The device operates to close the operating motor circuit to raise or lower regulator tap settings, thereby maintaining output voltage at a predetermined level.

So that the regulator does not operate excessively and attempt to correct minor momentary system voltage disturbances, a time delay or holding action is built into the control unit. This part of the control unit integrates the system voltage variation by summing the lengths of time that the system voltage is abnormal. If the voltage is either too high or low for a predetermined length of time, the control unit acts to correct the output voltage. The amount of time delay is adjustable and may be preset on the control panel.

Regulators normally are equipped with *line-drop compensators.* The compensator consists of an adjustable resistance and a reactance inserted in series with the voltage-control relay. Current is supplied to the line-drop compensator from a current transformer in the load circuit of the regulator. This current produces a voltage drop in the control circuit that is proportional to the load current (and therefore proportional to the line-impedance drop). In this way the voltage regulator automatically increases the voltage as the load increases to compensate for the increasing drop in the line. Regulators equipped with line-drop compensators may be adjusted to maintain a constant voltage at a remote point on a feeder circuit, rather than at the point at which the regulator is installed.

Other control equipment and accessories used with voltage regulators include bypass surge arresters for the series winding, a control switch to change the regulator from manual to automatic operation, a manual raise-and-lower switch, test terminals for connecting a portable voltmeter, an auxiliary relay for the operating motor, limit switches to prevent overtravel of the regulator, a position indicator, and protective and switching equipment for the auxiliary operating power.

12-22 TAP-CHANGING-UNDER-LOAD TRANSFORMERS

Tap-changing-under-load transformers, or load-ratio-control transformers, as they are sometimes called, are essentially combined transformers and voltage regulators. One winding of the transformer is tapped at various points, and these taps are connected to a tap-changing switch as in the step-type voltage regulator. These transformers are nearly always constructed as three-phase units.

One application of the tap-changing-under-load transformer is in a step-down station used to supply a load from a transmission line. In this application a constant voltage may be maintained for the load even though the transmission-line voltage varies considerably. One other common application of this transformer is in tie lines between two generating stations. A tap-changing-under-load transformer in such a tie line permits each station to maintain correct voltage at its own bus for supplying local loads, yet maintains the tie-line voltage at the correct value for interchange of power between the two stations. On many such interconnections it has been found that there has been a heavy interchange of reactive or wattless current over the tie line when the tie line is operated without tap-changing-under-load equipment. In effect, the variable-ratio transformer provides these interconnections with reactive voltampere control.

REVIEW QUESTIONS

1 What is the function of a transformer?
2 How does the efficiency of a transformer compare with rotating electric machinery such as motors and generators?
3 What is meant by (*a*) a primary winding; (*b*) a secondary winding?
4 How are transformer terminals marked?
5 What relationship exists between the voltages and the number of turns on the two windings of a transformer?
6 What relationship exists between the ratio of the voltages and the ratio of the currents in the high- and low-voltage windings?
7 Explain how a transformer regulates the amount of primary current required to supply a given secondary current.
8 What information is commonly found on a transformer nameplate?
9 What are two types of core construction used in transformers?

10 What is meant by leakage flux? How is it kept to a minimum?

11 What transformer losses vary with the amount of load on the transformer? Which losses are practically constant?

12 Why is oil placed in transformers?

13 Why are air-cooled transformers desirable in some locations?

14 Instrument transformers perform what two functions?

15 Are potential transformers connected in parallel or series with the load? How are current transformers connected?

16 What are some of the insulating materials used in transformer windings?

17 Why is the capacity of a transformer rated in kilovoltamperes?

18 What is the main difference between distribution and power transformers?

19 What are two general classifications of transformers based on the method of cooling?

20 Describe what is meant by a self-cooled, forced-air-cooled transformer.

21 What might be a typical application of a sealed dry-type transformer?

22 Why is it important to know the value of a transformer's impedance?

23 What precaution must be observed when working with current transformers?

24 What is an autotransformer? What are some of the advantages and disadvantages of the autotransformer?

25 Why must the transformer polarities be known when transformers are being connected for parallel operation?

26 What is meant by (a) additive polarity; (b) subtractive polarity?

27 What is a tap-changing switch? When might a transformer equipped with taps be desirable?

28 Why are single-phase distribution transformers made with the low-voltage winding in two sections?

29 What are four types of three-phase transformer connections?

30 What voltages are commonly provided by four-wire delta secondaries for light and power loads? By four-wire wye secondaries?

31 What are some advantages of using three-phase transformers in place of three single-phase transformers? What are some disadvantages?

32 Why is it necessary to maintain substantially constant voltage on an electrical system?

33 Describe how a transformer may be used to raise or lower voltage in a circuit.

34 How is the kilovoltampere rating determined for (a) a single-phase voltage regulator; (b) a three-phase voltage regulator?

35 Describe briefly the basic principle of operation of the step-voltage regulator.

36 What is the purpose of the switching reactor in a step-voltage regulator?

37 Describe briefly the operation of the voltage-control equipment used with a voltage regulator.

38 What are some of the usual voltage-regulator accessories?

PROBLEMS

12-1 The high-voltage winding of a transformer has a voltage rating of 480 V and has 400 turns. How many turns are there on the low-voltage winding if its voltage rating is 120 V?

12-2 A 1000-kVA 24,000/2400-V transformer operates at 40 V per turn. Find the number of turns on each winding and the current rating of each winding.

12-3 A 5-Ω resistance is connected across the 120-V secondary winding of a step-down transformer. If the primary current is 6 A, what is the primary voltage?

12-4 The low-voltage winding of a step-down welding transformer has one turn and the high-voltage winding 600 turns. When the transformer delivers 100 A, what is the current in the high-voltage winding?

12-5 Find the current ratings of each winding of a 100-kVA 2400/120V 60-Hz transformer.

12-6 A transformer is rated at 500 kVA, 60 Hz, and 2400/480 V. There are 200 turns on the 2400-V winding. When the transformer supplies rated load, find (*a*) the ampere-turns of each winding and (*b*) the current in each winding.

12-7 Find the size of wire in circular mils needed in each winding of a 10-kVA 2400/240-V 60-Hz transformer if the number of circular mils per ampere to be used in each winding is 1000. What is the closest AWG size in each case?

12-8 Following are data from a test of a 3-kVA 240/120-V 60-Hz transformer:

Resistance of 120-V winding . . . . 0.055 Ω
Resistance of 240-V winding . . . . 0.184 Ω
Core loss. . . . . 28 W

Find the efficiency of the transformer when it delivers rated kilo voltamperes at unity power factor.

12-9 A voltmeter, ammeter, and wattmeter are connected with suitable instrument transformers into a single-phase circuit and the following data taken:

Current transformer ratio.	5 to 1
Voltage transformer ratio	20 to 1
Voltmeter reading	110 V
Ammeter reading.	4 A
Wattmeter reading	360 W

Find the voltage, current, voltamperes, power factor, and power of the primary circuit.

12-10 A step-down 600/480-V autotransformer supplies a 10-kVA load. Find the primary and secondary line currents and the current in the winding common to both primary and secondary circuits.

12-11 An autotransformer starter used to start an induction motor on a 440-V line applies 65 percent of line voltage to the motor during the starting period. If the motor current is 140 A at starting, what is the current drawn from the line?

12-12 A 5-kVA 480/120-V transformer is equipped with high-voltage taps so that it may be operated at 480, 456, or 432 V depending on the tap setting. Find the current in the high-voltage winding for each tap setting, the transformer supplying rated kilovoltampere load at 120 V in each case.

12-13 A distribution transformer has two 120-V secondary windings that may be connected either in series or in parallel. The current rating of each winding is 41.7 A. What is the kilovoltampere rating of the transformer with the coils connected (*a*) in series and (*b*) in parallel?

12-14 Three single-phase transformers having a 20-to-1 ratio are delta-connected to a 2400-V three-phase line as step-down transformers. Find the voltage between the secondary terminals of each transformer. Find the secondary voltage of each transformer if the transformers are wye-connected to the same line.

12-15 The secondaries of a three-phase transformer bank supplying a three-phase 240-V motor are delta-connected. When the motor current is 10 A per terminal, find the current in each transformer winding.

12-16 A three-phase motor requires 80 kW at 80 percent power factor when operated at 240 V. If the motor is supplied from a 4160/240-V wye-delta step-down transformer bank, find the line currents on both the high- and low-voltage sides of the transformers.

12-17 The primary windings of three single-phase transformers are wye-connected, and the secondary windings are delta-connected. The voltage applied to each primary winding is 2300 V, and the secondary line voltage is 230 V. Find the secondary line current when the primary line current is 20 A.

12-18 What is the kilovoltampere rating of a single-phase voltage regulator that has a rated voltage of 2400 V, a current rating of 200 A, and a total regulation range of 20 percent (10 percent raise and 10 percent lower)?

12-19 A 12,000-V three-phase step regulator has a rating of 208 kVA and has a range of regulation of 10 percent raise and 10 percent lower. What is the full-load current rating of this regulator? How much kilovoltampere load can be carried by a circuit in which this regulator is installed without overloading the regulator?

13

Semiconductor Devices

The general topic of electronics, especially solid-state electronics using semiconductors, is too broad a subject to be treated in detail in a power-oriented text. Industrial and power engineers and technicians are increasingly involved with electronic circuits and equipment. As the semiconductor industry has evolved, it has developed devices capable of handling larger and larger current levels that are characteristic of power systems. Similarly, protection and control systems are utilizing electronic devices instead of electromechanical or hydraulic devices. These devices are discussed in Chaps. 19 and 21. The purpose of this chapter is to give a very brief introduction to some fundamentals and potential effects of the electronic circuits and devices in industry today.

The reader may pursue this subject in more depth by referring to the many excellent texts available.

13-1 SEMI-CONDUCTORS Materials such as copper and aluminum have many free electrons and are good conductors of electricity. Materials such as quartz or porcelain have few free electrons and are called insulators. Some materials having a conductivity that falls between that of good conductors and good insulators are called *semiconductors*. Some of these semiconductor materials can, by careful processing, be made suitable for use in solid-state electronic devices. Silicon, germanium, selenium, and certain oxides have been used in the manufacturing of solid-state electronic devices, but silicon is by far the most popular material.

To make a material having properties suitable for use in solid-state electronic devices, very minute amounts of other elements are added to pure silicon (or germanium) crystals during the manufacturing process. The process of adding other elements to the silicon is called *doping*. These other elements or *impurities* have atomic structures that differ from the silicon atomic structure. When these impurities are added in closely controlled amounts, they modify the electrical properties of the silicon, particularly its resistivity.

When silicon crystals are doped with impurity atoms of materials such as phosphorus or arsenic, the resulting material has an excess of loosely bound electrons. Since electrons are negative charge carriers, the material is called an *N-type semiconductor* (N for negative).

Pure silicon may, on the other hand, be doped with another type of impurity atom that has a deficiency of electrons with respect to silicon. Impurity elements of this type that have been used include boron, aluminum, and gallium. Since the resulting material has gaps or *holes* in its atomic structure, electrons, under certain conditions, can move into these holes. This leaves new holes in the material for other electrons to move into, and so on. The effect is just as if the holes were moving in a direction opposite to the electron movement. Thus the holes may be considered to have a positive charge since they represent an absence of electrons. It is convenient then to speak of holes in this type of semiconductor material as positive charge carriers and to refer to the material as a *P-type semiconductor* (P for positive).

In a conductor, the flow of electric current is considered to be the movement of free electrons. In a semiconductor, the current may be considered to be either the movement of electrons or the movement of holes. If the material is P-type material, that is, deficient in electrons, conduction is by holes. If the material is N-type material, which has an excess of electrons, conduction is by electrons.

The conventions used in the discussion of current flow in solid-state electric devices may thus be stated briefly as follows:

1. In N-type semiconductors, current is carried by negative charge carriers or electrons, which move in a direction opposite to the conventional current flow.
2. In P-type semiconductors, current is carried by positive charge carriers or holes, which move in the same direction as the conventional current flow.

13-2 SEMICONDUCTOR DIODES

When a section of P-type semiconductor is joined to a section of N-type semiconductor, it has been found that the junction so formed will pass current easily in one direction but will offer considerable resistance to the flow of current in the reverse direction. A two-electrode semiconductor device

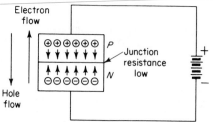

FIGURE 13-1 *Semiconductor diode with forward bias.*

having a P-type electrode and an N-type electrode is called a *semiconductor diode,* a *junction diode,* or a *semiconductor rectifier.*

A semiconductor diode is in a state of equilibrium when there is no external voltage applied to the diode. When the junction is formed initially, some of the free charges from both the P-type and N-type materials diffuse across the junction and recombine. This forms a neutral barrier called the *depletion layer* between the two regions. This depletion layer prevents a further exchange of charge carriers between the two regions of the diode. Thus, the diode is said to be in equilibrium.

When a diode is connected to an external battery as in Fig. 13-1 with the battery positive terminal connected to the P terminal of the diode and the negative terminal to the N terminal of the diode, the diode is said to be *forward-biased.* The bias voltage overcomes the small potential barrier of the depletion layer, allowing electrons to be attracted across the diode junction by the positive terminal of the battery. Likewise, holes are attracted in the opposite direction across the junction by the negative terminal of the battery. Electrons constantly enter the N terminal of the diode and leave the P terminal, and current flows in the circuit.

When the battery polarity is reversed as in Fig. 13-2, the diode is said to be *reverse-biased* and the excess electrons in the N-type material are attracted away from the junction by the positive terminal of the battery. Likewise, the holes are attracted away from the junction. This leaves the section of the diode near the junction practically void of charge carriers, and only an extremely small current called the *leakage current* can flow through the junction or in the external circuit. Since a forward-biased junction offers a low resistance to the flow of current and a reverse-biased

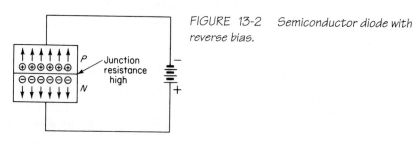

FIGURE 13-2 *Semiconductor diode with reverse bias.*

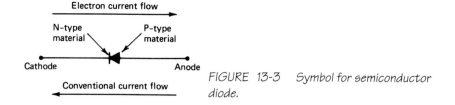

FIGURE 13-3   *Symbol for semiconductor diode.*

junction offers a high resistance to the flow of current, the junction has the properties of a rectifier.

A semiconductor diode is represented by the symbol shown in Fig. 13-3. Note that the terminal connected to the N-type material is identified as the cathode. The arrow in the symbol points in the direction of conventional current flow. Other solid-state device symbols use a similar approach (see Appendix A).

13-3 SILICON RECTIFIERS

A rectifier is a device for changing an alternating current to pulses of direct current. As indicated in the previous section, the semiconductor diode has the properties of a rectifier.

When an alternating voltage is applied to a silicon rectifier, as shown in Fig. 13-4, the diode is alternately forward- and reverse-biased during each cycle of the applied voltage. When the diode is forward-biased during the positive half-cycle, electron current flows freely. During the next half-cycle when the diode is reverse-biased, the diode is nonconducting. The diode conducts current only during the positive half-cycles of the applied voltage, resulting in a pulsating dc output with a waveshape like that

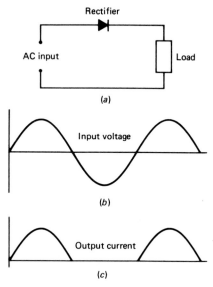

FIGURE 13-4   *Half-wave rectifier: (a) circuit; (b) input voltage wave; (c) output current wave.*

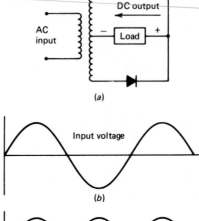

(a)

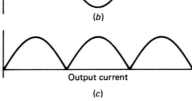

Input voltage

(b)

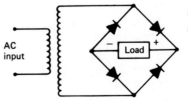

Output current

(c)

FIGURE 13-5   Full-wave rectifier with center-tapped transformer: (a) circuit; (b) input voltage wave; (c) output current wave.

shown in Fig. 13-4c. A rectifier producing an output current of this type is called a *half-wave rectifier*.

To provide a smoother current output, two diodes and a center-tapped transformer may be connected as shown in Fig. 13-5 to form what is known as a *full-wave rectifier*. With this arrangement, one diode conducts during each half-cycle, resulting in the output current waveform shown in Fig. 13-5c.

Another commonly used rectifier connection, called the *full-wave bridge rectifier*, is shown in Fig. 13-6. This rectifier circuit avoids the need for a center-tapped transformer and provides twice the output voltage obtained from the basic two-diode rectifier.

To provide the large amounts of direct current often required in industrial plants, three-phase rectifiers are usually used. A connection diagram of a three-phase, full-wave, bridge-type rectifier is shown in Fig. 13-7.

The pulsating output produced by rectifiers generates harmonics in the output (see Sec. 13-15). This output is suitable for some applications. But, in most cases, the pulsation and its harmonics cannot be tolerated by the equipment connected to the output. The pulsations may be reduced by

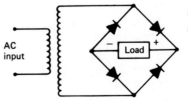

FIGURE 13-6   Single-phase full-wave bridge rectifier.

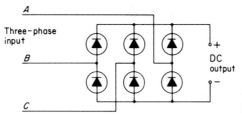

FIGURE 13-7 *Three-phase full-wave bridge rectifier.*

filter circuits that "absorb" certain of the harmonics. Filter circuits are combinations of inductance coils and capacitors arranged so that the pulsating energy is stored alternately in the coils and capacitors, resulting in a much smoother flow of current from the rectifier. Two typical filter circuits are shown in Fig. 13-8.

Silicon rectifiers are available commercially with current ratings ranging from a few milliamperes to several hundred amperes and with voltage ratings as high as 1000 V or more. They are highly efficient, achieving rectification efficiencies as high as 99 percent. Because of their many desirable characteristics, silicon rectifiers find wide application as industrial dc supplies for battery charging, electrochemical processes, variable-speed motor drives, and other similar applications where formerly dc generators were used.

13-4 ZENER DIODES

The Zener diode is a special type of silicon junction diode often used as a voltage regulator or voltage stabilizer. Like the silicon rectifier diode, the Zener diode has a very low resistance to the flow of current when it is forward-biased. When it is reverse-biased at low voltages, it allows only a very minute current to flow. However, when an applied reverse-bias voltage is gradually increased, a point is reached at which the Zener diode breaks down and suddenly begins to conduct. The sharp break from nonconductance to conductance is called the Zener effect, so named for Dr. Zener, who first proposed a theory explaining the effect.

The voltage level at which the breakdown occurs in a Zener diode may be controlled to within close limits during the manufacturing process. Thus these devices may be designed to have a wide range of breakdown voltages with ratings as low as about 2 V to as high as several hundred volts.

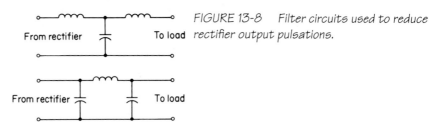

FIGURE 13-8 *Filter circuits used to reduce rectifier output pulsations.*

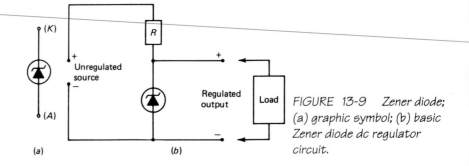

FIGURE 13-9 Zener diode;
(a) graphic symbol; (b) basic
Zener diode dc regulator
circuit.

At applied voltages greater than the breakdown point, the voltage drop across a conducting Zener diode remains essentially constant even though the current through the diode increases with the increased applied voltage. This makes the device well suited for use as a constant-voltage reference or control element.

The schematic symbol for a Zener diode is shown in Fig. 13-9*a*, and a simple voltage-regulating circuit using a Zener diode is illustrated in Fig. 13-9*b*. In this circuit, the constant-voltage breakdown characteristic is used to provide a regulated dc output. The cathode of the diode is connected to the positive side of the supply to provide the reverse bias, and the current through the diode is limited by the series resistor *R*. Since the output voltage is the same as the voltage drop across the diode, the output voltage remains constant even though the input voltage and hence the diode current may vary.

The voltage-limiting characteristic is very useful to protect one circuit from excessive voltages caused by another circuit. For example, some computers may require that input voltages be held within certain ranges to prevent damage to the computer electronics. These are also discussed in Chap. 21.

In addition to a breakdown voltage rating, Zener diodes have a power rating. This is the power that the diode can dissipate without being destroyed.

In addition to their use as voltage references and voltage regulators, Zener diodes find use in what are known as intrinsically safe barriers. These devices are used in industrial applications where a field device such as a process transmitter must be located in an explosive atmosphere. The barrier limits voltage and current so that sufficient power is not available to ignite the surrounding atmosphere. A simplified diagram of such a barrier is shown in Fig. 13-10. The two diodes as shown provide functional redundancy and, in conjunction with the resistor network, perform both voltage and current limiting. The barrier itself is located in a nonexplosive atmosphere.

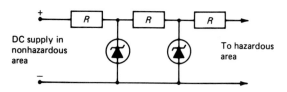

FIGURE 13-10   *Safety barrier between haz-*
ardous and nonhazardous areas.

13-5
TRANSISTORS

As described previously, a single-junction semiconductor device is capable of rectifying current. By adding a second junction, a semiconductor device is formed that is capable of amplification. Such a device is called a *transistor.*

The transistor consists of a sandwich of semiconductor metals containing alternate layers of N-type and P-type material. The transistor may be the PNP type consisting of two layers of P-type material separated by a thin layer of N-type material or the NPN type consisting of two layers of N-type material separated by a thin layer of P-type material. The arrangement of both types is shown in Fig. 13-11. In each type, one outer layer is called the *emitter,* the center layer the *base,* and the other outer layer the *collector.* The function of the emitter is to supply charge carriers to its junction with the base. The function of the collector is to remove charge carriers from its junction with the base.

Transistor action may be described as the control of currents and voltages in one junction by the currents and voltages in another junction. In the normal operation of a transistor, the emitter-base junction is forward-biased and the collector-base junction is reverse-biased. When biased in this manner, the transistor has a low-resistance emitter-base junction and a high-resistance collector-base junction. When the transistor is used as an amplifier, the emitter-base circuit is used as the input circuit and the collector-base circuit as the output circuit. The ability of the transistor to transfer current from a low-resistance input to a high-resistance output

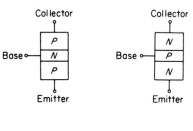

FIGURE 13-11   *Pictorial and graphical representation of transistors; (a) PNP type; (b) NPN type.*

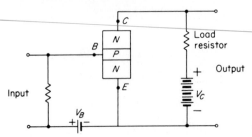

FIGURE 13-12 *Basic transistor amplifier circuit, common-emitter configuration.*

gives it its name, *transfer resistor,* which, when contracted, forms the word *transistor.*

A simplified circuit diagram of a basic NPN transistor amplifier circuit is shown in Fig. 13-12. The circuit shown is called a common-emitter circuit since the emitter is common to both the input and output circuits.

In the circuit shown, electrons move easily from the N-type emitter material to the P-type material in the base because of the forward bias. The electrons, upon entering the base, diffuse through the thin base and are attracted to the collector by the positive potential of battery V_C.

The characteristic of the transistor is such that nearly all the electrons entering the base from the emitter flow into the collector and out through its load resistance. The physical size, thickness, and arrangement of the materials forming the transistor junctions and the impurity concentrations in each are designed so that only a very small part of the emitter current flows in the base circuit. Hence, the collector current is almost equal to the emitter current. In practical transistor circuits, the collector current may be as much as 98 percent of the emitter current. Furthermore, the characteristics of the transistor are such that the ratio of collector current to the base current remains nearly constant over a wide range of currents.

If the voltage of the battery V_B in the emitter-base circuit of Fig. 13-12 is raised or lowered, the result is an increase or decrease in base current. The increase or decrease in base current causes a corresponding increase or decrease in collector current.

In the above description it has been assumed that the only currents flowing in the circuit were those caused by the batteries V_B and V_C; that is, it has been assumed that the input signal has been zero.

If in the circuit of Fig. 13-12, an ac voltage signal is applied to the circuit input, this signal alternately adds to and subtracts from the bias voltage V_B. As indicated previously, a change or variation in voltage in the emitter-base circuit results in a variation in base current. The variation in base current causes a corresponding variation in collector current. The collector current flowing in the load resistor results in an *IR* drop in this resistor, the magnitude of which is proportional to the current through it. Thus, the variation of input or signal voltage appears in amplified form across the load resistor or output terminals of the transistor amplifier circuit. Because of the polarities required for the proper biasing of the common-emitter circuit shown, there is a 180° phase shift between the input and the output voltages.

The above is a very much simplified description of the operation of an NPN transistor. The operation of the PNP transistor is similar except that bias-voltage polarities are reversed. A detailed analysis of transistor theory and operation involves consideration of advanced theories of physics and mathematics and is beyond the scope of this book.

Transistor amplifiers find a multitude of applications in communication equipment and in industrial control systems. Transistor amplifiers in communication systems may be used as radio and audio frequency amplifiers or as dc amplifiers.

13-6 TRANSISTORS AS SWITCHING DEVICES

As has been shown, a transistor conducts or amplifies when the emitter-base junction is forward-biased and the collector-base junction is reverse-biased. When both junctions are reverse-biased, the transistor will not conduct. Use is made of this characteristic of transistors in switching circuits.

In the simplified circuit shown in Fig. 13-13, it can be seen that when the switch S is operated to connect battery A into the base-emitter circuit, the emitter-base junction will be forward-biased. For this condition the transistor will conduct and current will flow through the load. If, however, the switch is operated to connect battery B into the circuit, the emitter-base junction is reverse-biased and the transistor will not conduct and no current will flow through the load. Used in this way, then, the transistor performs the function of a switch.

The basic concept in any switching circuit is that a definite change of state, usually a voltage change, causes operation of the switching device. The circuit described above operates in this manner; that is, a change in voltage state in the emitter-base circuit in effect turns on or off the output or load current of the transistor.

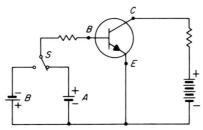

FIGURE 13-13 Basic circuit to illustrate the use of a transistor as a switch.

Transistors used as switches offer several advantages over electromechanical types of switches or relays. They are small, have no moving parts, and are easily actuated from various electric inputs. Probably more important are their exceptionally fast operating times as compared with their electromechanical counterparts.

Probably the most important use of transistor switching is in the logic circuitry used in digital computers. Other uses are in pulse and switching circuits in radar, television, and other communication circuits and in commercial power rectifiers, inverters, and frequency-converting equipment control circuitry.

13-7 THYRISTORS

Thyristors are multilayer semiconductor devices that can be switched from the OFF state to the ON state by a pulse of control current. Once conduction commences, it continues until the voltage across the thyristor is decreased to zero, as occurs, for example, every cycle in ac circuits.

There are several kinds of thyristors, but only three will be considered here, namely, the reverse-blocking triode thyristor called the *silicon controlled rectifier,* the bidirectional triode thyristor called the *triac,* and the bidirectional diode called the *diac.*

13-8 SILICON CONTROLLED RECTIFIERS (SCR)

The silicon controlled rectifier (SCR), like a diode, blocks reverse current. Unlike the diode, however, it is capable of blocking current under forward-voltage conditions also unless it has been switched electrically to the forward-current-conduction mode by a pulse of current flowing through its gate. As shown in Fig. 13-14, the SCR has three terminals: an anode, a cathode, and a gate. Also as shown, the SCR is composed of alternate layers of P- and N-type materials with three junctions. Junctions 1 and 3 are equivalent to those in ordinary PN semiconductor diodes. Both of these junctions will conduct current when they are biased in the forward direction, but conduction is blocked when they are reverse-biased. The NP junction 2 blocks conduction when the complete four-layer device is biased in the forward direction.

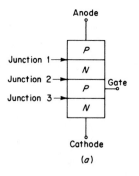
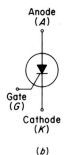

FIGURE 13-14 SCR thyristor: (a) junction diagram and (b) graphical symbol.

If, when an SCR is biased in the forward direction, a current pulse is injected into the P region of junction 2 by momentarily applying a positive voltage to the gate, junction 2 becomes forward-biased and conduction through the SCR begins. Once conduction has started it will continue even though the gate current is reduced to zero. Conduction can be interrupted only by means external to the SCR, such as by reducing the anode-to-cathode voltage to zero or by reducing the current flow to zero.

When the SCR is used as a rectifier, the average value of dc supplied by the rectifier to the load may be controlled by controlling the point in each cycle at which the gate voltage is applied to the SCR, or in other words, by controlling the time in each cycle at which the SCR is triggered. If zero gate signal is applied, the SCR blocks current flow for both positive and negative half-cycles of the sine wave and no current flows to the load. If the gate voltage is continuously applied, then the SCR performs as a conventional half-wave diode rectifier and it conducts for each positive half-cycle. If the SCR is triggered by application of the gate voltage at some point during each positive half-cycle, current will flow for the remainder of that half-cycle until the sine-wave voltage goes to zero. Thus, by adjusting the time in each positive half-cycle at which the gate voltage is applied to trigger the SCR, the average value of dc to the load may be controlled. The process of varying the point within the cycle at which forward conduction is permitted to begin is called *phase control* of the SCR.

13-9 THE TRIAC

The bidirectional triode thyristor or *triac,* like the SCR, has three terminals, a main terminal no. 1, a main terminal no. 2, and a gate. The triac can block current flow in either direction between main terminals no. 1 and no. 2, or it can be triggered into conduction in either direction in response to positive or negative signals applied to the gate terminal. Once conduction has started in the triac it continues until it is stopped by external means, just as in the SCR. In effect, the triac behaves like two SCRs oriented one in each direction as indicated in the equivalent circuit diagram of Fig. 13-15a. The standard symbol for the triac, shown in Fig. 13-15b, also reflects the bidirectional property of the triac.

Because triacs can conduct in both directions, they are used basically as ac power devices for controlling ac loads, virtually superseding the rheostat for ac power control. Triacs are also used as static switches in ac circuits.

As an aid in visualizing the similarities and differences of diodes, SCRs, and triacs, the output waveforms (idealized) of these devices are shown in Fig. 13-16. As shown, the diode conducts for a full half of each cycle and blocks current flow for the other half-cycle. The SCR can conduct for a full half-cycle or for any part of that half-cycle depending upon the

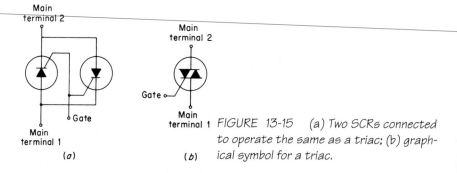

FIGURE 13-15   (a) Two SCRs connected to operate the same as a triac; (b) graphical symbol for a triac.

triggering time of its gate. It blocks current flow for the other half-cycle. The triac can be made to conduct in both halves of each cycle, the amount of conduction depending upon the time in each half-cycle that it is triggered by its gate. In summary, the output of diodes and SCRs is a form of dc; the output of triacs is a form of ac.

13-10 THE DIAC The diac, sometimes called a bilateral trigger or a bidirectional trigger diode, as the names imply, is applied primarily as a trigger device for other thyristors. The diac blocks current flow in either direction until the voltage across it exceeds the breakover voltage. Typical breakover voltages of the diac are 28 to 36 V. When this breakover voltage is exceeded, the diac triggers or fires and conduction through it begins. Conduction continues, as in the SCR and triac, until the voltage across it is reduced to zero. A symbol for the diac is shown in Fig. 13-17. Examples of the application of the diac are included in a later section.

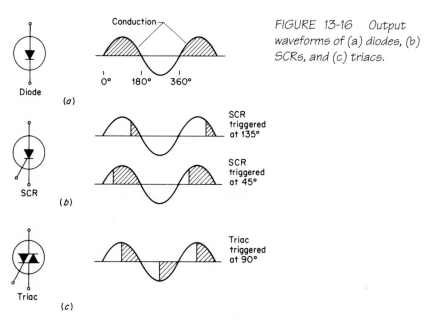

FIGURE 13-16   Output waveforms of (a) diodes, (b) SCRs, and (c) triacs.

FIGURE 13-17   *Standard graphical symbol for a diac.*

13-11 VARISTORS

Devices discussed immediately above require external pulses applied to gate elements to break down and initiate conduction. There are times when it is desirable to have circuit elements that can change their characteristics based upon the value of the voltage across them in the circuit. These characteristics are incorporated into semiconductor devices known as *varistors*. These devices change their resistance in a nonlinear manner depending upon voltage.

These devices have very high resistances (essentially, an open circuit) up to a given voltage. At that point, the varistor begins to conduct. Figure 13-18 compares the voltage-current characteristics for a "normal" linear resistor (see Chap. 2) and a varistor. Note that the voltage and current for a linear resistor has a constant relationship given by Ohm's law. A varistor has the ability to change this relationship. The voltage where conduction begins and the amount of current it can withstand over a period of time depend upon the varistor's chemical components.

These devices have application in a variety of circuits for protection against overvoltages such as lightning (see Sec. 19-14). The most common chemical composition for these varistors is a metal oxide such as zinc oxide.

13-12 OPTOELECTRONIC DEVICES

Optoelectronic semiconductor devices are devices that interact with light. There are two general classifications of these devices: *light sensors,* those devices that convert radiant energy (light) into electric energy, and *light emitters,* those devices that convert electric energy into light. These devices, together with other semiconductor devices, are being used increasingly in many industrial control applications. Symbols used for several optoelectronic devices are shown in Fig. 13-19.

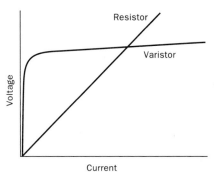

FIGURE 13-18   *A comparison of the voltage-current characteristics of a resistor and a varistor. The constant relationship between voltage and current based on Ohm's law is shown for a linear resistor.*

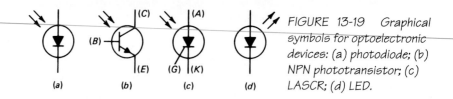

FIGURE 13-19 *Graphical symbols for optoelectronic devices: (a) photodiode; (b) NPN phototransistor; (c) LASCR; (d) LED.*

Three important optoelectronic devices, classified as light-sensing or light-detecting devices, are the photodiode, the phototransistor, and the light-activated SCR (LASCR).

The *photodiode,* as the name implies, is a diode designed specifically to make use of the inherent characteristic of a PN junction to interact with light. As indicated earlier in this chapter, a PN junction without external bias has a very thin depletion layer. However, when the junction is reverse-biased, the depletion layer is increased in width in proportion to the value of the bias voltage. Only a minute leakage current can flow across the junction. However, when light is allowed to fall on a properly designed reverse-biased junction, a current in addition to the leakage current flows across the junction. Moreover, the amount of current that flows is proportional to the intensity of the light. The resulting current is very small, and for many control applications it must be amplified to make it useful.

Phototransistors combine in one device the ability to detect light and provide amplification of the resulting current as well. When light strikes the collector-base junction of a phototransistor, reverse leakage current flows as in the photodiode. The small emitter-base current is used to control a much larger emitter-collector current in the usual circuit arrangement. The effect is to switch on or off or to regulate and amplify a current in proportion to the amount of light striking the collector-base junction.

The *light-activated SCR* or *LASCR* is a light-sensing optoelectronic device used to detect light, convert that light to a current, and use the current to trigger an SCR section within the device.

Like the SCR, the LASCR has three junctions. It also has three terminals: an anode, a cathode, and a gate, all as indicated by the symbol shown in Fig. 13-19. In addition, the LASCR has a window through which light is admitted to activate the light-sensitive area.

As normally used, the LASCR gate is biased to near the triggering value. Then light energy, when added to the gate bias, triggers the unit into conduction. Provisions must be made in the circuit to turn off the current as in the control of the SCR. Thus large currents can be switched on or off in the LASCR with small amounts of light energy.

The *light-emitting diode* or *LED* is a semiconductor diode that emits light when it is forward-biased. Different materials have been used in the formation of the P- and N-type materials for the LED junction. One producer of these devices uses gallium arsenide and gallium aluminum arsenide for the junction materials. A junction made of these materials emits an infrared light. Other materials used in differing proportions are combined to emit different light colors such as green or yellow.

When the LED is forward-biased, the bias current causes electrons to be injected into the P-type material and holes to be injected into the N-type material. In terms of energy levels, free electrons are at a higher level than are the holes. As the free electrons move through the region near the junction, they recombine with the holes. In the process of recombination, energy is released, some in the form of light and the remainder in the form of heat. The light output power efficiency is, however, very low—less than one percent.

Light-emitting diodes have several advantages over other kinds of light sources. They operate at low voltages, making them compatible with other semiconductor devices. They have very fast response times, they are small, and they have a long life. These attributes make the LED particularly well suited for displaying information obtained from electronic equipment.

Probably the most familiar use of the LED is its use in an array form used to display numbers. For example, an array of seven rectangular LEDs, properly arranged, can be used to form any digit from zero to nine. Other arrays can display alphanumerics, or both letters and numbers. Such arrays are used as readouts for calculators, digital clocks, measuring instruments, and other similar equipment.

An increasingly important use of the LED is its use as a light source in optically-coupled systems as described in the next paragraphs.

Often it is necessary to isolate one electric circuit from another or one part of a circuit from another part of the circuit. Electrical isolation may be necessary to ensure personal safety, to prevent equipment damage, to ensure proper equipment functioning, or for other reasons. Electrical isolation of two ac circuits can be accomplished relatively easily by the use of transformers. This is not so, however, for dc circuits. Optically coupled systems provide an excellent solution for this problem.

A *semiconductor optically coupled system* may be used to provide complete electrical isolation between two circuits. Such a system consists of a light source and a light detector often assembled within the same package. One combination used is an LED light source and a photodiode detector as shown diagrammatically in Fig. 13-20. When greater amplification is required, phototransistors or photo SCRs are used in the coupler output.

A typical example of the electrical isolation of circuits is the isolation used between functional modules of electronic control systems. This

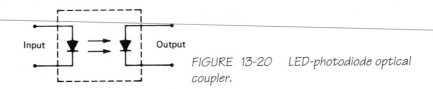

FIGURE 13-20   LED-photodiode optical coupler.

isolation prevents a minor electrical failure in one module or part of the system from cascading into adjacent parts and causing a catastrophic system failure. Signal information can still be transmitted between the various parts of the system even though the parts are electrically isolated by the optocouplers.

Optically coupled systems are not limited to use as electrical isolators. Many systems operate with the light source and the light detector in separate locations. These systems depend upon the interruption of the light beam to transmit information. The punched-card reader in which the data card is inserted between the light source and detector is an example of this type of system. Other applications include the sensing of the position of an object, fluid-level sensing, color determination, and intrusion alarms.

13-13 INVERTERS

An inverter is used to change dc power to ac power, as contrasted with a rectifier, which changes ac power to dc power. Solid-state inverters provide an efficient and economical method for accomplishing the conversion of dc to ac.

An inverter circuit is basically a type of *chopper.* In a chopper circuit, a dc supply is alternately opened and closed or "chopped" by a switching device such as a transistor or SCR. By varying either the ratio of on to off time of the switching device, or the repetition rate of the switching, the average voltage and frequency of the chopper output may be controlled.

The basic principle of operation of the inverter may be illustrated in a simplified way by the circuit shown in Fig. 13-21a, which shows two manually operated switches, A and B, in the circuit. With both switches open, no current flows in the circuit. When switch A is closed, current I_A flows in the direction shown, and when switch B is closed, I_B flows as shown. If switches A and B are alternately opened and closed in sequence, it can be seen that an alternating output can be generated at the load terminals of the circuit. In practical inverter circuits, SCRs or transistors are used to replace the mechanically operated switches shown in Fig. 13-21a.

One of the simplest inverter circuits is the center-tapped transformer circuit using two SCRs as shown in Fig. 13-21b. In this circuit the two SCRs alternately connect the output transformer to the dc supply, first with one polarity and then the other, to produce a square-wave output similar to that

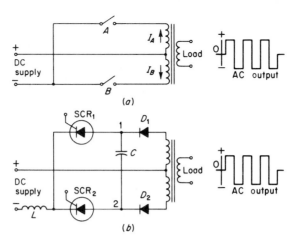

FIGURE 13-21 (a) Circuit to illustrate the inverter principle, and (b) basic inverter circuit.

shown. In the interest of simplicity, SCR gate trigger circuits are not shown in Fig. 13-21.

It must be remembered that after an SCR has begun to conduct, the gate loses control of the anode current. Thus, in order to use SCRs on a dc system, special turn-off control circuitry must be used to stop the SCR conduction at the right times. Circuits that are used to turn off an SCR are called *commutating* circuits. In the circuit shown in Fig. 13-21*b,* turn-off is accomplished by the action of the capacitor C. When SCR_1 is triggered and begins to conduct, the voltage across it drops very rapidly because of the voltage drop across the reactor L. This allows the capacitor to charge to a positive polarity at point 2 and to a negative polarity at point 1. Because of the autotransformer action of the output transformer, the capacitor charge approaches twice the battery voltage. At the beginning of the second half-cycle, when SCR_2 is triggered, capacitor C is then connected across SCR_1 with a polarity to reverse-bias it and thus cause conduction in SCR_1 to cease. With conduction in SCR_2 and with SCR_1 turned off, capacitor C charges to the opposite polarity and the cycle repeats, resulting in an alternating current in the output transformer. The diodes D_1 and D_2 prevent capacitor discharge current from flowing through the transformer.

There are several configurations of inverter circuits often used. Inverter configurations are somewhat analogous to rectifiers; that is, they may be half-wave, full-wave, single-phase bridge, or three-phase bridge circuits. Inverter circuits are also classified by the commutation method used. The inverter shown in Fig. 13-22 is a bridge-type inverter. The commutation is accomplished by the inductor L and the capacitor C. Turn-on is by

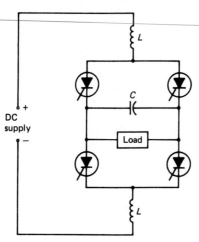

FIGURE 13-22 *Single-phase bridge-type SCR inverter.*

SCR triggering circuitry, which is not shown. For high-capacity applications, three-phase bridge inverters are used. An elementary diagram of a three-phase bridge inverter is shown in Fig. 13-23. The commutating capacitors and related components are not shown.

As shown in Fig. 13-21, the inverter output is not a sine wave. Even the output of the three-phase inverter of Fig. 13-23 is not a sine wave. This is due to the fact that the SCRs are digital devices and turn on and off at specific times. These nonperfect sine waves contain harmonics, which are discussed in Sec. 13-14.

If the ac output of an inverter is rectified to provide dc again, the overall circuit is called a *converter.* The usual use of the converter is to change the magnitude of a dc voltage. Applications of the converter include its use in power supplies for computers and telephone equipment.

Static inverters are used extensively in adjustable-speed or variable frequency ac motor drive systems (see Sec. 16-18) and in standby power supplies. Other applications include power supplies for control and instrumentation circuits and in automotive electrical systems.

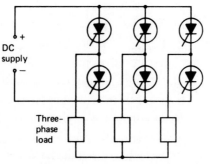

FIGURE 13-23 *Three-phase bridge-type SCR inverter.*

Devices and systems such as computers, airline and hotel reservation systems, process controllers, and boiler flame safeguard systems all require essentially uninterruptible power supplies for safety or for economic reasons. To supply backup power for these applications, various forms of *uninterruptible power supplies* (UPS) have been developed.

Static UPS systems have three basic parts; namely, a combination rectifier and battery charger, a battery, and an inverter. These basic parts or units may be combined in several different configurations. The simplest form of UPS is the continuous system shown in block-diagram form in Fig. 13-24. With this system the critical ac load is normally supplied through the rectifier, inverter, and the transfer switch. The rectifier converts commercial ac to dc, and the inverter changes the dc to closely regulated ac for the load. If there are power outages or transient disturbances in the commercial supply, the battery automatically picks up the load without any switching. The battery is maintained in a fully charged condition by the rectifier and is recharged by the rectifier after a power outage. The inverter output is locked into synchronism with the commercial ac supply as indicated by the dotted line in Fig. 13-24. This maintains exact clock timing for the critical load and permits operation of the manual transfer switch to take the UPS out of service if necessary.

An often used variation of the system shown in Fig. 13-24 is called the static switching transfer system. In this system, a static transfer switch is added in the output section of the UPS. This system supplies the critical ac load in the same manner as the continuous system does under normal conditions. However, if the UPS system fails or if the critical load demand becomes larger than the UPS can deliver, the critical load is transferred to the commercial power supply automatically by the static transfer switch. Manual transfer is also normally provided for this system to permit the removal of the UPS for routine maintenance.

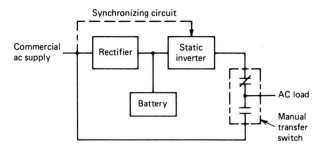

FIGURE 13-24   *Block diagram of a continuous-type static UPS.*

13-15
HARMONICS

The generation of an alternating emf was discussed in Chap. 9. As shown in Fig. 9-1 the emf created by a generator has a waveform in the shape of a sine wave. The waveform repeats itself 60 times per second on a 60 Hz power system. However, under some circumstances, frequencies other than 60 Hz are generated by other equipment on the same system. These frequencies are whole number multiples of the fundamental frequency and are called *harmonics*. At 60 Hz, these frequencies include the following:

Harmonic	Frequency
1	60
2	120
3	180
4	240
5	300
6	360
7	420

In three-phase power systems, only odd-numbered frequencies are usually present.

When various waveforms of differing frequencies combine, the resulting wave is distorted and is no longer a perfect sine wave. Figure 13-25 shows the effects on a waveform of combining various harmonics. This particular wave shows the unique characteristic result of energizing a transformer that creates a rare, even second harmonic (120 Hz). Note that the resulting wave is greatly distorted.

Generators are designed to produce as close to a perfect sine wave as possible. As electronic devices can change the wave shape, they may be a source of harmonics. These harmonics then flow to other areas of the power system. For example, the inverter described above has historically been a harmonic source. The six-pulse inverter (see Fig. 13-23) produces harmonics according to the following formula:

$$h = np \pm 1$$

where h is the harmonic, n is the whole number (1, 2, 3, etc.), and p is the number of pulses (6). Figure 13-26 illustrates some of the harmonic frequency waveshapes and the simplified resulting output wave.

Harmonics are of increasing concern in modern power systems as they can cause abnormal heating in electrical equipment, very high overvoltages, and interference with electronic operations due to distorted voltage inputs. Filters and newer type circuits are used to make equipment less susceptible to these problems.

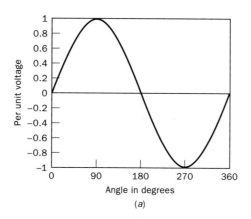

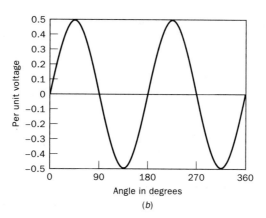

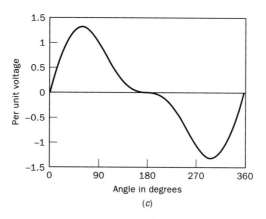

FIGURE 13-25 The current fundamental wave (a) is combined with the second harmonic (b) to form the characteristic (c).

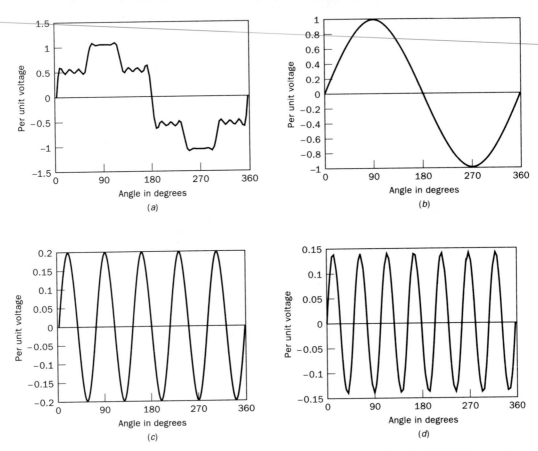

FIGURE 13-26   The unfiltered current output of a six-pulse inverter (see Fig 13-23) is a series of harmonic waves that are combined to form the total output waveform. The combined simplified current waveform is shown in (a). Representative current harmonics are as follows: (b) fundamental, (c) fifth, and (d) seventh. Frequencies up to the nineteenth were used to construct the (a) waveform.

REVIEW QUESTIONS

1   What are some materials used in the manufacture of semiconductor devices?
2   What is the effect of adding P-type and N-type impurities to semiconductor crystals?
3   Describe the conventions used when considering electric charge movement in semiconductor materials.
4   What important function can be performed by a semiconductor diode?
5   What is the depletion layer in a P-N junction?
6   What is meant by forward-biasing and reverse-biasing of a semiconductor junction?
7   What is a half-wave rectifier?

8 What advantage does a full-wave rectifier have over a half-wave rectifier?

9 How are the pulsations of the output of a rectifier "smoothed out"?

10 Give some applications for the Zener diode.

11 Name the three electrodes of a transistor.

12 How is the emitter-base junction biased in a transistor?

13 What characteristics of the transistor make it suitable for use as a switch?

14 Describe how a transistor functions as an amplifier.

15 Name three kinds of thyristors.

16 Give the names of the three terminals of the SCR.

17 Describe what is meant by phase control of an SCR.

18 Compare an SCR with a triac. What are their differences? Their similarities?

19 How many terminals does a triac have?

20 Describe the differences in the output waveforms obtainable with diodes, SCRs, and triacs.

21 What is a diac?

22 What are the two general classifications of optoelectronic devices?

23 What is a photodiode? A phototransistor? Describe briefly the action of each.

24 How many junctions does the LASCR have?

25 What is an LED?

26 Give an application of the optical coupler.

27 Define an inverter. What are some inverter configurations?

28 What are some of the harmful effects of harmonic currents?

14

Alternating-Current Generators

Alternating-current generators, sometimes called synchronous generators or alternators, provide a vital link in the long process of converting the energy in coal, oil, gas, or uranium into a useful form for use in industry and home.

Large generators used for supplying modern national electric power grids are driven by steam turbines or water wheels. Generators used to supply smaller, isolated power systems or to furnish supplemental peak-load power to larger grids are often combustion-turbine or diesel-engine driven. An interior view of a generating station in which two 66-MW steam turbine-driven generators are installed is shown in Fig. 14-1. The direct-connected exciter can be seen on the unit in the foreground. A 3750-kW generator coupled directly to a diesel engine is shown in Fig. 14-2.

14-1
CONSTRUCTION

Armature windings are placed on the rotating part of dc generators to provide a way of converting the alternating voltage generated in the windings to a direct voltage at the terminals through the use of a rotating commutator. The field poles are placed on the stationary part of the machine. In all but small, low-voltage, alternating-current generators, the field is placed on the rotating part, or *rotor,* and the armature winding on the stationary part, or *stator,* of the machine.

The revolving-field and stationary-armature construction simplifies the problems of insulating an ac generator. Since voltages as high as

FIGURE 14-1 Installation of two 66-MW steam-turbine-driven generators. (Stanley Consultants, Inc.)

18,000 to 24,000 V are commonly generated, this high voltage need not be brought out through slip rings and sliding contacts but can be brought directly to the system switching equipment through insulated leads from the stationary armature. This construction also has mechanical advantages

FIGURE 14-2 Slow-speed ac generator and its exciter connected to diesel engine. (Dresser Rand-Electric Machinery.)

in that armature windings are subjected to less vibration and centrifugal forces and can be braced better. The rotating field is supplied with direct current at 125, 250, or 375 V through slip rings and brushes, or through a direct cable connection between the field and the rotating rectifier if the brushless excitation system is used.

The armature or stator winding may be any one of many types, a widely used type being an open-circuited winding formed from separately insulated form-wound coils similar to the lap winding of the dc generator. Actually, such a winding is composed of three separate windings (on a three-phase generator), each displaced from the other two by 120 electrical degrees. The three windings may be either wye- or delta-connected. The wye connection is the more common since it lends itself well to the generation of high voltages directly, and a neutral wire can be brought out with the three lines to form a three-phase four-wire system as was shown in Chap. 11. An ac generator stator with windings in place is shown in Fig. 14-3.

A developed view of a simple three-phase winding is shown in Fig. 14-4. The winding shown in Fig. 14-4a is wye-connected; the method of connecting the terminals for a delta connection is shown in Fig. 14-4b. The winding pictured is called a *concentrated winding* since all the conductors of each phase are included in one slot under each pole. Commercial windings such as that shown in Fig. 14-3 are distributed windings, with the conductors of each phase group occupying two or more slots under each pole. The distributed winding provides a more uniform heat distribution and results in the generation of a better emf wave.

There are two distinct types of synchronous-generator field structures: the *salient-pole type* and the *cylindrical type*.

Slow-speed generators such as those driven by diesel engines or water turbines have rotors with projecting or salient field poles like the rotor

FIGURE 14-3   AC generator stator. (Kato Engineering/Division of Reliance Electric.)

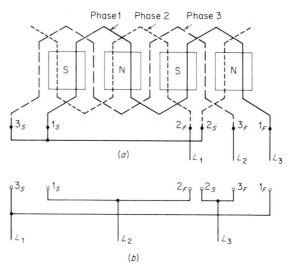

Phase 1   Phase 2   Phase 3

(a)

(b)

FIGURE 14-4   (a) Developed view of a simple three-phase stator winding, wye-connected. (b) Method of reconnecting the terminals for a delta connection.

shown in Fig. 14-5. The laminated pole pieces with their field coils are mounted on the rim of the steel spider, which is in turn keyed to the shaft.

High-speed or turbo-type generators have cylindrical rotors such as that shown in Fig. 14-6. The rotor shown will be wound for two poles and is designed to operate at 3600 revolutions per minute (rpm). The cylindrical construction is essential in high-speed machines because the salient-pole type is difficult to build to withstand the stresses at high speeds. Furthermore, salient-pole rotors have high windage losses at high speeds. Alternating-current generators with the cylindrical-rotor construction are driven by steam or gas turbines. The steam turbine-driven generators shown in Fig. 14-1 have two-pole cylindrical rotors.

FIGURE 14-5   Salient-pole rotor for slow-speed ac generator. (MagneTec-Louis Allis-Motors & Generators.)

FIGURE 14-6   Rotor for 3600 rpm ac generator. (General Electric Company.)

14-2 AC GENERATOR EXCITATION

The conventional excitation system prior to about 1960 consisted of a direct-current source connected to the ac generator field through two slip rings and brushes. The dc source was normally a motor-driven dc generator or a dc generator driven by the same prime mover that powered the ac generator.

Following the advent of solid-state devices, several different excitation systems using these devices have been developed and applied. In one system, power is taken from the ac generator terminals, converted to dc power by a stationary solid-state rectifier, and then supplied to the ac generator field by means of conventional slip rings and brushes. In a similar system used in large steam turbine-driven generators, power is supplied to a solid-state rectifier from a separate three-phase winding located in the top of the generator stator slots. The sole function of this winding is to provide the excitation power for the generator.

Still another excitation system used with both salient-pole- and cylindrical-rotor-type synchronous generators is the brushless system, in which a small ac generator mounted on the same shaft as the main generator is used as an exciter. The ac exciter has a rotating armature, the output of which is rectified by silicon diode rectifiers, which are also mounted on the main shaft. The rectified output of the ac exciter is fed directly by means of insulated connections along the shaft to the rotating synchronous generator field. The field of the ac exciter is stationary and is supplied from a separate dc source. The output of the ac exciter, and consequently the generated voltage of the synchronous generator, may thus be controlled by varying the field strength of the ac exciter. The brushless excitation system therefore has no commutator, slip rings, or brushes, which greatly improves reliability and simplifies maintenance of the machines.

14-3 VOLTAGE GENERATION

After an ac generator is brought up to its proper speed by its prime mover, its field is excited from a dc supply. As the poles move under the armature conductors on the stator, the field flux cutting across the conductors induces an emf in them. This is an alternating emf, since poles of alternate polarity successively pass by a given conductor. Since no commutator is used, the alternating emf generated appears at the stator winding terminals.

The amount of any generated emf depends on the rate at which magnetic flux lines are cut; or in the case of the ac generator, the amount of emf depends on the field strength and the speed of the rotor. Since most generators are operated at a constant speed, the amount of emf generated becomes dependent on the field excitation. This means that the amount of emf generated can be controlled by adjusting the amount of field excitation supplied to the generator. The field excitation may be readily controlled by varying the amount of excitation voltage applied to the generator field.

The power factor at which the generator operates is determined by the characteristics of the load being supplied (unless the generator is operating in parallel with other generators as explained in Sec. 14-7).

The frequency of the generated emf depends on the number of field poles and on the speed at which the generator is operated. In a given coil, one complete cycle of emf is generated when a pair of rotor poles (a north and a south pole) is moved past the coil. Thus the number of cycles generated in one revolution of the rotor will be equal to the number of pairs of poles on the rotor, or $p/2$, where p is the total number of poles. If n is the rotor speed in revolutions per minute, then $n/60$ is equal to the revolutions per second. The frequency in hertz or cycles per second is therefore

$$f = \frac{p}{2} \times \frac{n}{60} = \frac{pn}{120} \qquad (14\text{-}1)$$

By far the most common power frequency used in America is 60 Hz, with 25 Hz being used to a much lesser extent. The common frequency used in Europe is 50 Hz.

14-4 GENERATOR REGULATION

As load is added to an ac generator operating at a constant speed and with a constant field excitation, the terminal voltage will change. The amount of change will depend on the machine design and on the power factor of the load. The effects of different load power factors on the change in the terminal voltage with changes of load on an ac generator are shown in Fig. 14-7.

Regulation of an ac generator is defined as the percentage rise in terminal voltage as load is reduced from the rated full-load current to zero, the speed and excitation being constant, or

Percent regulation (at a stated power factor)

$$= \frac{\text{no-load voltage} - \text{full-load voltage}}{\text{full-load voltage}} \times 100$$

Factors affecting generator regulation are as follows:

1. IR drop in the armature winding
2. IX_L drop in the armature winding
3. Armature reaction (magnetizing effect of the armature currents)

In a dc generator, the generated emf E is the sum of the terminal voltage V_t and the IR drop in the armature circuit. In an ac generator, the voltage drop due to the inductive reactance of the winding must also be taken into account. Thus the generated emf of an ac generator is equal to the terminal voltage plus both the IR and IX_L drops in the armature winding.

A simplified phasor diagram of an ac generator operating at unity power factor is shown in Fig. 14-8a. The generated emf E is the phasor sum of the terminal voltage V_t, the IR drop that is in phase with the current I, and the IX_L drop, which leads I by 90°.

The phasor diagram of Fig. 14-8b represents the generator with the same load current as in Fig. 14-8a but with the current lagging the terminal voltage by 36.9° (power factor = 0.8 lag). As before, the generated emf is the phasor sum of V_t, the IR drop, and the IX_L drop in the armature winding. An inspection of Figs. 14-8a and 14-8b shows that for a given generated

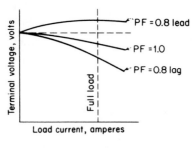

FIGURE 14-7 *Regulation curves of an ac generator at different power factors.*

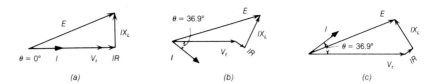

FIGURE 14-8   Simplified phasor diagrams of an ac generator operating at
(a) unity power factor; (b) 0.8 lagging power factor; (c) 0.8 leading power
factor.

emf the terminal voltage is less in the case of the 0.8 lagging power factor. At lower lagging power factors, the IR and IX_L drops lower the terminal voltage still further.

Figure 14-8c represents a condition in which a generator supplies a load with a leading power factor. When the IR and IX_L drops are added as phasors to the terminal voltage, it is found that the generated emf is less than the terminal voltage. This is possible because of the phase relations of the IR and IX_L drops with respect to the terminal voltage.

The phasor diagrams of Fig. 14-8a, b, and c show the effect of the IR and IX_L drops on the terminal voltage for a given E. However, E is *not* constant in an ac generator but varies with the amount of load and the power factor of the load due to the effect of armature reaction. At unity power factor, the effect of armature reaction is at a minimum, its effect being merely a distorting action on the main field flux. However, at lagging power factors, the magnetizing effect of the armature currents opposes the main field mmf, resulting in a weakened field flux and a lowered E. The lower the power factor in the lagging direction, the more the armature mmf demagnetizes the field.

At leading power factors, the armature mmf aids or strengthens the main field mmf, resulting in a higher generated emf with increases in load. This magnetizing effect increases as the power factor becomes more leading.

14-5 GENERATOR VOLTAGE REGULATORS

Since the terminal voltage of an ac generator varies considerably with changes in load, some means must be provided to maintain the constant voltage required for the operation of most electric equipment. A common way of doing this is to use an auxiliary device called a *voltage regulator* to control the amount of dc field excitation supplied to the generator. When the generator terminal voltage drops because of changes in load, the voltage regulator automatically increases the field excitation, which restores normal rated voltage. Similarly, when the terminal voltage increases because of load changes, the regulator restores normal rated voltage by decreasing the field excitation.

Most voltage regulators control the field excitation of the generator indirectly by operating in the exciter field circuit. Much less current need be handled by the regulator in the exciter field circuit than in the generator field circuit.

One type of generator voltage regulator is the direct-acting rheostatic type. Essentially this regulator consists of an automatically controlled variable resistance connected into the exciter field circuit. This regulator has been largely superseded by the static type in which moving mechanical parts are replaced by solid-state devices. The principle of operation of these regulators is the same, however; the voltage of the ac generator is regulated by varying the effective resistance in its exciter field circuit, which in turn varies the voltage output of the exciter and of the ac generator.

The simplified elementary diagram in Fig. 14-9 shows a static voltage regulator and its connection in an ac generator excitation system. In this system, the effective resistance of the exciter field circuit is varied by power transistors connected in parallel with the exciter field rheostat. The transistors are switched from a conducting to a nonconducting state a varying rate depending upon the amount of correction required in the ac generator voltage. This, in effect, alternately bypasses and inserts the rheostat into the exciter field circuit. The rate of on-off switching of the power transistors, which thereby regulates the exciter field current, is controlled by solid-state auxiliary and sensing devices connected to the generator current and potential transformers.

Auxiliary apparatus includes stabilizing equipment to damp out undue oscillations in the regulator and to prevent overshooting of the regulator on rapid changes in voltage. The current transformer connections are made to auxiliary apparatus used in connection with a cross-current compensation scheme to regulate automatically the division of kilovar load between generators operating in parallel.

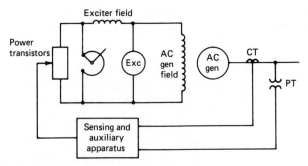

FIGURE 14-9   Simplified diagram of a static voltage regulator.

If the load on a generating station becomes so large that the rating of the running generators is exceeded, it becomes necessary to add another generator in parallel to increase the power available from the generating station.

Before two synchronous generators may be paralleled, the following conditions must be fulfilled:

1. Their phase sequence must be the same.
2. Their terminal voltages must be equal.
3. Their voltages must be in phase.
4. Their frequencies must be equal.

When two generators are operating so that these requirements are satisfied, they are said to be in *synchronism*. The operation of getting the machines into synchronism is called *synchronizing*.

Before a generator is synchronized with other generators for the first time, its phase sequence must be checked to determine that it has the same phase sequence as that of the other generators in the station. This normally is done with a test instrument called a phase-sequence indicator. One commonly used phase-sequence indicator is essentially a small three-phase motor that rotates in one direction for one phase sequence and in the opposite direction for the other phase sequence. The method of making the phase-sequence test and other synchronizing operations will be explained by making reference to the connection diagram shown in Fig. 14-10.

When a new generator has been installed and is ready for test, the generator is operated at approximately rated speed and rated voltage with its circuit breaker open. The phase-sequence indicator is connected temporarily to the system bus potential transformers at points *a, b,* and *c,* and the system phase sequence is noted on the indicator. The phase-sequence indicator connections are then transferred to the generator potential transformers, with temporary connections being made at points *a', b',* and *c',* and the phase sequence of the incoming generator is noted. If the sequences of the generator and of the system are the same, the generator is ready for the remaining steps in the synchronizing process. If, however, the phase sequence of the generator is opposite to that of the system, then any two of the three main leads of the generator must be interchanged to obtain the correct generator sequence. After it has been determined that the phase sequence of the generator matches that of the system and permanent connections have been made between the generator and its circuit breaker, it is not necessary to check the phase sequence each time the generator is synchronized.

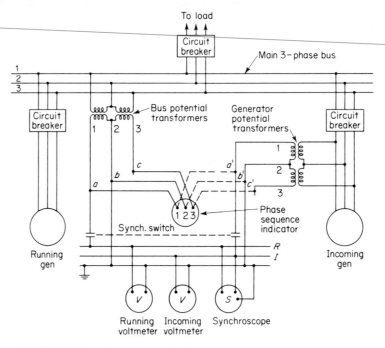

FIGURE 14-10 *Generator synchronizing connections.*

To determine that the remaining conditions for successful parallel operation are fulfilled, two voltmeters and a synchroscope are connected as shown in Fig. 14-10. A synchronizing switch is provided between both the bus and generator potential transformers and the synchronizing equipment, so that the synchronizing equipment may be switched on during the synchronizing operation.

With the incoming generator running at approximately rated speed, with its circuit breaker open, and with the synchronizing switch closed, the voltage of the generator, as indicated by the incoming voltmeter, is adjusted to match the bus voltage, as indicated by the running voltmeter. Generator voltage is raised or lowered by adjusting generator field excitation.

The synchroscope is then used to make certain that the generator voltage is in phase with the bus voltage and that the generator and bus frequencies are the same. The synchroscope is an instrument for indicating differences of phase and frequency between two voltages. It is essentially a split-phase motor in which torque is developed if the two applied voltages differ in frequency. Voltages from corresponding phases of the bus and the incoming generator are applied to the synchroscope. A pointer, which is attached to the rotor of the instrument, moves over the dial face in either a clockwise or a counterclockwise direction, depending on whether the incoming generator frequency is high or low with respect to the bus frequency. When the pointer stops in a vertical (12-o'clock) position, the synchroscope indicates that the frequencies are equal and the voltages are in phase. The dial markings of a synchroscope are as shown in Fig. 14-11.

In the synchronizing operation, the frequency of the incoming generator is raised or lowered to match that of the running generator or bus. Generator frequency is varied by increasing or decreasing the speed of the generator prime mover. When the synchroscope pointer stops in the vertical position and the two voltmeters read the same, the generator circuit breaker may be closed to parallel the running and incoming generators.

In practice, it is sometimes difficult to adjust the speed of the incoming generator closely enough to stop the synchroscope pointer in the correct position. If this is the case, the frequency of the incoming generator should be adjusted as closely as possible to that of the bus, and the circuit breaker closed just before the pointer reaches the vertical position when traveling very slowly in the fast direction. This causes the incoming generator to take a small amount of load immediately after the closing of the circuit breaker and results in stable operation of the paralleled generators.

The process of synchronizing an incoming generator to a generating station bus is usually accomplished by automatic equipment rather than by the manual method just described. A typical complement of equipment required for automatic synchronizing includes a speed matching relay, a voltage matching relay, a synchronizing relay, auxiliary relays, and transfer relays or switches.

The automatic synchronizing equipment is energized as a part of a generating unit startup sequence as in manual synchronizing. As the incoming unit approaches its rated speed, the speed matching relay provides "raise" or "lower" impulses to the generator prime mover to match the generator frequency to the bus frequency. Likewise, the voltage matching relay matches the generator and bus voltages by sending "raise" and "lower" impulses to the generator excitation system. When the frequencies and voltages are matched within predetermined limits and the voltages are

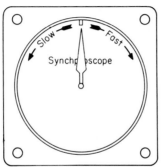

FIGURE 14-11   Synchroscope dial.

in phase, all as determined by the synchronizing relay, the closing impulse is given automatically to the generator circuit breaker. The closing impulse is given at a phase angle slightly in advance of synchronism to take into account the time required for the generator breaker to close its contacts.

Each generator in a generating station may be equipped with a complete complement of automatic synchronizing equipment or one common set of equipment may be switched from generator to generator by suitable transfer switches and equipment. Manual synchronizing equipment is usually provided as backup equipment for the automatic equipment.

14-7 PARALLEL OPERATION

After two synchronous generators have been paralleled, the load is usually divided between them in proportion to their ratings. That is, the larger the machine, the larger the share of the load it carries.

The proper division of load between generators is made by adjusting the governors of the prime movers on the generators. One prime-mover governor is opened while the other is closed slightly. In this way the system frequency is maintained at a constant value while the load is shifted from one machine to the other. Governor-control switches are mounted on the switchboard so that the operator is able to watch the switchboard instruments while making adjustments in load division.

The power factor of any ac distribution system depends on the load. A generator operating singly, then, must operate at the power factor of the load it supplies. However, when two or more generators operate in parallel, the power factor at which each operates is determined by its field excitation.

In general, the proper amount of field excitation for generators operating in parallel is the amount of excitation each generator would require if it were carrying its load alone at the same voltage and frequency.

If the excitation of a generator operating in parallel with other generators is increased above this normal value of excitation, its power factor changes in the lagging direction and its current output increases with no appreciable change in kilowatt load. Likewise, if the generator is underexcited, its power factor becomes more leading and its current output increases with no change in kilowatt output. The increased current in either case is not supplied to the load but circulates between the generators connected to the system, thereby increasing their losses and decreasing their useful capacity. It is desirable in most cases, therefore, to operate each generator at the same power factor, keeping the circulating current to a minimum.

Thus a change in field excitation causes a change in *ampere* load but not in *kilowatt* load. *Kilowatt-load division between synchronous gen-*

erators must be made by adjusting the prime-mover governor controls.

The voltage of a system supplied by several paralleled generators may be raised or lowered by increasing or decreasing, simultaneously, the field excitation of all generators. Likewise, system frequency can be raised or lowered by increasing or decreasing the speed of the several prime movers.

14-8 HUNTING

Synchronous generators operating in parallel sometimes have a tendency to *hunt*. If the driving torque applied to a generator is pulsating, such as that produced by a diesel engine, the generator rotor may be pulled periodically ahead of or behind its normal position as it rotates. This oscillating action is called *hunting*.

Hunting causes generators to shift load from one to another. In some cases this oscillation of power becomes cumulative and violent enough to cause the generators to fall out of synchronism.

The tendency of diesel-driven generators to hunt is reduced by using a heavy flywheel to reduce torque variations. Damper windings called *amortisseur* windings are placed on the surface of some generator rotors to decrease the tendency to hunt. The rotor shown in Fig. 14-5 is equipped with a damper winding consisting of short-circuited conductors embedded in the pole faces. When hunting occurs, there is a shifting of the armature flux across the pole faces, thereby inducing currents in the damper winding. Since any induced current opposes the action that produces it, the hunting action is opposed by the flow of the induced currents. Generators driven by steam turbines generally do not have a tendency to hunt since the torque applied does not pulsate.

14-9 RATINGS

Generator capacity is rated in kilovoltamperes and usually in kilowatts at a specified power factor. Other generator nameplate data includes voltage, current, frequency, number of phases, and speed ratings. Maximum temperature rise is stated together with the method of temperature measurement to be used. Excitation requirements are also stated, including rated field voltage and field amperes.

14-10 LOSSES AND EFFICIENCY

Losses of a synchronous generator are similar to those of a dc generator and include mechanical and iron losses, armature copper loss, and field-excitation loss.

The efficiency of an ac generator may be determined by a direct measurement of input and output or by calculation after the losses have been determined. Because of the difficulty of providing suitable power supply and loading devices required for a direct measurement, the efficiencies of ac generators, especially of larger ratings, are usually determined from their losses.

Alternating-current generator efficiency varies with the power factor at which the generator operates, since the kilowatt output decreases with a decrease in power factor while the field excitation (and field loss) must be changed to maintain rated voltage.

REVIEW QUESTIONS

1 Why are ac generators built with a stationary armature and revolving field?
2 Name two types of ac generator construction.
3 How is the field of an ac generator excited?
4 Upon what does the amount of emf generated in an ac generator depend? In actual operation which of these is variable?
5 Write the formula for frequency of an ac generator. What is the common frequency in use in America? In Europe?
6 What is meant by generator regulation?
7 What factors affect generator regulation?
8 How does the load power factor change the effect of armature reaction?
9 What is the purpose of a voltage regulator?
10 What requirements must be satisfied before ac generators may be paralleled?
11 What is meant when it is said that two ac generators are being synchronized?
12 Show how a phase-sequence indicator may be used to check the phase sequence of a generator.
13 When a generator is being synchronized, if the synchroscope pointer revolves once per second, what is the difference in frequency between the bus and the incoming generator?
14 How often must the phase sequence of an ac generator be checked?
15 What two things are indicated by a synchroscope? Is this the only instrument needed to indicate synchronism?
16 How is the kilowatt load divided in the desired proportion between two ac generators operating in parallel?
17 How is it possible to adjust the power factor of an ac generator that is operating in parallel with other generators?
18 What is meant by the term *hunting* as applied to generator operation?
19 What information is found on an ac generator nameplate?
20 Name three types of generator losses. Are ac generator losses the same at all power factors?

PROBLEMS

14-1 A diesel-driven synchronous generator is operated at 200 rpm and has a frequency rating of 60 Hz. How many poles does it have?

14-2 How many cycles are generated in one revolution of a 24-pole ac generator? How many revolutions per second must it make to generate a frequency of 60 Hz? How many rpm?

14-3 What must be the speed of a two-pole 25-Hz synchronous generator?

14-4 A three-phase generator has a load of 100 kW per phase at a power factor of 80 percent, lagging. (*a*) What is the three-phase kilowatt load; (*b*) the three-phase kilovoltampere load; (*c*) the current per terminal if the terminal voltage is 7200 V?

14-5 A wye-connected 2400-V 600-kVA unity power factor generator is reconnected for delta operation. What is the new voltage, current, and kilovoltampere rating?

14-6 Find the regulation of an ac generator that has a full-load voltage of 2400 V and no-load voltage of 3240 V at 80 percent lagging power factor. Will the percent regulation at unity power factor be larger than, smaller than, or the same as at 80 percent lagging power factor?

14-7 What is the kilowatt rating of a three-phase 875-kVA generator at (*a*) unity power factor; (*b*) 80 percent power factor; (*c*) 50 percent power factor? If the voltage rating is 2400 V, what is the current rating at each power factor?

14-8 A three-phase 480-V 1200-kVA synchronous generator supplies rated load at a lagging power factor of 75 percent. If the efficiency of the generator is 90 percent, what horsepower is the prime mover delivering to the generator?

15

Direct-Current Motors and Controls

A motor is a machine that converts electric energy into mechanical energy. The dc motor is very similar to a dc generator in construction. In fact, a machine that runs well as a generator will operate satisfactorily as a motor.

One common constructional difference between motors and generators should be noted, however. Since motors often are operated in locations in which they may be exposed to mechanical damage, dust, moisture, or corrosive fumes, they are normally more completely enclosed than are generators.

Two general classifications for dc motor enclosures have been established by motor manufacturers: *open motors* and *totally enclosed motors*. Open motors have ventilating openings that permit the passage of external cooling air over and around the motor windings. Although external air is admitted into the open motor, the ventilating openings are constructed to restrict the entrance of liquids or solid objects. Such open motors are classified according to their construction as being dripproof, splashproof, weather-protected, or guarded. For example, the motor shown in Fig. 15-1 is classified by its manufacturer as a dripproof, fully guarded motor. It is considered to be dripproof because liquids or solids falling from above cannot interfere with the operation of the motor. It is called a guarded motor because openings in the motor are arranged so that accidental contact cannot be made with live or rotating parts.

Totally enclosed motors, as the name indicates, are completely enclosed so that no ventilating air can enter the motor. The heat developed by the motor must be dissipated entirely by radiation from the motor enclosure.

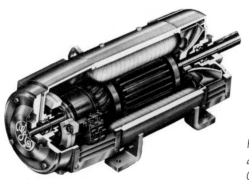

FIGURE 15-1 *Cutaway view of a dc motor. (General Electric Company.)*

Although the mechanical construction of dc motors and generators is similar, their functions are different. The function of a generator is to *generate a voltage* when conductors are moved through a field, while the function of a motor is to develop a *turning effort* or a *torque*. DC motor and generator characteristics are summarized in Appendix E.

In this chapter, the motor principle and the characteristics of the several types of dc motors are discussed. An introduction to motor control circuits and equipment is also included.

15-1 MOTOR PRINCIPLE

Every conductor carrying current has a magnetic field around it, the direction of which may be established by the right-hand rule. The strength of the field depends on the amount of current flowing in the conductor.

If a wire carrying current away from the reader is placed in a uniform magnetic field, the combined fields will be similar to that shown in Fig. 15-2. Above the conductor, the field due to the conductor is from left to right, or in the same direction as the main field. Below the conductor, the magnetic lines from the conductor and the main field magnetic lines are in opposite directions. The result is to strengthen the field or increase the flux density above the conductor and to weaken the field or decrease the flux density below the conductor.

It is convenient to think of magnetic lines as elastic bands under tension that are always trying to shorten themselves. Thus, the magnetic lines above the conductor exert a downward force on the conductor. Likewise, if the current in the conductor is reversed, the increased number of magnetic lines below the conductor will exert an upward force on the conductor.

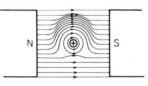

FIGURE 15-2 *Effect of placing a current-carrying conductor in a magnetic field.*

While the above explanation of how a force is developed on a conductor is convenient, it is somewhat artificial. However, it is based on a fundamental principle of physics, which may be proved experimentally and which may be stated as follows:

A conductor carrying current in a magnetic field tends to move at right angles to the field.

15-2 TORQUE DEVELOPMENT IN A MOTOR

Torque is defined as the action of a force on a body that tends to cause that body to rotate. Thus, the measure of the tendency of a motor armature to rotate is called the *torque* of the motor.

Motor-armature windings are wound in the same manner as are generator windings. When a voltage is applied to the brushes of a motor, current flows into the positive brush, through the commutator and armature windings, and out the negative brush. Armature conductors are wound so that all conductors under the south field poles carry current in one direction, while all the conductors under the north field poles carry current in the opposite direction. Figure 15-3 shows the distribution of armature currents in a four-pole motor for a given polarity of applied terminal voltage.

When voltage is applied to a motor, such as is shown in Fig. 15-3, current flows through the field winding, establishing the magnetic field. Current also flows through the armature winding from the positive brushes to the negative brushes. Since each armature conductor under the four pole faces is carrying current in a magnetic field, each of these conductors has a force exerted on it, tending to move it at right angles to that field.

An application of the right-hand rule to the armature conductors under the north pole *D* in Fig. 15-3 shows the magnetic field to be

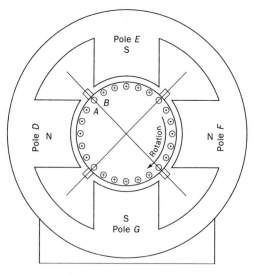

FIGURE 15-3 Armature-current directions in a four-pole motor for clockwise rotation.

strengthened under the conductors, resulting in an upward force on the conductors. Similarly, force is exerted to the right on the conductors under south pole *E,* downward on the conductors under north pole *F,* and to the left on the conductors under south pole *G.* Thus, there is a force developed on all the active armature conductors tending to turn the armature in a clockwise direction. The sum of these forces in pounds multiplied by the radius of the armature in feet is equal to the total torque developed by the motor in *pound-feet* (lb-ft). In SI units, the force in newtons multiplied by the radius of the armature in meters is equal to the torque in *newton-meters* (N-m). If the armature is free to turn, that is, if the connected load is not too great, the armature will begin to rotate in a clockwise direction.

As the armature rotates and the conductors move from under a pole into the neutral plane, the current is reversed in them by the action of the commutator. For example, in Fig. 15-3, as the conductor at *A* moves from under the north pole and approaches the neutral plane, the current is outward. At the neutral plane the current is reversed so that, as it moves under the south pole as at *B,* it carries current inward. Thus, the conductors under a given pole carry current in the same direction at all times.

It should be evident from Fig. 15-3 that if the armature current were reversed by reversing the armature leads, but leaving the field polarity the same, torque would be developed in a counterclockwise direction. Likewise, if the field polarity were reversed leaving the armature current as shown, torque would be developed in a counterclockwise direction. However, if both armature-current direction and field polarity were changed, torque would be developed in a clockwise direction. *The direction of rotation of a dc motor may be reversed by reversing either the field or the armature connections. If both are reversed, the direction of rotation remains unchanged.*

The force developed on each conductor of a motor armature is due to the combined action of the main field and the field around the conductor. It follows, then, that the force developed is directly proportional to the strength of the main field flux and to the strength of the field around each conductor. The field around each armature conductor depends on the amount of armature current flowing in that conductor. Therefore, the torque developed by a motor may be shown to be

$$T = K' \phi I_a \qquad (15\text{-}1)$$

where T = torque
K' = a constant depending on physical dimensions of motor
ϕ = flux per pole
I_a = armature current

This important equation is of use in analyzing a motor's performance under various operating conditions.

15-3 MEASUREMENT OF TORQUE

The torque developed by a motor may be measured by means of a *prony brake* attached to the motor pulley as shown in Fig. 15-4. Any desired load may be placed on the motor by adjusting the pressure of the brake shoes. The energy output of the motor is dissipated as heat, which is produced as the pulley turns in the brake shoes. For any given load, the torque output of the motor is the length of the brake arm r multiplied by the force F as registered on the spring scales. In USCS units, r is in feet and F is in pounds.

EXAMPLE 15-1

A prony brake arm is 42 in. long. The pressure on the motor pulley is adjusted so that a net force of 18 lb is measured on the scale. What torque is the motor developing?

$$\text{Torque} = \text{force} \times \text{brake arm}$$

$$T = Fr = 18 \times \frac{42}{12} = 63 \text{ lb-ft}$$

In SI units, r is measured in meters and F in newtons. If the spring scale is calibrated in kilograms, its reading must be multiplied by the gravitational constant 9.81 to obtain the force in newtons.

The turning or twisting effort that a motor develops at the instant of starting is called its *starting torque*. It may be measured with a prony brake by clamping the brake arm securely so that the pulley cannot turn. With the motor connected to the supply lines and the armature remaining stationary, the force is measured with the scale. The starting torque is then equal to the force multiplied by the brake-arm length.

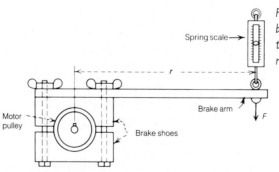

FIGURE 15-4 *Prony brake for measuring the torque developed by a motor.*

15-4 GENERATOR ACTION OF A MOTOR

As the armature of a generator is rotated by a prime mover, a torque is developed in the generator opposing the action of the prime mover. This countertorque may be considered to be a motor action in a generator. Likewise, a generator action is developed in every motor. Whenever a conductor cuts magnetic flux lines, an emf is induced in that conductor. The driving force is immaterial; if a conductor is moved through a magnetic field, an emf will be induced in that conductor. The direction of that emf is in accordance with *Fleming's right-hand rule* for generator action.

The conductor of Fig. 15-5 represents one of many current-carrying conductors of a motor armature. For the field polarity shown, the current in the conductor is such that by motor action it will be moved downward. As the conductor is moved downward, it cuts magnetic flux lines and an emf is induced in it. The direction of this emf by Fleming's right-hand rule for a generated emf is found to be toward the reader. Since this emf is in a direction opposite to the flow of current in the conductor, it is called a *counter electromotive force*.

Since the counter emf of a motor is generated by the action of armature conductors cutting magnetic flux lines, its value will depend on the field strength and the armature speed. The value of the counter emf E that is generated in a motor is given by the relation

$$E = K\phi n \qquad (15\text{-}2)$$

where K is a constant depending on the physical properties of the motor. Remember that Eq. (15-2) was used to determine the generated emf of a generator.

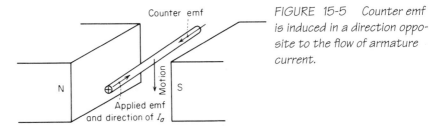

FIGURE 15-5 *Counter emf is induced in a direction opposite to the flow of armature current.*

The effective voltage acting in the armature circuit of a motor is the applied or terminal voltage minus the counter emf. The armature current by Ohm's law is

$$I_a = \frac{V_t - E}{R_a} \tag{15-3}$$

where I_a = armature current
V_t = motor terminal voltage
E = counter emf
R_a = armature-circuit resistance

Multiplying both sides of Eq. (15-3) by R_a and transposing results in

$$V_t = E + I_a R_a \tag{15-4}$$

which is the fundamental motor equation. Note that this is the same as the generator equation with the sign of the $I_a R_a$ term changed. The counter emf of a motor is always less than its terminal voltage.

EXAMPLE 15-3   Find the counter emf of a motor when the terminal voltage is 240 V and the armature current is 60 A. The armature resistance is 0.08 Ω.

$$V_t = E + I_a R_a$$

or $E = V_t - I_a R_a$

$$= 240 - (60 \times 0.08) = 240 - 4.8 = 235.2 \text{ V}$$

**15-5
EQUIVALENT
CIRCUIT OF A
MOTOR**

It is convenient when dealing with dc motors to represent the armature as an equivalent circuit. As in the generator, the armature circuit is equivalent to a source of emf in series with a resistance as shown in Fig. 15-6. However, in the motor, current flows from the line into the armature against the generated emf. It is evident from Fig. 15-6 that the terminal voltage must be balanced by both the IR drop in the armature circuit and the counter emf at all times. This is apparent also from Eq. (15-4).

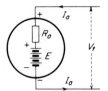

FIGURE 15-6   Equivalent circuit of a motor armature.

A comparison of the equivalent circuit of a generator (Fig. 8-17 see p. 104) and the equivalent circuit of a motor (Fig. 15-6) shows that the only difference is the direction of flow of armature current. In the generator, the generated emf is larger than the terminal voltage, resulting in a flow of current *from* the generator. In the motor, the generated emf is smaller than the terminal voltage, resulting in a flow of current *into* the motor. This again brings out the fact that the dc machine is a reversible machine. When it is supplied with mechanical energy, it delivers electric energy; when it is supplied with electric energy, it delivers mechanical energy.

15-6 POWER RELATIONSHIPS IN A MOTOR

If each term of the fundamental motor equation is multiplied by I_a, the resulting equation is

$$V_t I_a = E I_a + I_a^2 R_a \tag{15-5}$$

The term $V_t I_a$ is the power supplied to the armature of the motor. The power lost as heat in the armature circuit is represented by the term $I_a^2 R_a$. Thus, the term $E I_a$ must represent the power developed by the armature. This power is not all available at the pulley since some of this developed power must be used to overcome the mechanical or rotational losses of the motor.

What is the power developed by the motor in the preceding example (*a*) in watts and (*b*) in horsepower?   **EXAMPLE 15-4**

(*a*)   Developed power $= E I_a$ watts

$$= 235.2 + 60 = 14,112 \text{ W}$$

(*b*)   Horsepower $= \dfrac{\text{watts}}{746}$

$$= \dfrac{14,112}{746} = 18.92 \text{ hp}$$

It may also be shown that the power output of a motor calculated by using USCS units is

$$\text{Power output in horsepower} = \dfrac{2\pi n T}{33,000} = \dfrac{nT}{5252} \tag{15-6}$$

where $n =$ speed of motor, revolutions per minute (rpm)
$T =$ torque at motor pulley, pound-feet

When SI units are used, the power output of a motor is

$$\text{Power output in watts} = \frac{2\pi nT}{60} = \frac{nT}{9.55} \tag{15-7}$$

where n = speed of motor, rpm

T = torque at motor pulley, newton-meters

Thus the power output of a motor may be readily obtained by measuring its useful torque with a prony brake and measuring its speed with a tachometer. The power output may then be calculated by means of Eq. (15-6) or Eq. (15-7).

EXAMPLE 15-5   The measured speed of a motor is 1100 rpm. The net force registered on the scale used with a prony brake is 16 lb. If the brake arm measures 28 in., what is the horsepower output of the motor?

$$T = Fr$$

$$= 16 \times \frac{28}{12} = 37.3 \text{ lb-ft}$$

$$\text{Horsepower output} = \frac{nT}{5252}$$

$$= \frac{1100 \times 37.3}{5252} = 7.82 \text{ hp}$$

EXAMPLE 15-6   The measured speed of a motor is 1100 rpm. The net force registered on the scale used with a prony brake is 5 kg. If the brake arm measures 0.8 m, what is the power output of the motor?

$$\text{Force} = 5 \times 9.81 = 49.1 \text{ N}$$

$$T = Fr = 49.1 \times 0.8 = 39.24 \text{ N-}m$$

$$\text{Power output in watts} = \frac{nT}{9.55} = \frac{1100 \times 39.24}{9.55}$$

$$= 4520 \text{ W or } 4.52 \text{ kW}$$

15-7 INTERPOLES

Interpoles perform the same function in a motor as they do in a generator; that is, they provide a commutating flux that generates the emf necessary to neutralize the emf of self-induction in the armature coils undergoing commutation. Since the motor field flux is distorted in a direction opposite the field distortion in a generator, it follows that motor interpoles must have a polarity opposite that of generator interpoles; that is, *motor interpoles must have a polarity opposite that of the following main pole in the direction of armature rotation.* It should be noted that for given external connections, a dc machine that has correct interpole polarity when operated as a generator will have the correct interpole polarity when operated as a motor, since a reversal of armature current changes the polarity of the interpoles.

15-8 MOTOR SPEED REGULATION

Speed regulation of a motor is a term that describes the variation of motor speed as load on the motor is changed. If a motor is able to maintain a nearly constant speed for varying loads, the motor is said to have good speed regulation. The speed regulation of a given motor is dependent upon that motor's inherent characteristics and is usually expressed in percent as in the following:

$$\text{Percent regulation} = \frac{\text{no-load speed } - \text{ full-load speed}}{\text{full-load speed}} \times 100 \quad (15\text{-}8)$$

EXAMPLE 15-7

The no-load speed of a dc shunt motor is 1200 rpm. When the motor carries its rated load, the speed drops to 1120 rpm. What is the speed regulation in percent?

$$\text{Percent regulation} = \frac{NL \text{ speed } - FL \text{ speed}}{FL \text{ speed}} \times 100$$

$$= \frac{1200 - 1120}{1120} \times 100 = 7.14 \text{ percent}$$

15-9 THE SHUNT MOTOR

This is the most common type of dc motor. It is connected in the same way as the shunt generator, that is, with the shunt field directly across the terminals in parallel with the armature circuit. A field rheostat is usually connected in series with the field.

A shunt motor has good speed regulation and is classed as a constant-speed motor even though its speed does decrease slightly with an increase in load.

As load is added to a shunt motor, the motor immediately tends to slow down. The counter emf immediately decreases since it is dependent on the speed and the practically constant field flux. The decrease in the counter emf permits the flow of an increased armature current, thus providing more torque for the increased load. The increased armature current causes a larger I_aR_a drop, which means that the counter emf does not return to its former value but remains at some lower value. This is shown to be true by the fundamental motor equation

$$V_t = E + I_aR_a \qquad (15\text{-}9)$$

Since V_t is constant, the sum of the counter emf and the I_aR_a drop must remain constant. If I_aR_a becomes larger owing to an increase in load, E must decrease, thus causing a corresponding decrease in speed.

The basic speed of a shunt motor is the full-load full-field speed. Usual speed adjustment is made by inserting resistance in the field circuit with a field rheostat, thereby weakening the field flux. This method of speed control provides a smooth and efficient means of varying the motor speed from basic speed to a maximum speed that is set by both electrical and mechanical limitations of the motor. *Care must be taken never to open the field circuit of a shunt motor that is running unloaded.* The loss of the field flux causes the motor speed to increase to dangerously high values.

The speed of a shunt motor may also be changed by means of an adjustable resistance in the armature circuit, but this method is less efficient than the shunt-field control. The method is also objectionable because it causes the motor to have a very poor speed regulation.

The field flux of a shunt motor is very nearly constant. Since the torque of a motor is equal to $K'\phi I_a$, the torque is directly proportional to the armature current. Figure 15-7 shows both the torque-load and the speed-load characteristics of a typical shunt motor. Note that the torque increases in a practically straight-line relationship with an increase in armature current, while the speed drops slightly as armature current is increased.

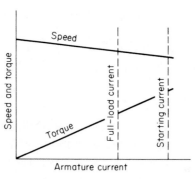

FIGURE 15-7   *Speed-load and torque-load curves of a shunt motor.*

Starters used with dc motors are usually designed to limit the armature starting current to 125 to 200 percent of full-load current. Since the torque of a shunt motor is proportional to the armature current, the starting torque will be 125 to 200 percent of full-load torque, depending on the value of the starting resistance. For example, if the starting current of a given shunt motor is 150 percent of full-load current, then the starting torque is 150 percent of full-load value (note Fig. 15-7).

Because the speed of a shunt motor may be fixed at any value between a maximum and minimum value, it is often used for driving such loads as machine tools. The operator may select any speed within the rating of the motor by adjusting the field rheostat. But at any one setting of the rheostat, the motor speed remains practically constant for all loads.

15-10 THE SERIES MOTOR

In the series motor, the field is connected in series with the armature. As the series field must carry full armature current, it is wound with a few turns of comparatively large wire. Any change in load causes a change in armature current and also a change in field flux. Therefore as the load changes, the speed changes.

It was shown that the speed of a shunt motor is inversely proportional to the field flux. This is true also in a series motor. The armature-circuit IR drop also varies with the load, but its effect is very small compared with the effect of the field flux. Therefore, the speed of a series motor depends almost entirely on the flux; the stronger the field flux, the lower the speed. Likewise, a decrease in load current and therefore in field current and field flux causes an increase in speed. Thus, the speed varies from a very high speed at light load to a low speed at full load.

A series motor does not have a definite no-load speed. As load is removed from the motor, the field flux decreases. If all the load is removed, the flux drops to practically zero and the motor speed may become dangerously high. For this reason, *the load should never be removed completely from a series motor.* A series motor is used only where load is directly connected to the shaft or geared to the shaft. Very small series motors usually have enough friction and other losses to keep the no-load speed down to a safe limit. The speed-load curve of a typical series motor is shown in Fig. 15-8.

In a shunt motor, the torque is practically proportional to the armature current since the field flux is almost constant. In the series motor the field flux varies with the armature current, and for small loads the field flux is almost directly proportional to the armature current. Since torque is equal to $K' \phi I_a$, this means that for small loads the torque is proportional to I_a^2. However, as the armature current approaches full-load value, the sat-

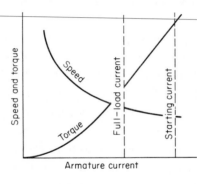

FIGURE 15-8   Speed-load and torque-load curves of a series motor.

uration of the magnetic circuit and armature reaction both prevent the torque from increasing as rapidly as the square of the current. This causes the torque-load curve to straighten out for the heavier loads as shown in Fig. 15-8.

It has been pointed out that the series motor develops a large torque for large armature currents and also operates at low speed for large armature currents. It is therefore a suitable motor for starting heavy loads. It is often used for driving cranes and winches where heavy loads must be moved slowly but where the lighter loads may be moved with greater speed. Another common application of the series motor is in electric-railway service.

15-11 THE COMPOUND MOTOR

The cumulative-compound motor combines the operating characteristics of the shunt and series motors. It has a definite no-load speed and may be safely operated at no load. As load is added, the increased amount of field flux causes the speed to decrease more than does the speed of a shunt motor. Thus, the speed regulation of a cumulative-compound motor is poorer than that of a shunt motor.

The torque of the cumulative-compound motor is greater than that of the shunt motor for a given armature current owing to the series field flux. The torque-load and speed-load curves are shown in Fig. 15-9.

Cumulative-compound motors are used when a fairly constant speed

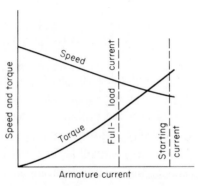

FIGURE 15-9   Speed-load and torque-load curves of a cumulative-compound motor.

is required with irregular loads or suddenly applied heavy loads. Such loads as presses, shears, and reciprocating machines are often driven by compound motors.

Figure 15-10 shows a comparison of the characteristics of series, shunt, and cumulative-compound motors with the same full-load horsepower output and the same light-load speed. A representative value of starting current is shown, being 150 percent of full-load current.

The speed regulation of the shunt motor is much better than that of the series or compound motor. Thus, the shunt motor is suitable for constant-speed applications. The compound and series motors are used where good speed regulation is not essential or where a varying speed may be desirable.

The exceptionally high starting torque of the series motor makes it an ideal motor for starting heavy loads at a reduced speed. Note from Fig. 15-10 that the series motor develops much more torque than the compound or shunt motors for a given armature current, at all but the lighter loads.

Thus, the different types of dc motors have widely varying characteristics. The character of the load, of course, determines the type of motor to be used in each case.

15-12 PERMANENT-MAGNET DC MOTORS

Permanent magnets rather than electromagnets are used to provide the magnetic field flux in permanent-magnet motors. The armature and commutator assemblies of these motors are similar to those of wound-field motors. Since no electric power is required for supplying the field in the permanent-magnet motor, the attendant field copper losses are eliminated. More importantly, the size and weight of the permanent-magnet motor is reduced substantially below that of the wound-field counterpart with a similar power rating. The development of higher-performance magnets is expected to reduce still further the size and weight of these motors. Permanent-magnet motors are most often used in the smaller sizes, usually fractional horsepower sizes.

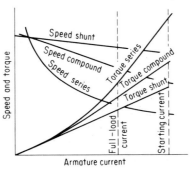

FIGURE 15-10   Comparative speed-load and torque-load curves for series, shunt, and compound motors.

Operationally, permanent-magnet motors are characterized by their high starting torque, their relatively high torque at low speeds, and their high efficiency. Typical applications include their use in line-printer drives, in precision positioning systems, and as reel motors in tape transports.

15-13 MOTOR-STARTING REQUIREMENTS

Two requirements must be met in starting dc motors, especially if they are to be started under load:

1. Both the motor and the supply lines must be protected from the flow of excessive current during the starting period.
2. The motor-starting torque should be made as large as possible, to bring the motor up to full speed in a minimum period of time.

At starting, when a motor armature is stationary, no counter emf is being generated. This means that the only thing to limit the amount of current being drawn from the supply is the armature-circuit resistance, which in most motors is very low, being of the order of 1 Ω or less. To meet the first starting requirement, an external resistance is placed in series with the motor-armature circuit during the starting period. To show why this is necessary, consider a 10-hp motor with an armature resistance of 0.5 Ω. If this motor were connected directly to a 230-V supply line, the resulting current would be

$$I = \frac{V}{R} = \frac{230}{0.5} = 460 \text{ A}$$

which is roughly 12 times normal full-load armature current for a motor of this size. This large inrush of current would in all probability damage the brushes, commutator, or windings. Starting resistances are usually designed to limit the starting current of a motor to 125 to 200 percent of the full-load current.

The amount of starting resistance required to limit the armature starting current to the desired values is

$$R_s = \frac{V_t}{I_s} - R_a \tag{15-10}$$

where R_s = starting resistance, ohms
V_t = motor terminal voltage, volts
I_s = desired armature starting current, amperes
R_a = armature-circuit resistance, ohms

Starting resistors are variable resistances, with the value of the resistance in the circuit at any time being manually or automatically controlled. The maximum value of the resistance is inserted in the armature

If the full-load armature current of the above motor is 40 A and it is desired to limit the starting current to 150 percent of this value, find the starting resistance that must be added in series with the armature.

$$R_S = \frac{V_t}{I_S} - R_a$$

$$= \frac{230}{40 \times 1.5} - 0.5$$

$$= \frac{230}{60} - 0.5 = 3.33 \ \Omega$$

circuit when the motor is first connected to the supply. As the motor speed increases, the counter emf increases, thereby decreasing the armature current. The starting resistance may then be cut out by successive steps until the motor has reached its full speed.

The second motor-starting requirement is met by providing a maximum value of field flux and by allowing a maximum safe value of armature current to flow during the starting period. In shunt and compound motors, a maximum field flux is obtained by cutting out the shunt-field rheostat, thereby applying full line voltage to the shunt field. In the series motor, the field flux is at a maximum by virtue of the heavy starting current flowing through the field winding. With a maximum field flux and a maximum allowable armature current, the starting torque (equal to $K'\phi I_a$) developed is at a maximum value, thereby bringing the motor to full speed in a short time.

15-14 CONTROLLERS AND STARTERS

A controller is a device for regulating the operation of the apparatus to which it is connected. A dc motor controller performs several basic functions such as starting, stopping, reversing, controlling speed, and providing some measure of protection for the motor that it governs.

A starter is a controller whose main function is to start and accelerate a motor.

15-15 DRUM CONTROLLERS

The drum controller is a manually operated controller formerly used extensively for starting, reversing, and controlling the speed of series motors and, to a lesser extent, dc shunt motors.

A drum controller consists essentially of a cylinder insulated from a central shaft on which is mounted a series of copper segments. As the cylinder is rotated by a handle, the various contact segments come in contact with stationary contact fingers. By designing the controller with different sequences for the making and breaking of the contacts as the drum is rotated, a variety of switching operations may be performed. The drum controller is no longer widely used since it has been largely replaced by automatic control equipment.

15-16 MAGNETIC CONTACTORS

Automatic starting, stopping, and reversing of motors are commonly accomplished by means of magnetically operated switches called *magnetic contactors*. Controllers using magnetic contactors to control the operation of the motors are called *magnetic controllers*.

Magnetic controllers may be actuated by means of a push button, which is often located at some point remote from the motor and controller. Controllers are also made entirely automatic by actuating them by various automatic devices. If, for example, a motor drives an air compressor that supplies air to a storage tank, the starting and stopping "signals" to the motor may be given by a pressure switch connected to the tank. When the pressure falls below a certain value, the motor is started; it is stopped again when the pressure has risen to the desired value.

The basic part of all magnetic controllers is the magnetic contactor. An operating coil is placed on an iron core so that when current flows through the coil the iron core becomes magnetized. This attracts a movable iron armature that carries one or more insulated electric contacts. As the armature is moved toward the core, the moving contacts are moved against stationary contacts. The contacts are connected in series with the controlled device, such as a motor-armature circuit, thereby completing the circuit when the operating coil is energized. When the circuit of the operating coil is opened, the iron core becomes demagnetized and the armature is released, being returned to its open position by means of a spring or by gravity. Since the contacts of a magnetic contactor may be designed to carry heavy currents, a means is provided whereby large amounts of current can be controlled by means of the relatively feeble current needed for energizing the operating coil.

To aid in extinguishing the arc formed when heavy currents are interrupted by electric contacts, *magnetic blowouts* are used. A magnetic blowout is an electromagnet so constructed that its field is set up across the arc. When the contacts open, the arc is moved at right angles to the field in the same way that a current-carrying conductor is acted upon. The field and contacts are so arranged that the arc is drawn away from the contacts, breaking it in a short time. To make the field strength proportional to the current being interrupted, the blowout coil is connected in series with the circuit being opened. A single-pole magnetic contactor equipped with a magnetic blowout coil is shown in Fig. 15-11.

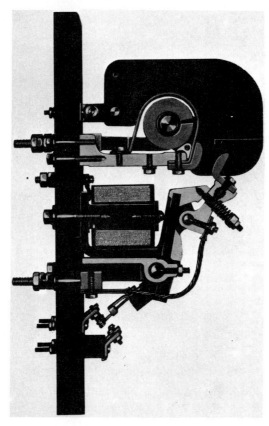

FIGURE 15-11 *Sectional view of a one-pole magnetic contactor equipped with magnetic blowout coil. (Square D Company.)*

Magnetic contactors are used when it is desired to control a motor or other device from a remote point. It is not necessary to run the main supply lines to the point of control. Only small control wires need to be used.

If a switch is to be opened and closed frequently, magnetic control is also desirable. An operator is easily able to operate a push button to actuate a contactor several times a minute, whereas the operation of heavy manual controls might cause undue fatigue.

Undervoltage release, undervoltage protection, and overload protection are easily provided by the use of magnetic contactors as described in the following sections.

15-17 CONTROL OF MAGNETIC CONTACTORS: MOTOR PROTECTION

Figure 15-12 shows a wiring diagram and a schematic control diagram of a two-wire control circuit using a magnetic contactor for connecting a motor to the line. It will be noted that the wiring diagram identifies the electric devices, terminals, and interconnecting wiring. The control schematic shows by means of graphic symbols the electrical connections and functions of the control circuits. The schematic diagram facilitates tracing the

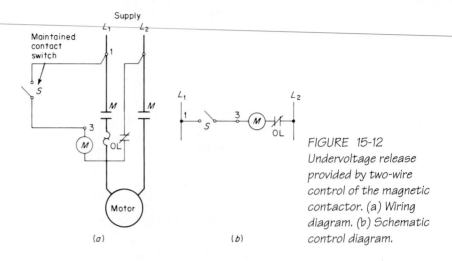

FIGURE 15-12
Undervoltage release
provided by two-wire
control of the magnetic
contactor. (a) Wiring
diagram. (b) Schematic
control diagram.

circuit and its functions without regard to the physical location of the component parts. Commonly used graphic symbols are shown in Appendix A.

When the switch S in Fig. 15-12 is closed, the operating-coil circuit is completed from point 1 on L_1 through the switch, coil M, and the normally closed overload relay contacts to L_2. This closes the main contacts M connecting the motor to the supply. The method for inserting and removing the starting resistance is not shown as this will be considered in Sec. 15-18. When the switch S is opened, contactor coil M is deenergized, the main contacts M open, and the motor is disconnected from the line.

If, while the motor in Fig. 15-12 is in operation, the supply voltage falls below a certain value or fails completely, the contactor disconnects the motor from the line. When the supply voltage is restored, the contactor M is reenergized and the motor restarts. This type of control is said to provide the motor with *undervoltage release*. This is desirable where it is necessary that the motor be restarted automatically after a voltage failure.

The switch S in Fig. 15-12 may be replaced by any of a number of automatic devices such as a thermostat, float switch, photoelectric cell, or limit switch.

A three-wire push-button control circuit for a contactor is shown in Fig. 15-13. The START button is a momentary contact switch that is held normally open by a spring. The STOP button is held normally closed by a spring. When an operator depresses the START button, the operating-coil circuit is completed from point 1 on L_1, through the STOP button, the START button, the coil and the normally closed overload relay contacts to L_2. This energizes contactor M, closing the main contacts M that connect the motor to the supply. At the same time, contactor M auxiliary or interlock contacts M-1 are closed.

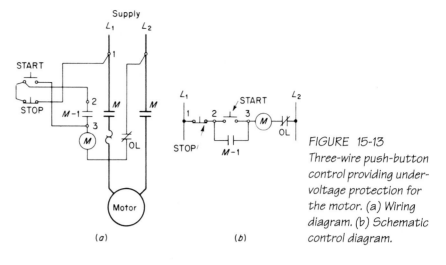

FIGURE 15-13
Three-wire push-button control providing under-voltage protection for the motor. (a) Wiring diagram. (b) Schematic control diagram.

When contacts M-1 close, a new circuit is established from point 1 through the STOP button to point 2, through contacts M-1 to point 3, through coil M and the overload relay to L_2. Since the operating-coil circuit is now maintained by the M-1 contacts, the operator may release the START button. When the STOP button is depressed, the coil is deenergized, thereby opening both the main and auxiliary contacts and stopping the motor.

If the supply voltage drops below a certain value, or fails while the motor is running, the main contacts and the auxiliary contacts are both opened. Upon return of the supply voltage, the contactor cannot close until the START button is again closed. Because a contactor that is controlled by a three-wire control circuit maintains the interruption of the circuit even after the line voltage is restored, it is said to provide *undervoltage protection* for the motor. This protection is used when it is desired to prevent the unexpected starting of a motor.

Overload protection for motors is used to protect the motor and control apparatus from excessive heating due to motor overloads. *Thermal-overload* relays are commonly used for motor-overload protection. There are two principal types of thermal-overload relays. Both types are operated from the heat generated in a heating element through which the motor current passes. In one type the heat bends a bimetallic strip, and in the other the heat melts a film of solder. Both act to open the motor control circuit.

A bimetallic strip is made of two different metals whose surfaces have been welded together. One of the metals expands rapidly when heated, while the other is not affected greatly by the heat. When heat is applied, the strip is caused to deflect due to the different expansion of the two metals. When the motor current reaches a predetermined value, the heat generated deflects the strip far enough to trip a latch that opens the motor-control circuit. When the strip cools sufficiently, the relay may be reset and the motor restarted.

In the solder-film type of overload relay the heat generated at a given overload current melts the film, which releases a latch arrangement and opens the motor control circuit. When the solder has cooled enough to hold the latch, the relay may be reset.

In either type of thermal relay, the time required to operate is determined by the amount of current flowing in the heater. Thus, the relay operates slowly for light overloads but disconnects the motor in a short time for dangerously heavy overloads. Relays in which the time of operation is inversely proportional to the amount of current flowing are called *inverse-time relays*.

Overload relay heating elements and contacts are connected in the motor main and control circuits, respectively, as shown in Figs. 15-12 and 15-13.

15-18 MAGNETIC CONTROLLERS FOR STARTING DC MOTORS

It was shown in Sec. 15-13 that a resistance must be connected in series with a dc motor-armature circuit during the starting period. If magnetic control is to be used for starting a motor, some method must be used for automatically inserting this resistance at starting and removing it when the motor reaches normal speed. A device that is often used to accomplish this is the *time-limit acceleration starter*. This starter cuts out the starting resistor in steps with contactors that operate successively at definite time intervals as the motor accelerates.

A time-limit acceleration starter is shown in Fig. 15-14. Figure 15-15

FIGURE 15-14 Time-limit acceleration starter. (Square D Company.)

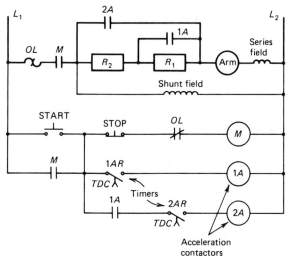

FIGURE 15-15
Schematic diagram of a
time-limit acceleration
starter. (Square D
Company.)

shows a schematic wiring diagram for this type of starter in which the
starting resistors are bypassed in two steps as the motor accelerates.

The time-limit acceleration starter shown in Fig. 15-15 is controlled
with a three-wire push-button control circuit. Depressing the START button
energizes the main contactor *M*, the contacts of which connect the motor
shunt-field and armature circuits to the supply lines and start acceleration
timer 1AR. The two acceleration timers are wired in series with the acceleration contactor coils and appear on the diagram as normally open time-
closing contacts. Immediately following the closing of main contactor *M*,
resistors R_1 and R_2 are both connected in series with the armature circuit
since the contacts of both acceleration contactors 1A and 2A are open. After
a time delay during which the motor begins to accelerate, timer 1AR closes
its contact to energize contactor 1A, which bypasses resistor R1 and also
starts timer 2AR. As the motor continues to accelerate, timer 2AR closes its
contact to energize contactor 2A, which bypasses resistor R_2. This applies
full line voltage to the armature circuit, and the motor accelerates to its
normal speed.

Time-limit acceleration starters used with motors of 10 hp and larger
are usually designed to have three, four, or five acceleration points rather
than the two points shown for the starter in Fig. 15-15. These starters may
also be designed to provide for dynamic braking and reversing service.

**15-19 DYNAMIC
BRAKING**

Often it is desirable to have a means of stopping a motor quickly. One
method of bringing a motor to a quick stop is by means of *dynamic braking*. If a running motor is disconnected from its supply lines and a resistance load (often the starting resistance) is placed across the armature

terminals, the motor then acts as a generator, sending a current through the resistance. This causes the energy possessed by the rotating armature to be expended quickly in the form of heat in the resistance, which brings the motor to a standstill quickly. The amount of braking effort available depends on the motor speed, the motor field strength, and the value of the resistance. This type of braking is used extensively in connection with the control of elevators and hoists and in other applications in which motors must be started, stopped, and reversed frequently.

15-20 POWER SUPPLIES FOR DC MOTORS

Since dc-power distribution systems are not readily available, dc motors usually have their own power supply in the form of a battery, an ac-to-dc motor-generator set, or a rectified ac supply.

Following the advent and development of semiconductor electronic components, static ac-to-dc rectifying systems became economically feasible. These systems, using silicon controlled rectifiers (SCR) or thyristors together with other solid-state components (see Chap. 13), provide a relatively low-cost method for supplying and controlling dc motors from commercial ac supply systems. Such systems find wide application not only in the control of general-purpose dc motor drives but also in highly automated industrial drives.

When a dc motor is operated from a rectified ac system, its performance may not be the same as when that motor is operated from a low-ripple dc supply such as a battery or a generator. The pulsating voltage and distorted wave forms present in the output of some rectified power supplies may raise motor temperature and increase noise and may adversely affect motor commutation and efficiency.

There are several types of rectified power supplies that can be used as a source of supply for dc motors, and each type of rectifier may have a different effect on motors connected to it. DC motors and their power supplies must be carefully matched to obtain optimum motor performance.

15-21 MOTOR SPEED CONTROL

The factors that may affect motor speed become evident when the equation for counter emf, Eq. (15-2), is combined with the fundamental motor equation, Eq. (15-4). This results in

$$V_t = K\phi n + I_a R_a \qquad (15\text{-}11)$$

or

$$n = \frac{V_t - I_a R_a}{K\phi}$$

Equation (15-11) is called the motor speed equation. From this equation it can be seen that motor speed n is dependent upon four variables, namely, the field flux ϕ, the armature-circuit resistance R_a, the terminal voltage V_t, and the armature current I_a.

Speed control of a motor can be effected by the manipulation of three of the variables in Eq. (15-11) by manual or automatic means. The armature current I_a is determined by the amount of load being supplied by the motor armature and hence cannot be used for motor speed control. Thus, the three basic speed-control methods are

1. Control of field flux ϕ
2. Control of armature-circuit resistance R_a
3. Control of terminal voltage V_t

Field Flux Control.   Field current and in turn field flux in a shunt or compound motor is readily varied by means of a field rheostat connected in series with the shunt field. Increasing resistance in the field circuit causes a decrease in field flux and therefore an increase in speed. Likewise, a decrease in field-circuit resistance causes a decrease in speed.

Armature-Circuit Resistance Control.   Motor armature-circuit resistance may be changed by introducing a variable resistance in series with the armature. As the series resistance is increased, the voltage across the motor armature is reduced and motor speed decreases. Likewise, motor speed is increased when the series resistance is decreased. This is the method of speed control normally used for series motors.

Terminal Voltage Control.   Motor speed can be controlled by varying the voltage applied to the armature circuit. Prior to the advent of relatively inexpensive solid-state components, the provision of a variable or multivoltage supply of dc voltage was very costly. This method of speed control was largely limited to the Ward-Leonard system or variations of it for the control of large dc motors. A dc rectifier drive system and the Ward-Leonard system of speed control are discussed in the following sections.

15-22 DC RECTIFIER DRIVE SYSTEM   A dc rectifier drive system is an adjustable-speed drive consisting of a rectifier, control equipment, and a dc motor, all of which is suitable for operating directly from a standard commercial ac supply system. Such drive systems may be very simple ones, or they may be complex industrial drive systems with ratings up to 100 hp, or even higher.

The basic elements of a dc rectifier drive for a shunt-wound dc motor are shown in Fig. 15-16. This is the configuration of a general-purpose

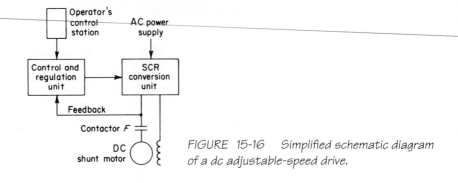

FIGURE 15-16   Simplified schematic diagram of a dc adjustable-speed drive.

drive suitable for powering conveyors, elevators, and machine tools, or for other loads requiring a continuously variable, closely regulated speed. As indicated by the diagram, the SCR unit converts the incoming ac supply voltage to an adjustable dc voltage. The contactor F connects the motor armature to the dc supply voltage.

The operator's control station contains a START-STOP button to control contactor F and a potentiometer used to set the speed level of the motor. During operation of the motor, a feedback voltage is compared with the reference voltage established by the setpoint on the potentiometer. The difference in these two voltages is used to regulate the triggering of the SCR unit, which in turn regulates the output voltage of the conversion unit and hence the speed of the motor. DC excitation for the motor field is supplied by a diode rectifier in the SCR conversion unit. A typical drive of this type may have a speed control range of 30 to 1 and a speed regulation of 2 percent. A more precise degree of speed regulation may be obtained when the feedback signal is obtained from a tachometer generator driven from the dc motor shaft. Speed regulation with this arrangement may be held to within 0.1 percent on some drives.

Many of the larger, more sophisticated dc adjustable-speed drives for industrial use provide for motor reversing, dynamic braking, and regeneration. Regeneration is the ability of a motor to return power to the ac source when the motor load becomes overhauling and the load attempts to drive the motor. A basic regenerative dc drive allows operation of the drive motor in any of four modes: forward motoring, forward regeneration, reverse motoring, and reverse regeneration. Other features often supplied with larger dc drives include special protective features such as motor field loss protection and protection against ac line disturbances. Special test and diagnostic equipment is also sometimes furnished as a part of the drive. A typical dc SCR regenerative drive unit that provides many of these features is shown in Fig. 15-17.

FIGURE 15-17 DC motor drive unit. (General Electric Company.)

15-23 WARD-LEONARD SYSTEM OF SPEED CONTROL

Some applications of dc motors require a wide and finely graduated range of speed control. The Ward-Leonard, or adjustable-voltage, system provides such control and involves a separate generator that drives an adjustable-speed motor. By varying the field excitation of the generator, the voltage applied to the motor may be varied over a wide range, resulting in a wide speed range on the motor.

A simplified diagram of the system is shown in Fig. 15-18. The driving motor is usually a constant-speed ac motor. Directly connected to the prime mover is the separately excited dc main generator G that drives the main motor M. To supply the field excitation of both G and M, a small self-excited dc generator or exciter E is also connected directly to the driving motor. Some systems use a static power supply for the field excitation in place of a rotating exciter.

To start the motor M, the field excitation of G is raised so that the voltage V_t is gradually increased across M, which causes the motor to accelerate. No starting resistance is needed since the voltage V_t is raised from a very low value, allowing the counter emf of the motor to increase with the applied voltage. The motor speed may be set at any speed between zero and its maximum speed by adjusting the field excitation of the generator G to the desired value. Since the field flux of motor M is constant, it has the

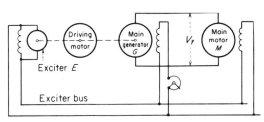

FIGURE 15-18 Simplified diagram of the Ward-Leonard system of speed control.

speed-load characteristics of a shunt motor and the speed regulation is good at a given armature voltage.

If, while the motor M is operating, the voltage V_t is suddenly decreased to a value below the counter emf of the motor, the armature current is reversed with the motor acting as a generator, driving G as a motor. Dynamic braking is thus established, which brings the motor to a quick stop. The motor M may be reversed by dropping V_t to zero and reversing the generator field current. As the voltage V_t is increased with the opposite polarity, the motor accelerates in the opposite direction.

An important requirement in many industrial motor control applications is the maintenance of a constant speed of a drive motor or the maintenance of a variable speed that is a function of some operating quantity. For example, the speed of the individual sections of a paper-making machine must be maintained within an accuracy of a fraction of one percent. Special control systems, called *closed-cycle feedback control systems*, have been developed for use with the conventional Ward-Leonard system to obtain extremely accurate speed control.

Some common applications of the Ward-Leonard system are in steel mills for reversing rolling mills, excavator drives, elevators for tall buildings, gun turrets, and special hoist motors.

The first cost of the Ward-Leonard system is high, and it is relatively inefficient since several energy transformations are involved. However, the speed control provided is very effective; that is, the response to changes in speed is fast, a wide range of speeds is available, and reversing and dynamic braking are provided. In applications where these factors are important, the advantages gained in using the Ward-Leonard system justify the increased cost and low efficiency.

REVIEW QUESTIONS

1 What is the function of a motor?
2 Upon what fundamental principle does motor operation depend?
3 Define torque. What is the unit?
4 Describe how torque is developed in a motor.
5 How may the direction of rotation of a dc motor be reversed?
6 What two things determine the amount of force developed on an armature conductor?
7 Give the torque equation of a motor.
8 If the mechanical output of a motor is increased, what must happen to the electrical input?
9 Explain how the torque of a motor is measured.
10 What is starting torque?

11 When a motor is in operation, why is the armature current *not* equal to the line voltage divided by the armature resistance? Give the correct equation for armature current.

12 What is the fundamental motor equation? How does it differ from the generator equation?

13 Given a prony brake and a tachometer, explain how the output of a motor may be measured. What other instruments would be needed to obtain data for finding the efficiency of a motor?

14 What is the unit of torque in USCS units? In SI units?

15 When the torque and speed are known, what is the formula for computing motor output?

16 What is the purpose of interpoles?

17 If the resistance of a motor shunt-field rheostat is increased, what happens to the speed of the motor?

18 What is a common method of controlling the speed of a shunt motor?

19 Define speed regulation.

20 Explain how a motor automatically adjusts its input to take care of changes in mechanical load.

21 What happens to the speed of a shunt motor as load is increased?

22 How is the basic speed of a shunt motor defined?

23 Is it possible to lower the speed below the basic speed by varying the shunt-field-rheostat resistance?

24 Why is it dangerous to open the field of a shunt motor running at no load?

25 If the torque of a shunt motor is doubled, what happens to the armature current, assuming a constant field flux?

26 Why is the speed regulation of a series motor poorer than that of a shunt motor?

27 Why should a series motor never be operated without load?

28 Why does a series motor have a higher starting torque than a shunt motor of similar capacity?

29 What are some common applications of (*a*) shunt motors, (*b*) series motors?

30 How does the speed regulation of a compound motor compare with that of (*a*) a series motor, (*b*) a shunt motor?

31 How does the starting torque of a compound motor compare with that of (*a*) a series motor, (*b*) a shunt motor?

32 Give some applications of compound motors.

33 What are some of the characteristics of permanent-magnet dc motors?

34 What two requirements must be met in starting dc motors?

35 What is the purpose of the starting resistance used with dc motors?

36 Give some of the functions of a controller.

37 Explain what is meant by undervoltage protection of a motor.

38 Where is the drum controller used?

39 What is a magnetic blowout? How does it operate?

40 What are some advantages of using magnetic contactors to control the operation of electric motors?

41 What is meant by undervoltage release? In what applications would it be desirable?

42 What kind of protection is provided for a motor by the three-wire push-button control of a magnetic contactor?

43 How is overload protection commonly provided for a motor?

44 How is time delay obtained for operation of the accelerating contacts in a time-limit acceleration starter?

45 Describe what is meant by dynamic braking.

46 What are some advantages of a rectified ac supply for a dc motor? Some disadvantages?

47 Name three methods by which speed of a dc motor may be controlled.

48 Describe briefly a dc adjustable-speed drive.

49 What are some advantages of the Ward-Leonard system of speed control? What is the major disadvantage?

PROBLEMS

15-1 A 120-V shunt motor develops a torque of 20 lb-ft when the armature current is 12 A. Assuming a constant value of field flux, what torque will the motor develop when the armature current is 20 A?

15-2 When developing a torque of 18 lb-ft, a 120-V shunt motor draws a current of 10 A from the source. The field current is 1.5 A. What current will the motor draw from the line when it is developing a torque of 60 lb-ft, assuming that the field flux does not change in value?

15-3 A prony brake attached to a motor pulley has a brake-arm length of 0.5 m. If, when the motor runs at 1150 rpm, the scale indicates a net force of 3.5 kg, what is the kilowatt output?

15-4 A prony brake is to be built to measure the torque output of a 10-hp motor that operates at a speed of 1200 rpm. If the scale available has a maximum capacity of 15 lb, what is the length of the shortest brake arm that could be used to measure any torque output of the motor up to 125 percent of its rated value?

15-5 A shunt motor on a 240-V line has an armature current of 75 A. If the field-circuit resistance is 100 Ω, find the field current, the line current, and the kilowatt input to the motor.

15-6 A shunt motor draws 6 kW from a 240-V line. If the field resistance is 96 Ω, find the line current, the field current, and the armature current of the motor.

15-7 A shunt motor connected to a 120-V line runs at a speed of 1200 rpm when the armature current is 20 A. The armature resistance is 0.1 Ω. Assuming constant field flux, what is the speed when the armature current is 50 A? (*Hint:* Speed is directly proportional to counter emf.)

15-8 Find the armature current of a shunt motor when the terminal voltage is 112 V, the counter emf is 107 V, and the armature-circuit resistance is 0.3 Ω.

15-9 The armature resistance of a 240-V shunt motor is 0.16 Ω. At no load, the line current is 6.4 A at rated voltage. The field current is 2.2 A, and the speed is 1280 rpm. Find the motor speed when the line current increases to 75 A, the field current remaining the same. Assume that armature reaction does not affect the field flux.

15-10 The counter emf of a shunt motor is 227 V, the field resistance is 160 Ω, and the field current is 1.5 A. The line current is 36.5 A. (*a*) Find the armature resistance. (*b*) If the line current during starting must be limited to 60 A, how much starter resistance must be added in series with the armature?

15-11 A shunt motor draws a current of 38 A from a 120-V source. The field-circuit resistance is 50 Ω, and the armature-circuit resistance is 0.25 Ω. Find (*a*) the field current, (*b*) the armature current, (*c*) the counter emf, and (*d*) the counter emf at starting (armature at a standstill).

15-12 Find the horsepower developed by the motor in the preceding problem. If the mechanical and iron losses total 550 W, what is the horsepower output?

15-13 What is the kilowatt output of a motor that is running at 1200 rpm and is supplying a torque of 24 N-m to the load?

15-14 What horsepower is developed by a motor when the armature current is 20 A, the applied voltage is 115 V, and the counter emf is 110 V?

15-15 A 10-hp motor has a shunt-field resistance of 115 Ω and a field current of 2 A. What is the applied voltage?

15-16 A certain shunt motor is connected to a 240-V line. The armature-circuit resistance is 0.05 Ω. When the armature current is 60 A, what is the counter emf?

15-17 A motor has a no-load speed of 900 rpm and a full-load speed of 850 rpm. What is the speed regulation?

15-18 The armature resistance of a shunt motor is 0.048 Ω. When the motor is connected across 120 V, it develops a counter emf of 113 V. Find (a) the IR drop in the armature circuit, (b) the armature current, (c) the armature current if the armature were stationary, and (d) the counter emf when the armature current is 160 A.

15-19 The efficiency, at rated load, of a 100-hp 600-V shunt motor is 85 percent. The field resistance is 190 Ω, and the armature resistance is 0.22 Ω. The full-load speed is 1200 rpm. Find (a) the rated line current, (b) the armature current at full load, (c) the counter emf at full load, and (d) the field current.

15-20 The power input to a shunt motor is 5600 W for a given load on the motor. The terminal voltage is 220 V, the I_aR_a drop is 6.4 V, and the armature resistance is 0.27 Ω. Find (a) the counter emf, (b) the power taken by the field, and (c) the field current.

15-21 At full load a 15-hp motor draws 54.8 A from a 240-V line. (a) What is the motor efficiency? (b) What is the motor efficiency at no load?

15-22 At rated load the rotational losses (iron losses plus mechanical losses) of a 240-V shunt motor are 865 W. The field resistance is 92 Ω, and the armature-circuit resistance is 0.12 Ω. The rated motor current is 140 A. Find (a) the field copper losses, (b) the armature copper losses, (c) the rated horsepower output, and (d) the efficiency.

15-23 A long-shunt compound motor requires an armature current of 12 A to carry a certain load. The armature resistance is 0.05 Ω, and the series-field resistance is 0.15 Ω. The motor is connected to a 115-V source. Find (a) the counter emf, and (b) the developed horsepower.

15-24 A 10-hp compound motor connected short-shunt is supplied by a 115-V source. The full-load current is 90 A. The shunt-field resistance is 90 Ω, the armature resistance is 0.08 Ω, and the series-field resistance is 0.04 Ω. Find (a) the shunt-field current, (b) the armature current, (c) the counter emf, (d) the efficiency at full load, (e) the full-load copper losses, and (f) the rotational losses (iron losses plus mechanical losses).

16

Polyphase Induction Motors and Controls

Because of its simple, rugged construction and good operating characteristics, the induction motor is the most commonly used type of ac motor. It consists of two parts: the stator, or stationary part, and the rotor, or rotating part. The stator is connected to the ac supply. The rotor is not connected electrically to the supply but has current induced in it by transformer action from the stator. Because of this, the stator is sometimes referred to as the primary and the rotor as the secondary of the motor.

Two types of three-phase induction motors are discussed in this chapter: squirrel-cage and wound-rotor motors. Both motors operate on the same basic principle and have the same stator construction but differ in rotor construction. Single-phase induction motors are discussed in Chap. 18.

16-1
CONSTRUCTION

A cutaway view of a typical three-phase squirrel-cage motor is shown in Fig. 16-1. Note the extremely simple construction as compared, for example, with a dc motor.

The stator core is built of slotted sheet-steel laminations that are supported in a stator frame of cast iron or fabricated steel plate. Windings, similar to the stator windings of a synchronous generator, are spaced in the stator slots 120 electrical degrees apart. The phase windings may be either wye- or delta-connected.

The rotor of a squirrel-cage motor is constructed of a laminated core with the conductors placed parallel, or approximately parallel, to the shaft

FIGURE 16-1   Cutaway view of a
squirrel-cage induction motor.
(MagneTek, St. Louis, MO.)

and embedded in the surface of the core. The conductors are not insulated
from the core, since the rotor currents naturally follow the path of least
resistance, that is, the rotor conductors. At each end of the rotor, the rotor
conductors are all short-circuited by continuous end rings. The rotor conductors and their end rings are similar to a revolving squirrel cage,
explaining the name.

The rotor bars and end rings of smaller squirrel-cage motors are cast
of copper or aluminum in one piece on the rotor core. In the larger motors,
the rotor bars, instead of being cast, are wedged into the rotor slots and are
then welded securely to the end rings. Squirrel-cage rotor bars are not
always placed parallel to the motor shaft but are often skewed. This results
in a more uniform torque and also reduces the magnetic humming noise
when the motor is running.

Wound-rotor or slip-ring motors differ from squirrel-cage motors in
rotor construction. As the name implies, the rotor is wound with an insulated winding similar to the stator winding. The rotor phase windings are wye-
connected and the open end of each phase is connected to a slip ring
mounted on the rotor shaft. Figure 16-2 shows a cutaway view of a wound
rotor motor. The three slip rings and brushes can be seen to the left of the
rotor winding. The rotor winding is not connected to a power supply; the
slip rings and brushes merely provide a means for connecting external
starting and control equipment into the rotor circuit. Information on
wound-rotor motor control is included in Section 16-18.

Motors have been classified according to environmental protection
and methods of cooling by the National Electrical Manufacturers
Association (NEMA). The three general classes are open motors, totally
enclosed motors, and motors with encapsulated or sealed windings.

Enclosures for the so-called open motors are constructed to provide
several degrees of protection. The open dripproof motor is constructed so

FIGURE 16-2 Cutaway view of a wound-rotor induction motor. (General Electric Company.)

that operation is not interfered with when drops of liquid or solid particles strike or enter the enclosure at any angle from 1 to 15° downward from the vertical. Guarded motors have protected openings to prevent accidental contact with live metal or rotating parts. Weather-protected motors have ventilating openings constructed so that entrance of rain, snow, or airborne particles is minimized.

Totally enclosed motors are enclosed to prevent the free exchange of air between the inside and the outside of the enclosure. They are classified as totally enclosed nonventilated, totally enclosed fan-cooled, explosion-proof, waterproof, or any of a number of other similar subclassifications.

The encapsulated or sealed-winding motor has its windings encapsulated or sealed with insulating resins or similar materials to seal and to protect the windings from severe environmental conditions.

16-2 THE ROTATING MAGNETIC FIELD

An induction motor depends for its operation on a rotating magnetic field that is established in the air gap of the motor by the stator currents. As mentioned above, a three-phase stator winding is wound with the phase windings spaced 120 electrical degrees apart.

A simplified stator winding layout of a two-pole wye-connected motor is shown in Fig. 16-3a. When the winding is energized from a three-phase supply, the phase currents vary in time phase as shown in Fig. 16-3b, and a pulsating flux is built up by each of the three windings. However, owing to the spacing of the windings and the phase difference of the currents in the

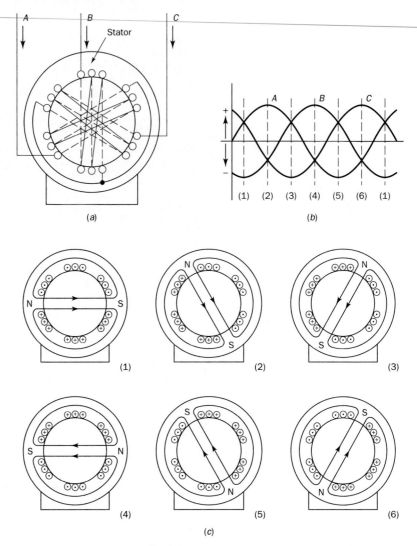

FIGURE 16-3 (a) Simplified two-pole stator winding layout. (b) Variation of current in the phase windings. (c) Resulting flux distribution in the motor at the successive instants indicated by the numbers on the curves in (b).

windings, the flux produced by each phase combines to form a resultant flux that moves around the stator surface at a constant speed. This resultant flux is called the *rotating magnetic field.*

The manner in which the rotating magnetic field is established by a three-phase stator winding may be illustrated by considering the direction of current flow through each of the three phases at successive time intervals. As shown in Fig. 16-3*b,* six different instants in time are marked off at

60° intervals for one complete cycle of the stator current waves. It will be assumed that, when currents have positive values in Fig. 16-3b, the current is flowing in the direction of the arrows in Fig. 16-3a. For example, at time 1 (Fig. 16-3b), currents in phases A and C are positive and the current in phase B is negative. Therefore, current is flowing toward the motor or with the arrows in lines A and C and away from the motor or against the arrow in line B.

Figure 16-3c shows the current directions in the stator conductors and the resulting field at each of the six different instants chosen in Fig. 16-3b. At time 1, the current is outward in the three coil sides on the upper half of the stator and is inward in the three coil sides on the lower half of the stator. The resulting field is established toward the right, creating a south pole on the stator surface at the right and a north pole at the left as shown. At time 2, the current is still in the same direction in phases A and B but has reversed in phase C. This establishes a field of the same strength as at time 1 but in a position 60° in a clockwise direction. The remaining diagrams in Fig. 16-3c show the resultant field at times 4, 5, and 6.

It is evident from Fig. 16-3 that the field resulting from the currents flowing in the three stator windings shifts around the stator surface, moving through a definite distance in each of the time intervals considered. Use is made of the principle of this rotating magnetic field in both induction and synchronous motors.

16-3 SPEED OF THE ROTATING MAGNETIC FIELD

In the two-pole winding of Fig. 16-3, the field makes one complete revolution in one cycle of the current. In a four-pole winding in which each phase has two separate groups of coils connected in series, it may be shown that the rotating field makes one revolution in two cycles of the current. In a six-pole winding, the field makes one revolution in three cycles of the current. In general, the field makes one revolution in $p/2$ cycles, or

$$\text{Cycles} = \frac{p}{2} \times \text{revolutions}$$

and

$$\text{Cycles per second} = \frac{p}{2} \times \text{revolutions per second}$$

Since revolutions per second equal revolutions per minute n divided by 60 and the number of cycles per second is frequency f,

$$f = \frac{p}{2} \times \frac{n}{60} = \frac{np}{120}$$

$$n = \frac{120\,f}{p} \tag{16-1}$$

The speed at which the rotating magnetic field revolves is called the *synchronous speed* of the motor. It will be noted that the same relation exists between the frequency, poles, and synchronous speed of a motor as exists between the frequency, poles, and speed of rotation of a synchronous generator. For a constant supply frequency, the synchronous speed is constant in any given motor.

If any two of the three supply lines to the stator winding in Fig. 16-3a are reversed, thereby reversing the phase sequence of the stator currents, an analysis similar to that of Sec. 16-2 will show the rotation of the magnetic field to be in the reverse direction, or counterclockwise. As will be shown in the following sections, the rotor turns in the same direction as the rotating flux. Therefore, *the direction of rotation of a three-phase motor may be reversed by interchanging any two of the three motor supply lines.*

16-4 PRINCIPLE OF OPERATION

In a dc motor, current is drawn from the supply and conducted into the armature conductors through the brushes and commutator. When the armature conductors carry current in the magnetic field established by the field circuit, a force is exerted on the conductors, which tends to move them at right angles to the field.

In an induction motor, there is no electric connection to the rotor, the rotor currents being induced currents. However, the same condition exists as in the dc motor; that is, the rotor conductors carry current in a magnetic field and thereby have a force exerted upon them tending to move them at right angles to the field.

When the stator winding is energized from a three-phase supply, a rotating magnetic field is established that rotates at synchronous speed. As the field sweeps across the rotor conductors, an emf is induced in these conductors just as an emf is induced in the secondary winding of a transformer by the flux of the primary currents. Since the rotor circuit is complete, either through end rings or an external resistance, the induced emf causes a current to flow in the rotor conductors. The rotor conductors carrying current in the rotating stator field thus have a force exerted upon them.

Figure 16-4 represents a section of an induction motor stator and rotor, with the magnetic field assumed to be rotating in a clockwise direction and with the rotor stationary, as at starting. For the flux direction and motion shown, an application of Fleming's right-hand rule shows the direction of the induced current in the rotor conductor to be toward the reader. At the instant being considered, with the current-carrying conductor in the magnetic field as shown, force is exerted upward upon the conductor since the magnetic field below the conductor is stronger than the field

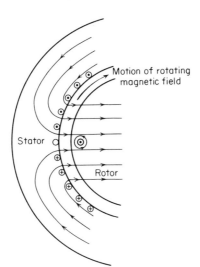

FIGURE 16-4 *Section of an induction motor rotor and stator showing the magnetic field in the air gap.*

above. For simplicity, only one rotor conductor is shown. However, other adjacent rotor conductors in the stator field likewise carry current in the same direction as the conductor shown and also have an upward force exerted upon them. One-half cycle later, the stator field direction will have reversed, but the rotor current will have also reversed, so that the force on the rotor is still upward. Likewise, rotor conductors under other stator field poles will have a force exerted upon them all tending to turn the rotor in the clockwise direction. If the developed torque is great enough to overcome the resisting torque of the load, the motor will accelerate in the clockwise direction, which is the same direction as the rotation of the stator magnetic field.

16-5 SPEED AND SLIP

An induction motor cannot run at synchronous speed. If it were possible, by some means, for the rotor to attain synchronous speed, the rotor would then be standing still with respect to the rotating flux. Then no emf would be induced in the rotor, no rotor current would flow, and therefore there would be no torque developed. The rotor speed even at no load must be slightly less than synchronous speed in order that current may be induced in the rotor, thereby producing a torque. This difference between rotor speed and synchronous speed is called *slip*. Slip may be expressed in revolutions per minute, but is more commonly expressed as percent of synchronous speed.

$$\text{Percent slip} = \frac{\text{synchronous speed } - \text{ rotor speed}}{\text{synchronous speed}} \times 100$$

Writing the above expression using symbols results in

$$\text{Percent } S = \frac{N_s - N_r}{N_s} \times 100 \qquad (16\text{-}2)$$

EXAMPLE 16-1   A four-pole 60-Hz squirrel-cage motor has a full-load speed of 1740 rpm. What is the percent slip at full load?

The synchronous speed from Eq. (16-1) is

$$N_s = \frac{120f}{p} = \frac{120 \times 60}{4} = 1800 \text{ rpm}$$

Slip in revolutions per minute $= 1800 - 1740 = 60$ rpm

$$\text{Percent slip} = \frac{60}{1800} \times 100 = 3.33 \text{ percent}$$

16-6 ROTOR FREQUENCY   If a two-pole 60-Hz induction motor (synchronous speed = 3600 rpm) operates at 5 percent slip, the slip in revolutions per minute is 3600 × 0.05 or 180 rpm. This means that a pair of stator poles passes a given rotor conductor 180 times every minute, or three times every second. When a pair of poles moves across a conductor, one cycle of emf is induced in that conductor. Thus the rotor conductor mentioned above will have induced in it an emf with a frequency of 3 Hz. If the slip is increased to 10 percent or 360 rpm, the frequency of the rotor emf and current is increased to 6 Hz. If the motor slip is increased to 100 percent, the rotor frequency will be 60 Hz.

It is obvious then that the rotor frequency depends on the slip—the greater the slip, the greater the rotor frequency. For any value of slip, the rotor frequency f_r is equal to the stator frequency f_s multiplied by the slip S expressed as a decimal, or

$$f_r = Sf_s \qquad (16\text{-}3)$$

Rotor frequency is significant because as it varies, the rotor reactance $(X_r = 2\pi f_r L_r)$ also varies, thus affecting both starting and running characteristics of the motor, as will be explained in following sections.

16-7 TORQUE AND SPEED OF AN INDUCTION MOTOR   Torque is produced in an induction motor by the interaction of the stator and rotor fluxes. The flux produced by the stator currents revolves at synchronous speed. In order that rotor currents may be induced, making possible the production of a torque, the rotor must turn at a speed less than synchronous speed. At no load, the rotor lags behind the stator flux only a small amount, since the only torque required is that needed to overcome the motor losses. As mechanical load is added, the rotor speed decreases. A decrease in rotor speed allows the constant-speed rotating field to sweep across the rotor conductors at a faster rate, thereby inducing larger rotor currents. This results in a larger torque output at a reduced speed.

Since the rotor impedance is low, a small decrease in speed produces a large increase in rotor current. For this reason the speed regulation of a standard squirrel-cage motor is low, the full-load slip being 3 to 5 percent. Although the motor speed does decrease slightly with increased load, the speed regulation is good enough that the induction motor is classed as a constant-speed motor.

With increasing loads, the increased rotor currents are in such a direction as to decrease the stator flux, thereby temporarily decreasing the counter emf in the stator windings. The decreased counter emf allows more stator current to flow, thereby increasing the power input to the motor. It will be noted that the action of the induction motor in adjusting its stator or primary current with changes of current in its rotor or secondary circuit is very similar to the changes occurring in a transformer with changes in load.

The torque of an induction motor is due to the interaction of the rotor and stator fields and is dependent on the strength of those fields and the phase relations between them. Mathematically,

$$T = K\phi I_r \cos \theta_r \tag{16-4}$$

where T = torque
K = a constant
ϕ = rotating stator flux
I_r = rotor current
$\cos \theta_r$ = rotor power factor

Throughout the normal range of operation, K, ϕ, and $\cos \theta_r$ are substantially constant, the torque increasing directly with the rotor current I_r. The rotor current in turn increases in almost direct proportion to the motor slip. The variation of torque with slip of a typical squirrel-cage motor of the standard type is shown in Fig. 16-5. Note that as slip increases from zero to about 10 percent the torque increases in an almost straight-line relationship with the slip.

As was explained in Sec. 16-6, an increase in slip causes an increase in rotor frequency and rotor reactance. Since the rotor-circuit resistance is constant, an increased rotor reactance means a decrease in rotor power factor (rotor power factor = R_r/Z_r). However, in the standard motor the change

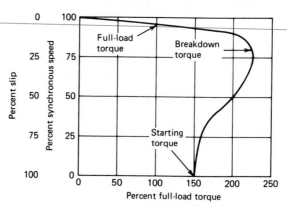

FIGURE 16-5   *Variation of torque with slip for a standard squirrel-cage motor.*

in slip is so small as load is increased from zero to full load, or even considerably above full load, that the change in rotor impedance is almost negligible.

However, as load and slip are increased much beyond the rated or full-load values, the increase in rotor reactance becomes appreciable. The increasing value of rotor impedance not only decreases the rotor power factor, but also lowers the rate of increase of rotor current so that the torque does not continue to increase directly with the slip. With the decreasing power factor and the lowered rate of increase in rotor current, the torque increase becomes less rapid and finally reaches maximum value at about 20 percent slip in the standard squirrel-cage motor. This maximum value of torque is called the *pull-out* or *breakdown torque* of the motor. If load is increased beyond the breakdown point, the decrease in rotor power factor is greater than the increase in rotor current, resulting in a decreasing torque, and the motor quickly comes to a stop.

The value of breakdown torque varies with the design of the motor but ranges from 200 to 300 percent of full-load torque in standard squirrel-cage motors. The value of slip at which the breakdown torque occurs depends on the value of rotor resistance; the higher the resistance, the higher the value of slip at breakdown. Some squirrel-cage motors for special applications are designed with high-resistance rotors so that the maximum torque occurs at 100 percent slip or at starting.

16-8 EFFECT OF LOAD ON POWER FACTOR

The current drawn by an induction motor running at no load is largely a magnetizing current, the no-load current lagging the applied voltage by a large angle. Thus the power factor of a lightly loaded induction motor is very low. Because of the air gap, the reluctance of the magnetic circuit is high, resulting in a relatively large value of no-load current compared with a transformer.

As load is added, the active or power component of the current increases, resulting in a higher power factor. However, because of the large value of magnetizing current, which is present regardless of load, the power factor of an induction motor even at full load seldom exceeds 90 percent.

The variation of the power factor of a typical motor with load is shown in Fig. 16-6. Other performance curves also shown are variations of speed, slip, efficiency, stator current, and torque for different values of load.

16-9 STARTING CONDITIONS

As stated in Sec. 16-6, the rotor frequency and reactance are high under starting conditions, that is, with 100 percent slip. Thus in the highly reactive rotor circuit, the rotor currents lag the rotor emf by a large angle. This means that the maximum current flow occurs in a rotor conductor at some time after the maximum density of the stator flux has passed that conductor. This results in a high starting current at low power factor, which results in a low value of starting torque.

As the rotor accelerates, the rotor frequency, and hence the rotor reactance, decreases, causing the torque to increase up to its maximum value. As the motor accelerates further, the torque decreases to the value required to carry the load on the motor at a constant speed.

The change in torque during the starting period is shown in Fig. 16-5, the starting torque being the value at the lower end of the curve at zero speed, or at 100 percent slip.

The high rotor reactance compared with the rotor resistance at start-

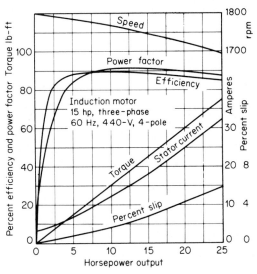

FIGURE 16-6 Performance curves of a typical squirrel-cage motor.

ing results in a poor starting torque with a high starting current. One method of correcting this condition is to design the rotor circuit with a high resistance. This increases the rotor power factor and improves the motor starting characteristics. But under the running conditions, the motor operates at a reduced efficiency because of the high I^2R loss in the rotor circuit. Another effect of a high rotor resistance, which may or may not be a disadvantage, is the increased speed regulation. Standard squirrel-cage motors are a compromise design; that is, they have a rotor resistance high enough to provide fair starting characteristics but that causes the motor to operate at a slightly reduced efficiency.

The standard or normal-torque normal-starting-current squirrel-cage motor has a starting torque of 80 to 200 percent of full-load value with a starting current of 500 to 700 percent of full-load current.

16-10 THE DOUBLE-SQUIRREL-CAGE MOTOR

The double-squirrel-cage motor is designed to provide a high starting torque with a low starting current. To accomplish this, the rotor is designed so that the motor operates with the advantages of a high-resistance rotor circuit during the starting period and a low-resistance rotor circuit under running conditions.

As the name denotes, the double-squirrel-cage motor has two rotor windings, one placed inside the other. A sectional view of this type of rotor winding is shown in Fig. 16-7. As shown, the outer cage winding is made with small high-resistance bars and, since it is near the rotor surface, it has a low inductance. The inner cage has a low resistance because of its larger conductors, but it has a high inductance because its bars are almost completely surrounded by iron.

During the starting period when the rotor frequency is high, the impedance of the outer winding is less than that of the inner winding, which causes a large proportion of the rotor currents to flow in the outer high-resistance winding. This provides the good starting characteristics of a high-resistance cage winding. As the motor accelerates, the rotor frequency decreases, lowering the reactance of the inner winding, allowing it to carry a larger proportion of the total rotor current. At the normal operating speed of the motor, the rotor frequency is so low that nearly all the rotor current flows in the low-resistance inner cage, resulting in good operating efficiency and speed regulation.

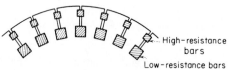

High-resistance bars

Low-resistance bars

FIGURE 16-7   Section of the rotor of a double-squirrel-cage motor.

Thus the outer winding produces the high starting and accelerating torque, while the inner winding provides the running torque at good efficiency.

The starting torque of the double-squirrel-cage motor ranges from 200 to 250 percent of full-load torque, with a starting current of 400 to 600 percent of full-load value. It is classed as a high-torque low-starting-current motor.

16-11 SPEED CONTROL OF SQUIRREL-CAGE MOTORS

From the formula for the synchronous speed of a motor, Eq. (16-1), it is evident that the supply frequency and the number of stator poles are the only variable factors determining synchronous speed. A variation of frequency is impossible when a motor is connected directly to ordinary commercial supply lines, since such supplies are constant-frequency systems. However, when an ac generator supplies but one motor, as in some ship propulsion systems, the supply frequency may be varied by varying the speed of the generator prime mover, thereby varying the speed of the motor. Likewise, if an adjustable-frequency supply is installed between a constant-frequency system and a motor, the motor speed can be varied by varying the output frequency of the adjustable-frequency supply. Following the advent of semiconductor devices, static variable-frequency supplies were developed and are commercially available for this use. Adjustable-speed ac drive systems are discussed in Sec. 16-18.

Multispeed squirrel-cage motors, designed to be operated on constant-frequency systems, are provided with stator windings that may be reconnected to form different numbers of poles. Two-speed motors usually have one winding that may be switched through suitable control equipment to provide two speeds, one of which is half of the other. Four-speed motors are equipped with two separate stator windings, each of which provides two speeds. While such motors do not have continuously adjustable speeds as do dc motors, or ac motors supplied from adjustable-frequency power supplies, they are suitable for such applications as ventilating fans, conveyors, machine tools, or other applications requiring several definite speeds but not necessarily a continuously adjustable speed drive.

16-12 STARTING INDUCTION MOTORS

In general, induction motors may be started either by connecting the motor directly across the supply circuit or by applying a reduced voltage to the motor during the starting period. Controllers used for starting motors by either method may be operated manually or magnetically.

Unlike dc motors, induction motors may be connected directly across the line without damage to the motor. However, because of the voltage disturbance created on the supply lines by their heavy starting currents, motors larger than 7½ to 10 hp are often started at a reduced voltage. The maximum allowable horsepower rating to be started on full voltage depends, however, on the motor design, the supply capacity, and the rules and regulations of the owners of the supply lines.

A greater starting torque is exerted by a motor when it is started on full voltage than when it is started on reduced voltage. In fact, it may be shown that the torque of an induction motor is proportional to the square of the applied voltage. Thus if the voltage is reduced to 80 percent of its rated value during starting, the starting torque is reduced to only 64 percent of that obtained by full-voltage starting. The reduced voltage applied to the motor during the starting period lowers the starting current but, at the same time, increases the accelerating time because of the reduced value of the starting torque. The type of load being started, then, also has a bearing on the method of starting to be used. If, for example, a particular load might be damaged by sudden starting and should be accelerated slowly, then reduced-voltage starting must be used.

Commonly used starters described here are

1. Full-voltage or across-the-line starters
2. Reduced-voltage starters
   (*a*) Primary-resistor starters
   (*b*) Autotransformer starters
   (*c*) Solid-state starters
3. Part-winding starters
4. Wye-delta starters

FIGURE 16-8   *Three-pole magnetic across-the-line starter. (Square D Company.)*

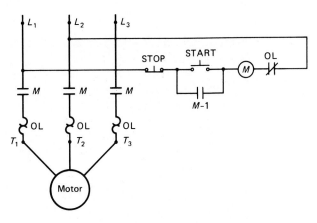

FIGURE 16-9 Circuit diagram of an across-the-line starter with a three-wire control circuit.

16-13 ACROSS-THE-LINE STARTERS

Motors are started on full line voltage by means of across-the-line starters. A magnetically operated across-the-line starter is shown in Fig. 16-8, and a connection diagram of the starter, using three-wire push-button control, is shown in Fig. 16-9. The operation is the same as that of the dc magnetic contactor with three-wire push-button control described in Sec. 15-17, with the exception that a three-pole contactor is used in the ac starter.

Across-the-line starters are generally equipped with thermal overload relays and when used with the three-wire push-button control as shown in Fig. 16-9 also provide undervoltage protection for the motor.

16-14 PRIMARY-RESISTOR STARTERS

Reduced voltage is obtained in the primary-resistor starter by means of resistances that are connected in series with each stator lead during the starting period. The voltage drop in the resistors produces a reduced voltage at the motor terminals. At a definite time after the motor is connected to the line through the resistors, accelerating contacts close; this short-circuits the starting resistors and applies full voltage to the motor.

A magnetically operated primary-resistor starter with the cover removed is shown in Fig. 16-10, and a typical elementary wiring diagram of this type of starter is shown in Fig. 16-11. When the START button is closed, the main contactor M is closed, connecting the motor to the line through the starting resistors. At the same time, a timing relay TR is started. After a definite time has elapsed during which the motor has accelerated, the time-delay contact TR closes. This energizes the accelerating contactor A, which short-circuits the starting resistors and applies full voltage to the motor.

In the starter just described, the starting resistance is cut out in one step. To obtain smoother acceleration with less line disturbance, starters are available in which the starting resistance is cut out in several steps.

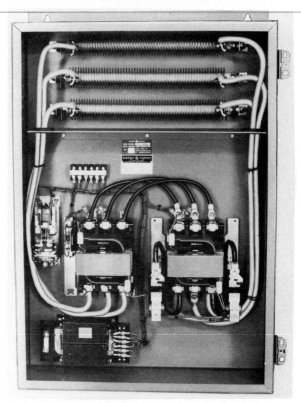

FIGURE 16-10 Primary-resistor starter. (General Electric Company.)

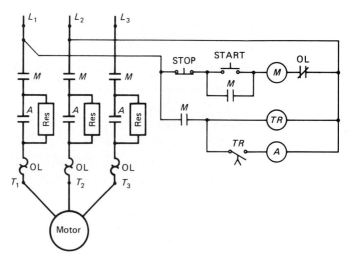

FIGURE 16-11 Circuit diagram of a primary-resistor starter.

Overload and undervoltage protection are provided in the same manner as in the across-the-line starter.

16-15 AUTOTRANSFOR- MER STARTERS

Autotransformer starters are used to reduce the voltage applied to the motor during the starting period through the use of a three-phase autotransformer. The autotransformer is provided with taps to provide a selection of 50, 65, or 80 percent of line voltage as a starting voltage with a corresponding reduction in supply line currents. Since motor starting torque varies as the square of the applied voltage, the resulting torque when using these taps will be 25, 42, and 64 percent, respectively, of full-voltage values. A tap can therefore be selected to match the starting torque required by a given motor and driven load.

Autotransformer starters may be either manually or magnetically operated. A magnetically operated starter is shown in Fig. 16-12, and an elementary diagram of a typical magnetic starter is shown in Fig. 16-13. Manually operated autotransformer starters are seldom used in modern motor control systems.

The starter shown in Fig. 16-13 has a two-pole starting contactor 1S,

FIGURE 16-12 *Magnetically operated auto-transformer starter. (Square D Company.)*

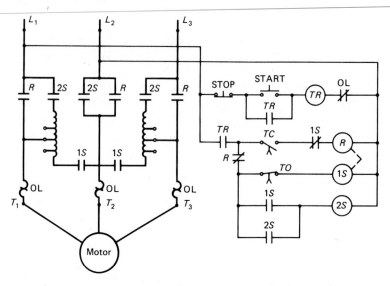

FIGURE 16-13   *Circuit diagram of an autotransformer starter.*

a three-pole starting contactor *2S,* and a three-pole running contactor *R.* Contactors *1S* and *R* are mechanically interlocked so that they cannot be closed simultaneously. Control relay *TR* has both time-opening and time-closing contacts *TO* and *TC.*

When the control relay *TR* is energized by closing the START button, contactors *1S* and *2S* are both energized through the *TR* contact connecting the motor to preselected taps on the autotransformer. After a time delay, during which the motor accelerates, timing relay contact *TO* opens and contact *TC* closes, which transfers the motor from the autotransformer to the line by dropping out starting contactors *1S* and *2S* and closing running contactor *R.* Starters of this type provide what is called closed-transition starting since the motor is not disconnected from the line at any time during the transition from reduced to full voltage.

Overload and undervoltage protection are provided by the thermal overload relays and the three-wire control circuit as shown.

16-16 SOLID-STATE REDUCED-VOLTAGE STARTERS

After solid-state power devices became readily available in the 1960s, solid-state motor starters were developed by a number of manufacturers. Although individual designs vary among manufacturers, all such starters in effect regulate motor starting current electronically to provide a smooth, stepless motor acceleration. A simplified block diagram of the power circuits of one such starter is shown in Fig. 16-14.

The basic solid-state power device used in this starter is a type of thyristor called the silicon controlled rectifier, or SCR. (SCRs are discussed

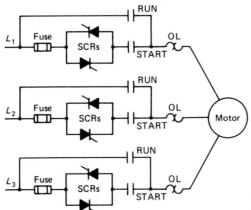

FIGURE 16-14   *Solid-state starter, simplified diagram.*

in more detail in Chap. 13.) The SCR blocks the flow of current in one direction but allows current to flow in the other direction after it has received a "turn on" signal called a gate pulse. Six SCRs are arranged in what is called a full-wave rectifier configuration as shown in Fig. 16-14. With the appropriate solid-state control circuitry, control of motor current or acceleration time may be achieved by applying the gate pulses to the SCRs at different times within each half-cycle of the power-supply voltage. If the gate pulse is applied early in the half-cycle, the SCR output is high. If the gate pulse is applied late in the half-cycle, the output is low. Input voltage to the motor can thus be varied from zero to full value during the starting period, causing the motor to accelerate smoothly from zero to full speed.

When the starter shown in Fig. 16-14 is energized, the START contacts are closed and the motor accelerates under the control of the SCRs. After the motor reaches full speed, the RUN contacts close, the SCRs are turned off, and the START contacts open.

Protective functions usually provided by the solid-state starter include detection and tripping for abnormal conditions such as a shorted SCR, phase-current unbalance, or motor overload. Current-limiting fuses or circuit breakers provide short-circuit protection for the starter and motor circuits.

16-17 PART-WINDING AND WYE-DELTA STARTERS

Two starters that are not strictly reduced-voltage starters but that are used, however, to limit motor starting currents are *part-winding* and *wye-delta starters*.

A requirement for the use of part-winding starting is that the motor must have two separate parallel windings with six leads brought out of the motor to the starter. The starter is designed to connect one of the three-phase motor windings to the supply when the starting impulse is received. After a time delay during which the motor accelerates, the starter automatically connects the second motor winding to the supply in parallel with the first. Although full voltage is applied to each of the motor windings, the initial motor starting current is considerably less than the starting current resulting from normal full-voltage starting of the standard motor. Care must be exercised in the application of this starting method, however, since the torque produced by the first half of the motor winding may be inadequate to accelerate some high-inertia loads.

Wye-delta starting, as the name implies, involves connecting a motor winding first in wye during the starting period and then in delta after the motor has begun to accelerate. Thus, during the starting period the supply phase voltage, which is only 57.7 percent of the line voltage, is applied to the motor winding and the starting inrush is reduced proportionally. Both ends of each phase of the motor winding must be brought out to the starter so that switching may be accomplished. This starter is often used for starting motors that drive loads having long acceleration times.

16-18 AC MOTOR DRIVES

DC motor drives were the first to employ solid-state components. As a result, the solid-state dc motor drive has dominated both the industrial and consumer markets in the past. However, solid-state ac drives have been developed to the point where they are competitive with the dc drives for many applications. In the solid-state ac drive, the induction or synchronous ac motor replaces the more complicated and less efficient dc motor used in the dc drives.

Four of the more generally used types of solid-state ac drive systems are discussed briefly here, namely, the primary-voltage control drive, the cycloconverter drive, the inverter drive, and the wound-rotor drive.

Primary-Voltage Control AC Drives. This drive system, sometimes called the phase-controlled converter drive, is the simplest and least costly solid-state ac drive system. In this system, the voltage applied to a squirrel-cage induction motor is varied by controlling the firing time of SCRs connected between the ac supply and the motor. This phase-control system in effect varies the voltage applied to the motor as would a rheostat or a variable-voltage transformer.

When the voltage applied to a running induction motor is reduced, the strength of the rotating magnetic field in the air gap is reduced. This results in less voltage being induced in the rotor, less rotor current developed, and less torque developed, and so the motor slows down. It can be shown that the torque reduces approximately as the square of the reduction in voltage. Standard squirrel-cage motors are not suited for primary-voltage control service because they overheat when they are operated at the lower voltages. Because of this, the usual motor used with this drive is the NEMA design D motor, which has a high resistance rotor and a slip of 8 to 13 percent.

Several thyristor circuit configurations have been used to vary the voltage applied to the motor. Tests conducted by one research organization indicate that two phase-controlled thyristors in each of the three phases supplying a wye-connected motor resulted in the best motor performance. This is the configuration often used for this drive.

The primary-voltage control ac drive system has speed-torque characteristics that suit it for pump and fan drives in ratings generally below 125 hp. This drive has been used in single-pump drives and also in multiple-pump drives. In these systems, a programmed control of several motors is used to maintain a constant liquid level in storage tanks or reservoirs.

Cycloconverter AC Motor Drive. The cycloconverter drive consists of a frequency converter with an adjustable-frequency output driving a synchronous or induction motor. Power conversion is accomplished by a set of thyristors in each phase of the three-phase input, a total of 36 thyristors being used in one system. Triggering of the thyristors is such that small increments of primary power are successively switched in a manner to simulate a lower-frequency sine wave in the output.

Advantages claimed for the cycloconverter are minimum heating, high efficiency, and the capability of motor regeneration and reversing. A major limitation is the low-frequency output when this system is operated from commercial 50- or 60-Hz systems. This drive is an economically attractive drive for large low-speed loads such as papermaking machines, mine hoists, ball mills, or other machines having large slow-speed revolving drums. For some applications, a completely gearless ac drive using a single low-speed synchronous motor supplied from a cycloconverter can be used to replace earlier, much more complicated conventional drives.

Inverter AC Motor Drive. The inverter drive, like the cycloconverter, is an adjustable-frequency ac drive. However, unlike the cycloconverter, which provides a direct ac-to-ac conversion, the inverter drive consists of a rectifier that provides dc power from the ac line. An inverter then converts the dc power to an adjustable-frequency power supply for the motor. The elements of an inverter drive system are shown in Fig. 16-15.

Several configurations are used for the components of the inverter drive, depending upon the manufacturer of the drive and the rating of the

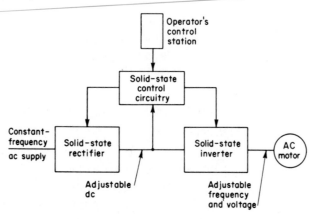

FIGURE 16-15   Block diagram of an inverter ac motor
drive.

drive. For example, the drive represented in Fig. 16-15 is supplied by one
manufacturer in a range of sizes from 1 to 5 hp. Incoming 230-V, single-
phase ac is rectified to a constant-voltage dc. The dc is then inverted by a
transistor-inverter section to an adjustable-frequency, three-phase output
with a frequency range of 6 to 120 Hz. Output is suitable for operating standard three-phase squirrel-cage induction motors. Special features that can
be provided with the drive include reversing, dynamic braking, and several
variations in the control devices supplied with the operator's station.

The inverter drive has a wider frequency and speed range than the
cycloconverter and can be operated up to the supply frequency, and as high
as 100 percent above that frequency in some systems. Applications of the
inverter ac drive, which is the most popular solid-state ac drive, include
drives for machine tools, conveyors, pumps, and air-handling systems. One
important application of this drive has been in synthetic-fiber-spinning
lines where a highly accurate power supply is required for variable-speed
synchronous motors.

The Wound-Rotor Drive.   As indicated in Sec. 16-1, the wound
rotor or slip-ring induction motor has an insulated three-phase rotor winding usually wye-connected and terminated in three slip rings. Three
external variable resistors are connected through brushes to the rotor
slip rings during the starting period, giving the motor a high starting
torque with a low starting current. Conventionally, the external resistance
is in the form of three rheostats, often manually operated. After the motor
is accelerated to near full speed, the rheostats are usually bypassed.
However, limited speed control can be provided by inserting rotor resistance

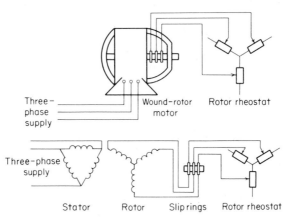

FIGURE 16-16 *Circuits of a wound-rotor induction motor showing the rheostats for speed control.*

during motor operation. This method of speed control is inefficient since power is wasted as heat in the rheostats. This method of control is illustrated in Fig. 16-16.

With the advent of solid-state power and control devices, developments have overcome some of the undesirable features of wound-rotor drives. Control systems are now available in which the external rotor resistance may be controlled electronically. This makes possible automatic control of both motor starting and motor speed. Slip power recovery systems have also been developed that improve wound-rotor-drive efficiency.

The components of one automatic control system in which rotor resistance is controlled electronically for motor starting and speed control are shown in Fig. 16-17. As shown, the three rotor currents are converted to a single direct current by a three-phase bridge rectifier. The dc is supplied to a single resistor through a smoothing reactor. Connected in parallel with the resistor is an electronic chopper consisting of one or more thyristors and solid-state circuitry. The chopper functions as an electronic switch that opens and closes at a rapid rate in response to the control circuitry. By varying the length of the ON and OFF periods, it is possible to obtain a variation

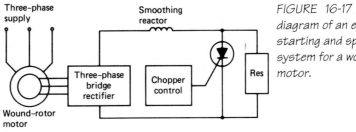

FIGURE 16-17 *Block diagram of an electronic starting and speed control system for a wound-rotor motor.*

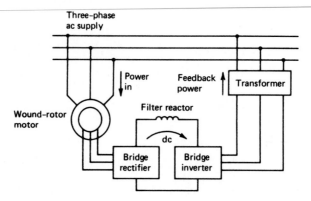

FIGURE 16-18   Slip power recovery system.

in the value of the effective resistance between zero and its full value. Feedback control (not shown) from a speed-sensing device on the motor shaft can be used to vary the effective resistance to maintain a constant pre-set speed on the motor.

The efficiency of a wound-rotor motor operating at reduced speed can be improved by an arrangement in which the rotor slip power is returned to the ac supply rather than being wasted as heat in the rotor rheostat. The essentials of such a circuit are shown in Fig. 16-18.

The slip power recovery system, when equipped with the necessary controls, is ideally suited for any application where a wound rotor must be run at less than its rated speed. Water pump and fan drives are typical applications for this system.

**16-19
INDUCTION-
MOTOR LOSSES,
EFFICIENCY, AND
RATINGS**
Induction-motor losses include stator and rotor copper losses, stator and rotor core losses, and friction and windage losses. As in the transformer, the core losses are practically constant for all loads. For practical purposes, the friction and windage losses are also considered to be constant. The copper losses, of course, vary with the load.

The efficiency curve of a typical 15-hp squirrel-cage motor is shown in Fig. 16-6. Note that the efficiency increases with load up to about three-fourths full load; above this point it begins to decrease because of the relatively larger values of I^2R loss.

In general, the larger the motor, the higher the full-load efficiency. Average values of full-load efficiencies of squirrel-cage motors are about as follows: 5 hp, 83 percent; 25 hp, 88 percent; and 100 hp, 90 percent. Wound-rotor motor efficiencies are slightly lower.

A large part of the total electric energy used by modern industry is used by ac polyphase motors in the size range of 1 to 100 hp. Any increase in the efficiency of this class of motors therefore has the potential for effecting large energy-cost savings. Recognizing this, motor manufacturers have designed a series of *energy-efficient* motors having higher-than-standard efficiencies. This improved efficiency is obtained by the use of special low-loss lamination steel, larger stator winding conductors, optimized rotor-winding design, longer stators and rotors, and by other similar design changes. These changes result in higher production costs and, in turn, a premium-priced motor. Motor users are thus offered the opportunity, when purchasing a motor, to weigh the economics of a higher first cost of an energy-efficient motor against the savings in energy cost over the life of the motor.

Motor nameplate data usually include the following in addition to the manufacturer's serial, frame, and type numbers: horsepower, speed, voltage, frequency, and temperature rise in a specified time. If a motor is designed to be used on more than one voltage or to operate at more than one speed by reconnecting the windings, a connection diagram is also often included on the nameplate.

A code letter that indicates the motor input in kilovoltamperes per horsepower with the rotor at a standstill is included on the nameplate of motors larger than $1/2$ hp. This information is used in determining the size of fuses or circuit breakers to be used in the motor supply circuit.

Another item found on most motor nameplates is a number called the service factor. When rated voltage and frequency are maintained on the motor under standard specified operating conditions, the motor may be overloaded up to the horsepower obtained by multiplying the nameplate horsepower by the service factor. This means, for example, that a motor rated at 10 hp and having a service factor of 1.15 may be loaded to 11.5 hp when rated voltage and frequency are maintained at the motor terminals. The service factor for most motors is 1.15.

Squirrel-cage motors are designed to have a wide variety of starting and operating characteristics, the different characteristics being attained largely by different designs of the rotor circuits. The motor referred to in this chapter as the standard motor is a normal-torque normal-starting-current motor and is a general-purpose motor classified by NEMA as a design B motor. Other types of motors designed for special applications include the following: the NEMA design A motor, which has a higher breakdown torque and higher starting current than the design B motor; the NEMA design C motor, a high-starting-torque and normal-breakdown-torque motor; and the NEMA design D motor, which is a high-starting-torque, high-slip motor. The operating characteristics of each are denoted by the name.

1 What are the two parts of an induction motor? How are the two circuits of an induction motor similar to the circuits of a transformer?

2 How does the construction of the squirrel-cage motor differ from that of the wound-rotor motor? Describe the rotor construction of each.

3 Describe briefly how the rotating magnetic field is produced by the three-phase stator currents. In a two-pole winding, how many revolutions does the rotating magnetic field make in one cycle of the current?

4 In general, what factors determine the synchronous speed of a motor?

5 How may the direction of rotation of a three-phase motor be reversed?

6 Upon what basic principle does the induction motor, like the dc motor, depend for its operation?

7 Describe briefly how torque is produced in an induction motor. Is the direction of rotation of the rotor the same as or opposite to that of the rotating field?

8 Why does an induction motor run at less than synchronous speed? What is the difference between synchronous and rotor speed called?

9 Show why the rotor frequency is equal to the slip multiplied by the stator frequency. What is the rotor frequency when the rotor is at a standstill with the stator excited?

10 What effect does slip have on rotor reactance?

11 What is the order of magnitude of the full-load slip of a standard squirrel-cage motor?

12 Describe the changes in speed, rotor current, and torque as load is added to an induction motor. How is the motor able to adjust its stator current with changes in mechanical load?

13 Upon what factors does the torque of an induction motor depend?

14 What is the effect of the change in rotor frequency as load is increased from no load to full load? What is its effect when the load is increased considerably above full load?

15 What is meant by pull-out or breakdown torque? If load is increased beyond the breakdown torque, what happens to the speed of the motor?

16 In the standard squirrel-cage motor, at about what value of slip does breakdown torque occur? How does the value of rotor resistance affect the point at which breakdown torque occurs?

17 Why is the power factor of a lightly loaded induction motor very low? What happens to the power factor as load is increased on the motor?

18 As load is increased from no load to full load on an induction motor, what happens to the slip, speed, torque, stator current, power factor, and efficiency of the motor?

19 Why does a squirrel-cage motor have a poor starting torque?

20 What are objections to building a squirrel-cage motor with high enough rotor resistance to provide good starting torque?

21 Describe the action of the double-squirrel-cage motor. How does its starting torque and current compare with the standard squirrel-cage motor?

22 In what way may a squirrel-cage motor be made to operate over a wide range of continuously adjustable speeds? What are the limitations of this method of speed control?

23 How may the same squirrel-cage motor be made to operate at two definite speeds?

24 What are two general methods of starting squirrel-cage motors?

25 What determines the maximum size of a squirrel-cage motor that can be started on full voltage?

26 It is desired that an air-compressor motor be started automatically by an across-the-line starter that is controlled by a pressure switch on the air tank. Draw a connection diagram showing the starter and control wires. Use a two-wire control and include a thermal overload relay in the diagram.

27 Name three different types of reduced-voltage starters.

28 Describe the operation of a magnetically operated primary-resistor starter.

29 Two motors are started at 50 percent of line voltage, one with a primary-resistor starter and the other with an autotransformer starter. The starting current taken by each motor is 100 A per terminal. What is the value of the current on the line side of the starters in each case?

30 What is a major advantage of solid-state reduced-voltage starters over primary-resistor and autotransformer starters?

31 What are the special motor-design requirements if the motor is to be used with a part-winding or wye-delta starter?

32 Name four types of solid-state ac motor drives.

33 Why is a standard squirrel-cage motor not suitable for primary-voltage control service? Which NEMA design type motor is usually used?

34 What is an advantage of the inverter ac-motor drive over the cycloconverter drive?

35 Draw a schematic diagram of a manually controlled wound-rotor drive.

36 Describe the essential features of a slip-power recovery system when used with a wound-rotor motor.

37 List the losses of an induction motor. Which are considered constant and which are variable?

38 What are some of the data usually found on a motor nameplate?

39 What is meant by the term *motor service factor?*

PROBLEMS

16-1 Find the synchronous speed of a 60-Hz motor that has an eight-pole stator winding.

16-2 Make a table showing the synchronous speeds of 2-, 4-, 6-, 8-, and 12-pole induction motors for frequencies of 25, 50, and 60 Hz.

16-3 A six-pole 60-Hz induction motor has a full-load slip of 5 percent. What is the full-load rotor speed?

16-4 A squirrel-cage-motor stator winding is wound for four poles. At full load, the motor operates at 1720 rpm with a slip speed of 80 rpm. What is the supply frequency?

16-5 What is the rotor frequency of an eight-pole 60-Hz squirrel-cage motor operating at 850 rpm?

16-6 How much larger is the rotor reactance of a squirrel-cage motor at starting (with the rotor at a standstill) than it is when the motor operates at 5 percent slip?

16-7 The three-phase induction propulsion motors for an aircraft carrier have stators that may be connected for either 22 or 44 poles. The frequency of the supply may be varied from 20 to 65 Hz. What are the maximum and minimum synchronous speeds obtainable from the motors?

16-8 A 50-hp 230-V three-phase motor requires a full-load current of 130 A per terminal at a power factor of 88 percent. What is its full-load efficiency?

16-9 A 10-hp 460-V three-phase squirrel-cage motor at three-fourths rated load operates at an efficiency of 85 percent and at a power factor of 85 percent. What is the line current per terminal?

16-10 What are the rated full-load speeds of a 60-Hz motor that may be connected for either four- or eight-pole operation, assuming the motor to operate with a full-load slip of 5 percent for either connection?

16-11 What is the percentage of full-voltage starting torque developed by a motor when it is started at 65 percent of full voltage?

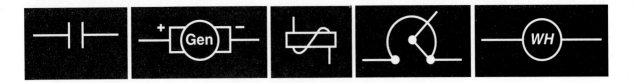

17
Synchronous Motors

Just as dc generators may be operated as motors, ac synchronous generators may be operated as motors. If, when two synchronous generators are operating in parallel, the prime mover is disconnected from one, it will continue to rotate, drawing power from the other to supply its losses. When mechanical load is added, the machine continues to operate at a constant speed. When operated in this manner, the machine is called a *synchronous motor*.

Synchronous motor construction is very similar to synchronous generator construction except that most motors are salient-pole machines. Synchronous motors, like generators, may be classified as brush type or brushless type. The brush type, as the name denotes, uses slip rings and brushes to connect the machine field to an external excitation supply. The brushless synchronous motor, like the brushless generator, has the field excitation and control systems mounted on the rotor. The rotor of a salient-pole brushless synchronous motor is shown in Fig. 17-1.

This chapter deals with three-phase synchronous motors. Single-phase synchronous motors are discussed in Chap. 18.

17-1 OPERATION OF THE SYNCHRONOUS MOTOR

When an induction motor is connected to a three-phase supply, a rotating magnetic field is established; this induces currents in the rotor windings, thereby producing a torque. The rotor can never turn at synchronous speed, since there must be relative motion between the rotating magnetic field and the rotor in order that currents can be induced in the rotor circuit.

When the stator windings of a synchronous motor are excited with three-phase voltages, a rotating magnetic field is established as in the

FIGURE 17-1 *Rotor of a slow-speed brushless synchronous motor. (Dresser Rand-Electric Machinery.)*

induction motor. However, in the synchronous motor, the rotor circuit is not excited by induction but by a source of direct current as in the ac generator. If the rotor is brought up to synchronous speed by some means, with the rotor poles excited, the poles of the rotor are attracted by the poles of the rotating magnetic field and the rotor continues to turn at synchronous speed. In other words, the rotor is locked into step magnetically with the rotating magnetic field. If for any reason the rotor is pulled out of step with the rotating stator flux, the attraction is lost and torque is no longer developed. Therefore, a synchronous motor develops torque only when it is running at synchronous speed. With the stator of a motor energized and the magnetic field revolving at synchronous speed, and with the rotor at a standstill, the rotor poles are first attracted in one direction and then the other, resulting in a net torque of zero. It is evident, then, that a synchronous motor is not self-starting.

17-2 STARTING SYNCHRONOUS MOTORS

Since a synchronous motor is not self-starting, some auxiliary means must be provided to start the motor. One starting method consists of using a small auxiliary induction motor to accelerate the rotor of the main motor to synchronous speed. This starting method is seldom used, however, and most synchronous motors are started by the use of a squirrel-cage winding embedded in the face of the rotor poles. The motor is then started as an induction motor and is accelerated to near synchronous speed by this means. At the proper instant, dc field excitation is applied and the motor pulls into synchronism. The squirrel-cage winding may be seen on the rotor shown in Fig. 17-1.

The amount of torque that a motor is able to exert when pulling into synchronism is called the *pull-in torque*. The value of pull-in torque developed by a synchronous motor varies widely with the application for which the motor is designed. For reciprocating compressors, a common application for synchronous-motor drive, the motor is designed to have a pull-in torque of about 50 percent of full-load value. Motors for other applications may be designed to have pull-in torques as high as 150 percent of full-load torque.

During the starting period, while the rotor is at a standstill, or is rotating at a speed much lower than synchronous speed, the rotating magnetic field rapidly cuts across the dc field coils, inducing an emf in them. Since each field coil is wound with many turns and the several coils are connected in series, the induced emf may become dangerously high. The usual method of limiting the value of field voltage during starting is to short-circuit the field winding through a field discharge resistor. This resistor is disconnected as the motor nears synchronous speed.

Low-speed synchronous motors equipped with squirrel-cage starting windings are usually started on full voltage. Reduced-voltage starting, using autotransformer or resistor-type starters, is very often used on high-speed motors, or on other motors that require large amounts of starting current or where voltage fluctuations on the supply circuits must be kept to a minimum.

A starter designed for full-voltage starting of a brush-type synchronous motor is shown in Fig. 17-2. Starters of the type shown perform several functions in starting a motor and protecting it from damage during both the starting and running periods. These functions are described briefly in the following.

In the synchronous-motor starter, a line contactor similar to that used in induction-motor starters is used to energize the stator windings. To prevent damage from the voltage induced in the field windings, a field contactor in the starter short-circuits the field winding through a field-discharge resistor during the starting period and opens when the motor nears synchronous speed. After the motor accelerates to approximately 95 percent of synchronous speed, synchronizing equipment in the starter energizes the motor field. This function is usually controlled by a field-frequency relay that ensures application of the field at the proper motor speed.

If the starting period of a synchronous motor is prolonged for any reason, the squirrel-cage winding may be overheated since it is designed for

FIGURE 17-2 *Full-voltage synchronous motor starter. (Square D Company.)*

starting duty only. Likewise, the cage winding is subject to damage if the motor should pull out of synchronism and continue to run as an induction motor. Protective devices in the motor starter are used to detect these conditions and open the motor power supply.

As in an induction-motor starter, thermal overload relays are normally a part of the protective equipment in a synchronous-motor starter. Other operating and protective equipment usually included are start-stop push buttons, a dc field ammeter, an ac ammeter, and a loss-of-field relay. In the starter shown in Fig. 17-2, a number of the synchronizing and protective functions are performed by a solid-state module that monitors motor operation during the starting and running periods.

17-3 SYNCHRONOUS-MOTOR EXCITATION

As indicated earlier, excitation systems for synchronous motors may be external to the motor and connected to the motor field through slip rings and brushes, or the motor may be the brushless type. The brushless type is largely supplanting the brush type primarily because many moving parts such as brushes, commutators, slip rings, and relays are replaced by static solid-state components mounted on the rotor of the brushless motor.

The exciter for the brushless motor is an ac generator. The exciter armature is mounted on the rotor of the main motor and the exciter field is

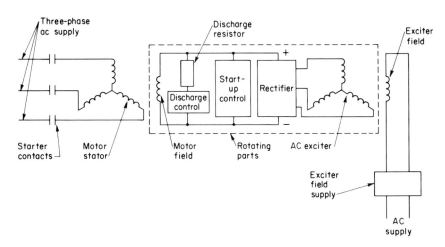

FIGURE 17-3   Simplified diagram of a brushless synchronous motor.

stationary. The output of the exciter is rectified by a silicon diode bridge-type rectifier, which is also mounted on the main rotor. This rectified output is connected directly to the motor field, thereby eliminating the need for any brushes or slip rings. Also mounted on the rotor are the field-discharge resistor and the solid-state control circuitry. This control equipment automatically connects the field-discharge resistor during the starting period and applies the field excitation to the motor at the proper instant to cause it to pull into synchronism. Protective equipment deenergizes the field circuit should the motor pull out of synchronism.

The armature of the ac exciter and some of the other excitation equipment may be seen in the photo of the brushless synchronous motor in Fig. 17-1. A simplified block diagram of a brushless motor is shown in Fig. 17-3.

17-4 EFFECT OF LOAD ON A SYNCHRONOUS MOTOR

In dc motors and induction motors, an addition of load causes the motor speed to decrease. The decrease in speed reduces the counter emf enough so that additional current is drawn from the source to carry the increased load at a reduced speed. In the synchronous motor, this action cannot take place, since the rotor is locked into step magnetically with the rotating magnetic field and must continue to rotate at synchronous speed for all loads.

The relative positions of a stator pole and a dc field pole of a synchronous motor running at no load are shown in Fig. 17-4a. The center lines of the two poles coincide. However, when load is added to the motor, there is a backward shift of the rotor pole relative to the stator pole, as shown in Fig. 17-4b. There is no change in speed; there is merely a shift

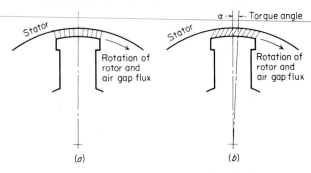

FIGURE 17-4 (a) Relative positions of a dc field pole and a pole of the rotating magnetic field of a synchronous motor at no load. (b) As load is added, the dc field pole drops behind the rotating magnetic field.

in relative positions of the two poles, both of which continue to rotate at synchronous speed. The angular displacement between the rotor and stator poles is called the *torque angle.*

When a synchronous motor operates at no load or with a torque angle of practically 0°, the counter emf of the motor is equal and opposite to the applied or terminal voltage (neglecting the motor losses). With increasing loads and torque angles, the phase position of the counter emf changes with respect to the applied voltage, which allows more stator current to flow to carry the additional load. This can best be shown by phasor diagrams.

The phasor diagram in Fig. 17-5*a* represents the condition in a synchronous motor at no load, the counter emf E being equal and opposite to the applied V_t. In Fig. 17-5*b*, enough load has been added to cause E to drop behind its no-load position by the angle α, corresponding to the change in torque angle similar to that shown in Fig. 17-4. The applied and counter voltages are no longer in direct opposition, their resultant voltage being the voltage V_r as shown. The resultant voltage V_r causes a current I to flow in the stator windings; I lags V_r by nearly 90° owing to the high inductance of the stator windings. The power input to the motor is thus $V_t I \cos \theta$ (for one phase), θ being the angle between the terminal voltage V_t and the stator current I.

A further increase in load results in a larger torque angle, which in turn increases the value of V_r and I as shown in Fig. 17-5*c*. Thus a synchronous motor is able to supply increasing mechanical loads, not by a decrease in speed, but by a shift in relative positions of the rotor and the rotating magnetic field. Note in Fig. 17-5 that for increasing loads with a constant value of E, the phase angle θ increases in the lagging direction.

If too great a mechanical load is placed on a synchronous motor, the

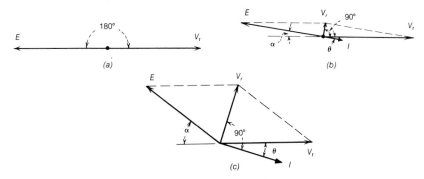

FIGURE 17-5   Phasor diagrams of a synchronous motor for three different load conditions with constant field excitation. (a) No load. (b) Small load. (c) Heavy load.

rotor is pulled out of synchronism, after which it comes to a stop; or if it has a squirrel-cage winding, it continues to operate as an induction motor until protective devices operate to stop the motor. The maximum value of torque that a motor can develop without losing its synchronism is called its *pull-out torque.* Pull-out torque, like pull-in torque, may be designed for widely varying values, depending on the application of the motor. These values range from 125 to 350 percent of full-load torque values.

17-5 POWER FACTOR; EFFECT OF CHANGING FIELD EXCITATION

One of the outstanding characteristics of a synchronous motor is the fact that it may be made to operate over a wide range of power factors by adjustment of its field excitation.

Figure 17-6a shows the phasor diagram of a synchronous motor with a mechanical load that results in the torque angle α and with the field excitation adjusted so that the motor operates at unity power factor. If, while the motor is carrying the same mechanical load, the field excitation

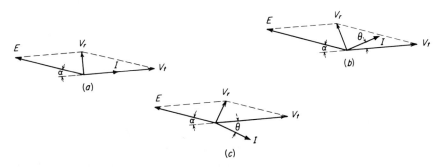

FIGURE 17-6   Phasor diagrams of a synchronous motor with a constant mechanical load but with different amounts of field excitation. (a) Field excitation adjusted for unity-power-factor operation. (b) Field overexcited. (c) Field underexcited.

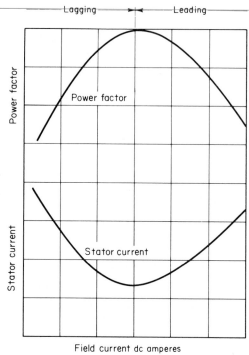

Power factor

Power factor

Stator current

Stator current

Field current dc amperes

FIGURE 17-7 Variation of stator current and motor power factor with a variation of field excitation at a constant load.

is increased, the counter emf E increases to the value shown in Fig. 17-6b. This results in a change in phase position of the stator current I with respect to the terminal voltage V_t so that the motor then operates at a leading power factor. Since the power output is unchanged, the value of I must increase with the decrease in power factor. Further increases in field excitation cause the motor to operate at a still more leading power factor with corresponding increases in stator current.

Figure 17-6c shows what happens if the field excitation is reduced below the value represented in Fig. 17-6a. A reduction in excitation changes the power factor in the lagging direction, and as the excitation is further reduced it becomes still more lagging with corresponding increases in stator current.

Hence for a given load, the power factor of a synchronous motor may be varied from a low lagging value to a low leading value. Figure 17-7 shows the variation of armature current and power factor of a synchronous motor with a variation of field excitation and for a constant load.

The normal value of field current for a given load is that value that results in unity power factor and a minimum stator current at that load. When the motor is overexcited, it operates at a leading power factor; when it is underexcited, it operates at a lagging power factor.

The efficiency of synchronous motors is in general higher than that of induction motors of the same horsepower and speed rating. The types of losses are identical with those of synchronous generators.

Synchronous motors are divided into two distinct classifications as to speed: motors with speeds above 500 rpm are classed as high-speed motors, and those with speeds less than 500 rpm are classed as low-speed motors.

Synchronous motors are rated as unity-power-factor, 90 percent leading-power-factor, or 80 percent leading-power-factor motors. Other power-factor ratings may be used for special applications. While leading-power-factor motors provide more power-factor correction, they are more expensive than unity-power-factor motors since they must have larger current-carrying capacities.

Unity-power-factor motors may be operated at leading power factors but with less than rated horsepower output. This is because more than rated stator current must flow to carry the rated horsepower load at a reduced power factor. Likewise, field current must be increased above normal value.

Synchronous-motor nameplate data include the same items found on ac generator nameplates with the kilovoltampere rating being replaced with a horsepower rating.

Synchronous motors are used for constant-speed power applications in sizes above 20 hp and more often in sizes larger than 100 hp. A very common application is for driving air or gas compressors. It is especially desirable that compressors be driven at a constant speed since their output and efficiency vary considerably with the speed at which they are operated.

Other common applications are in public-works pumping and for driving fans, blowers, and pulverizers. The advent of the brushless synchronous motor has increased the field of synchronous-motor application so that they are now used where formerly only induction motors were considered.

An outstanding advantage of synchronous motors is the fact that they may be operated at unity or leading power factors. When operated on the same electrical system with induction motors, or other devices that operate at lagging power factors, the leading reactive kilovoltamperes, or kilovars (kvar), supplied by the synchronous motors compensate for the lagging reactive kilovoltamperes, or kilovars, of other devices, resulting in an improvement in the overall system power factor.

Low power factors are undesirable for several reasons. Generators, transformers, and supply circuits are limited in ratings by their current-carrying capacities. This means that the kilowatt load that they can deliver is directly proportional to the power factor of the loads that they supply. For example, a system can deliver only 70 percent of the kilowatt load at 0.7 power factor that it can deliver at unity power factor.

The voltage regulation of generators, transformers, and supply lines is also poorer at low power factors than at unity power factor.

Since line and equipment losses are proportional to I^2R, these losses are higher in an electrical system for a given kilowatt load at low power factors than at unity power factor. Thus any improvement in power factor increases system efficiency and in general improves the operating characteristics of the system. For this reason electrical system operators often include power-factor clauses in industrial power contracts that place a penalty on a consumer for operating at a power factor below a certain value.

To gain some of the above-mentioned benefits from high power factors, very often synchronous motors are used to compensate for the lagging

---

EXAMPLE 17-1   The load of an industrial concern is 400 kVA at a power factor of 75 percent, lagging. An additional motor load of 100 kW is needed. Find the new kilovoltampere load and the power factor of the load if the motor to be added is (*a*) an induction motor with a power factor of 90 percent, lagging, and (*b*) an 80 percent power-factor (leading) synchronous motor.

Original load, 400 kVA at 75 percent power factor.

$$kW = 400 \times 0.75 = 300 \text{ kW}$$

$$(kVA)^2 = (kW)^2 + (kvar)^2$$

$$kvar = \sqrt{(kVA)^2 - (kW)^2}$$

$$= \sqrt{400^2 - 300^2} = 264.6 \text{ kvar, lagging}$$

(*a*) Induction motor, 100 kW at 90 percent lagging power factor.

$$kVA = \frac{kW}{\text{power factor}} = \frac{100}{0.9} = 111.1 \text{ kVA}$$

$$kvar = \sqrt{(111.1)^2 - (100)^2} = 48.4 \text{ kvar, lagging}$$

Resultant kilowatt load = 300 + 100 = 400 kW

Resultant kilovar load = 264.6 + 48.4 = 313 kvar

EXAMPLE 17-1
(CONTINUED)

$$\text{Resultant kilovoltampere load} = \sqrt{(kW)^2 + (kvar)^2}$$

$$= \sqrt{(400^2) + (313)^2} = 507.8 \text{ kVA}$$

$$\text{Resultant power factor} = \frac{kW}{kVA} = \frac{400}{507.8}$$

$$= 0.787 = 78.7 \text{ percent, lagging}$$

Since the relation between kilowatts, kilovars, and kilovoltamperes of any load may be represented by a right triangle, the solution of the above problem may be made graphically as shown in Fig. 17-8. The triangle OAE represents the original load; $OE = 300$ kW, $EA = 264.6$ kvar, and $OA = 400$ kVA. The angle θ_1 is the power-factor angle of the original load. Triangle ABC represents the added induction-motor load, sides AC, CB, and AB representing 100 kW, 48.4 kvar, and 111.1 kVA, respectively, with the phase angle represented as θ_2. Triangle OBD represents the resultant load with sides OD, DB, and OB representing 400 kW, 313 kvar, and 507.8 kVA, respectively. The resultant power-factor angle is θ_3.

(*b*) Synchronous motor, 100 kW at 80 percent leading power factor.

$$kVA = \frac{kW}{\text{power factor}} = \frac{100}{0.8} = 125 \text{ kVA}$$

$$kvar = \sqrt{(125)^2 - (100)^2} = 75 \text{ kvar (leading)}$$

$$\text{Resultant kilowatt load} = 300 + 100 = 400 \text{ kW}$$

$$\text{Resultant kilovar load} = 264.6 \text{ (lagging)} - 75 \text{ (leading)}$$

$$= 189.6 \text{ kvar (lagging)}$$

$$\text{Resultant kilovoltampere load} = \sqrt{(kW)^2 + (kvar)^2}$$

$$= \sqrt{(400)^2 + (189.6)^2} = 442.5 \text{ kVA}$$

$$\text{Resultant power factor} = \frac{kW}{kVA} = \frac{400}{442.5} = 0.903$$

$$= 90.3 \text{ percent, lagging}$$

The graphical solution is shown in Fig. 17-9, with triangles *OEA, ABC,* and *OCD* representing the original load, the added load, and the resultant load, respectively.

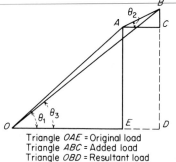

Triangle *OAE* = Original load
Triangle *ABC* = Added load
Triangle *OBD* = Resultant load

FIGURE 17-8   Graphical solution for Example 17-1a.

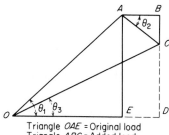

Triangle *OAE* = Original load
Triangle *ABC* = Added load
Triangle *OCD* = Resultant load

FIGURE 17-9   Graphical solution for Example 17-1b.

reactive kilovoltamperes taken by induction motors. The following example will be used to illustrate how a synchronous motor used on the same system with induction motors improves the overall power factor.

Synchronous motors, when operated without mechanical load and for the purpose of improving power factor, are called synchronous condensers. By considerably overexciting its field, a synchronous condenser is made to operate at a very low leading power factor, the only kilowatt input required being that necessary to supply its losses. When used at the end of a long transmission line, the synchronous condenser neutralizes the effects of lagging-power-factor loads, thereby improving the regulation of the transmission line.

1   How does the construction of a synchronous motor compare with that of a synchronous generator?

2   Describe the principle of operation of a synchronous motor. How does the principle of operation of a synchronous motor differ from that of an induction motor?

3   What determines the speed of a synchronous motor? How may the speed be varied?

4   How would it be possible to reverse the direction of rotation of a synchronous motor?

5 Why is a synchronous motor not self-starting? Name some ways in which they are started.

6 What is pull-in torque?

7 What precaution must be taken in regard to the field circuit of a synchronous motor during the starting period?

8 What types of controllers are used for starting synchronous motors? What forms of motor protection are generally provided?

9 How does a synchronous motor adjust its electric input with changes in mechanical output?

10 For a constant field excitation, what is the effect on the motor power factor of an increasing load?

11 What is meant by the pull-out torque of a synchronous motor?

12 For a constant mechanical output, how may the power factor at which a synchronous motor operates be changed?

13 Compare the method of adjusting the power factor of a synchronous motor with the method of adjusting the power factor of a synchronous generator operating in parallel with other synchronous generators.

14 When is a synchronous motor said to be (*a*) overexcited; (*b*) underexcited?

15 How are synchronous motors classified as to speed?

16 What are the standard power-factor ratings of synchronous motors?

17 What precaution must be taken in operating a synchronous motor at a power factor that is more leading than its rated power factor?

18 Under what operating condition would the field circuit be apt to heat abnormally?

19 What are some applications of synchronous motors?

20 Why is a low power factor undesirable?

21 Explain how a synchronous motor can improve the power factor of a load with a low lagging power factor.

22 What is a synchronous condenser?

PROBLEMS

17-1 What is the speed of a 30-pole 60-Hz 440-V synchronous motor? Is this motor classed as a high- or low-speed motor?

17-2 What is the percent speed regulation of the motor in Prob. 17-1.

17-3 A motor-generator set used for frequency conversion consists of a 10-pole 25-Hz synchronous motor and a direct-connected 24-pole synchronous generator. What is the generator frequency?

17-4 If a six-pole induction motor electrically connected to the generator-terminals in Prob. 17-3 has a full-load slip of 5 percent, what is its full-load speed in revolutions per minute?

17-5 The propulsion motors used on an aircraft carrier are rated 5900 hp, three-phase, 2400 V, 62.5 Hz, and 139 rpm. How many poles do they have? The speed of the above motors may be varied by a variation of the supply frequency between 16 and 62.5 Hz. What are the maximum and minimum speeds? Assuming a full-load efficiency of 85 percent and unity power factor, what is the full-load current input per terminal?

17-6 In Example 17-1, what must be the power factor of the added 100-kW load if it improves the overall plant power factor to 100 percent?

17-7 A transmission line delivers a load of 7500 kVA at a power factor of 70 percent, lagging. If a synchronous condenser is to be located at the end of the line, to improve the load power factor to 100 percent, how many kilovars must it supply to the system, assuming it to operate at 0 percent leading power factor?

17-8 A synchronous motor with an input of 480 kW is added to a system that has an existing load of 700 kW at 80 percent lagging power factor. What will be the new system kilowatt load, kilovoltampere load, and power factor if the new motor is operated at (*a*) 80 percent lagging power factor, (*b*) 100 percent power factor, and (*c*) 80 percent leading power factor?

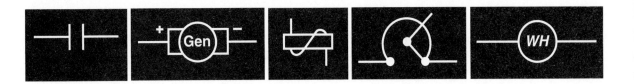

18

Single-Phase Motors

Single-phase motors may be classified by electrical type into three general classes: (1) series motors, (2) induction motors, and (3) synchronous motors. These three classes of motors are discussed in this chapter.

18-1 THE AC SERIES MOTOR

Direct-current shunt or series motors rotate in the same direction regardless of the polarity of the supply. Thus, it might be expected that either motor would operate on alternating current. It has been found, however, that the shunt motor develops little torque when it is connected to an ac supply. The high inductance of the field winding causes the field current to lag the armature current by such a large angle that a very low net torque results. However, in the series motor, the field and armature currents are the same, so that the main field and armature field are in phase. Therefore, about the same torque is developed with a given alternating current as with a like amount of direct current in a series motor.

However, some changes must be made in the design of a dc series motor if it is to operate satisfactorily on alternating current. When an ordinary dc series motor is connected to an ac supply, the current drawn by the motor is limited to a relatively low value by the high series-field impedance. Consequently, the power developed is small. To reduce the field reactance to a minimum, ac series motors are designed with as few field turns as possible, sufficient field flux being obtained by using a low-reluctance magnetic circuit. The effect of armature reaction in the series motor is overcome by the use of compensating windings in the pole faces. The compensating winding may be short-circuited on itself, in which case the motor is said to be *inductively* compensated, or it may be connected in series with the armature and field, in which case the motor is *conductively* compensated. All parts of the magnetic circuit of an ac series motor must be laminated to reduce the eddy-current losses.

The operating characteristics of the ac series motor are very similar to those of dc series motors. The speed increases to a high value with a decrease in load. In the very small series motors, the losses are usually large enough at no load to limit the speed to a definite value. The torque is high for high armature currents, thus giving the motor a good starting torque.

Since inductive reactance is directly proportional to frequency, ac series-motor operating characteristics are better at lower frequencies. Some series motors are built in large sizes for traction service and are designed to operate at low frequencies, 25 Hz or less. However, fractional horsepower sizes may be designed to operate satisfactorily at 50 or 60 Hz. For some applications, it is desirable to use a motor that will operate on either ac or dc circuits. By a compromise design, fractional-horsepower series motors may be built to operate satisfactorily either on 50 or 60 Hz or on direct current at 115 or 230 V. These motors are called *universal* motors. Common applications for the universal motor are in vacuum cleaners, sewing machines, and portable tools.

18-2 THE SINGLE-PHASE INDUCTION-MOTOR PRINCIPLE

In a two-phase induction motor, the phase coils are spaced 90 electrical degrees apart on the stator surface. When these coils are excited by two-phase voltages, that is, voltages that differ in time phase by 90°, the resulting phase currents are out of phase by 90°. Each of the two currents establishes a pulsating or alternating field, but these fields combine to produce a single rotating field as in the three-phase motor (Chap. 16). The rotating field sweeping across the rotor conductors induces a current in these conductors that causes a torque to be developed in the direction of the rotation of the field. This is the basic principle of operation of the two-phase motor. It is briefly reviewed here because conditions in a single-phase motor are similar to those in a two-phase motor.

If a single-phase voltage is applied to the stator winding of a single-phase induction motor, alternating current will flow in that winding. This stator current establishes a field similar to that shown in Fig. 18-1. During the half-cycle that the stator current is flowing in the direction indicated, a south pole is established on the stator surface at *A* and a north pole at *C*. During the next half-cycle, the stator poles are reversed. Although the stator field strength is varying and reversing its polarity periodically, its action is always along the line *AC*. This field, then, does not rotate but is a pulsating stationary field.

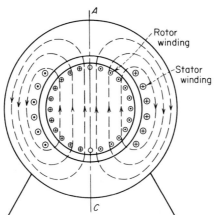

FIGURE 18-1 The stator field pulsates along the line AC. No torque is developed.

As in a transformer, voltages are induced in the secondary circuit, in this case the rotor. Since single-phase induction-motor rotors are of the squirrel-cage type similar to those of polyphase motors, rotor currents are caused to flow in the directions shown in Fig. 18-1. These rotor currents establish poles on the rotor surface, but since these poles are always in direct line (along the line *AC*) with the stator poles, no torque is developed in either direction. Therefore, a *single-phase induction motor is not self-starting* and requires some special starting means.

When, however, the single-phase stator winding in Fig. 18-1 is excited and the rotor is made to turn by some auxiliary device that will be described later, the rotor conductors cut across the stator field, causing an emf to be generated in them. This is illustrated in Fig. 18-2, which shows the rotor being turned in a clockwise direction. If the stator flux is upward at the instant shown, the direction of the generated rotor emfs, as determined by Fleming's right-hand rule, will be outward in the upper half of the rotor and inward in the lower half of the rotor, as indicated by the dots

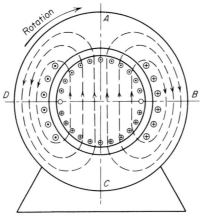

FIGURE 18-2 The motor of Fig. 18-1, with the rotor turning, generates an emf in the rotor conductors in the direction shown by the dots and crosses.

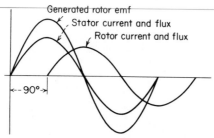

FIGURE 18-3 *Phase relations of the stator current and flux, the generated rotor emf, and the rotor current and flux. The rotor flux lags the stator flux by 90°.*

and crosses. One-half cycle later the direction of the generated emfs will be reversed. The generated rotor emfs vary *in phase* with the stator current and flux. Because of the low resistance and high inductance of the rotor, however, the resulting rotor current *lags* the generated rotor emf by nearly 90°. Figure 18-3 shows the phase relations of the stator current and flux, the rotor emf, and the rotor current and flux.

The maximum value of the field produced by the rotor currents, as shown in Fig. 18-4, occurs nearly one-fourth cycle after the generated rotor emf has reached its maximum value. Because this rotor field is at right angles to the stator field, it is called a *cross field*. Since the rotor currents are caused to flow by an alternating emf, the field resulting from these currents also alternates and its action is always along the line *DB*, Fig. 18-4. The cross field is very similar to a field that would be produced by another coil placed on the stator in the spaces at *A* and *C* in Fig. 18-4 and excited by a voltage 90° behind the voltage applied to the coil in place, as in a two-phase motor. Since the cross field acts at right angles to the stator field and also lags the stator field by 90° in *time phase,* the two fields combine to form a resultant rotating field that revolves at synchronous speed.

It should be kept in mind, however, that the cross field is produced by

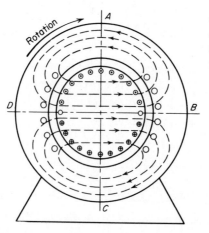

FIGURE 18-4 *Cross-field established by the rotor currents one-fourth cycle later than the stator field shown in Fig. 18-2.*

a *generator action* and therefore is present only when the rotor is turning. It also follows that the strength of the cross field must be proportional to the rotor speed. At synchronous speed the cross field is nearly the same strength as the stator field. But since an induction motor must operate at some speed lower than synchronous speed, the cross field is somewhat weaker than the stator field at actual operating speeds. This means that the rotating field in a single-phase induction motor is irregular and is not of constant strength, as is the field of a polyphase motor. Therefore, the torque developed by a single-phase induction motor is irregular or pulsating. This is the reason why many single-phase motors are set on rubber or spring mounts to reduce the vibration and noise inherent in such motors.

18-3 SPLIT-PHASE STARTING

It was shown in the preceding section that, when two windings are spaced 90 electrical degrees apart on the stator of a motor and are excited by two alternating emfs that are 90° displaced in time phase, a rotating magnetic field is produced. If two windings so spaced are connected in parallel to a single-phase source, the field produced will alternate but will not revolve since the two windings are equivalent to but one single-phase winding. If, however, an impedance is connected in series with one of these windings, the currents may be made to differ in time phase. By the proper selection of such an impedance, the currents may be made to differ by as much as 90°, thereby producing a rotating field very much like the field of a two-phase motor. This is the principle of *phase splitting*.

18-4 THE RESISTANCE-START SPLIT-PHASE MOTOR

When two similar motor stator windings spaced 90 electrical degrees apart are connected in parallel to a single-phase source, the currents through the two windings lag the applied voltage by the same angle. By connecting a resistance in series with one winding, the current through that winding is caused to be more nearly in phase with the applied voltage. Since the current in the first winding is not affected by the addition of the resistance, the currents through the two windings are displaced in time phase. This is the condition necessary to produce a revolving field. A motor using this method of phase splitting is called a *resistance-start motor*. The resistance-start split-phase motor is commonly known as the *split-phase motor* even though there are other forms of split-phase motors.

In practice, instead of using an external series resistor, one winding is designed to have a higher resistance and lower reactance than the other winding. The high-resistance winding is called the *starting* or *auxiliary winding,* and the low-resistance winding is called the *main winding.*

The phase relations of the applied voltage V, the main-winding current I_m, and the starting-winding current I_s at the instant of starting are

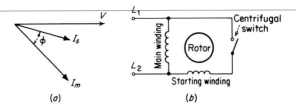

FIGURE 18-5 (a) Phase relations in and (b) connection diagram of the resistance-start motor.

shown in Fig. 18-5*a*. The angle φ between the main and starting winding currents is small, being of the order of 30°, but this is enough phase difference to provide a weak rotating magnetic field. Since the currents in the two windings are not equal in magnitude, the rotating field is not uniform and the starting torque so produced is small.

It was shown in the preceding section that a single-phase induction motor develops a torque once it has been started. For this reason, the starting winding of the resistance-start motor is disconnected when the motor reaches a predetermined speed, usually 70 to 80 percent of synchronous speed. This is accomplished by a centrifugally operated or a solid-state switch. A connection diagram showing the main and starting windings of a resistance-start motor is shown in Fig. 18-5*b*.

The resistance-start split-phase motor is very widely used for easily started loads. Common applications are for driving fans, woodworking tools, grinders, oil burners, and various other low-starting-torque applications. Because of its low starting torque, this motor is seldom used in sizes larger than 1/3 hp.

18-5 THE CAPACITOR-START MOTOR

An improvement can be made in the starting characteristics of a split-phase motor by connecting a capacitor in series with the starting winding as shown in the connection diagram in Fig. 18-6. A motor of this type is called a *capacitor-start motor.* The current through the main winding lags the applied voltage as in the resistance-start motor. By choosing a capacitor of the proper size, the current in the starting winding may be made to *lead* the voltage. Thus, the time-phase difference φ in the currents of the two windings at starting may be made nearly 90° as is shown in Fig. 18-6. This more nearly approximates the action of the two-phase revolving field than does the resistance-start motor, thereby resulting in a relatively higher starting torque.

To provide adequate starting winding current and optimum phase shift, the starting capacitor must have a large microfarad rating. The elec-

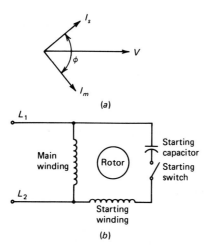

FIGURE 18-6 (a) Phase relations in and (b) connection diagram of the capacitor-start motor.

trolytic capacitor has a large capacity rating with a relatively small size and low cost and is therefore used extensively for motor-starting duty. This type of capacitor is not, however, suitable to be left permanently connected in the circuit. The capacitor-start motor, like the resistance-start motor, has a starting switch that disconnects the starting winding as the motor approaches synchronous speed.

Starting torque of the general-purpose capacitor-start motor is approximately 350 percent of full-load torque, compared with about 150 percent for the resistance-start split-phase motor. This motor is used for heavy-duty drives such as refrigerators, compressors, pumps, and air conditioners. The usual size range is from 1/8 to 3/4 hp, but sizes up to 2 hp are available.

18-6 THE PERMANENT-SPLIT CAPACITOR MOTOR

The permanent-split capacitor motor, as the name indicates, has a capacitor permanently connected in series with the starting or auxiliary winding and in parallel with the main winding. The connection diagram is, therefore, the same as that of a capacitor-start motor with the starting switch omitted. The operation of the motor is similar to that of a two-phase motor since a rotating magnetic field is established by the two windings for both the starting and running conditions.

It has been found that the size of the capacitor needed to provide the best running conditions for a motor is much smaller than that required to give good starting torque. The permanent-split capacitor motor is, therefore, a compromise design that uses a relatively small oil-filled capacitor rated for continuous duty. This results in a motor with a very low starting torque but that is quieter and more efficient than the capacitor-start motor.

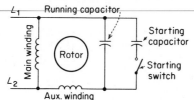

FIGURE 18-7 *Connection diagram of a two-value capacitor motor.*

The permanent-split capacitor motor is used primarily for direct-connected fans and blowers where the motor can be matched accurately with the driven load. Sizes used are generally smaller than $1/3$ hp.

18-7 THE TWO-VALUE CAPACITOR MOTOR

The two-value capacitor motor is a motor in which a different value of capacitance is used for starting and for running conditions. In the usual switching arrangement, a starting capacitor in series with a starting switch is connected in parallel with a running capacitor and the paralleled capacitors are in series with the auxiliary winding as shown in Fig. 18-7. A cutaway view of a two-value capacitor motor is shown in Fig. 18-8.

The use of separate starting and running capacitors permits the motor designer to select an optimum size for each, which results in excellent starting torque and good running performance. The starting and running capacitors normally used are electrolytic and oil-filled types, respectively. This motor design is used normally only for larger single-phase motor applications where particularly high starting torque is required.

18-8 THE SHADED-POLE MOTOR

A shading coil is a low-resistance short-circuited copper loop that is placed around a part of each pole of a motor. Figure 18-9 represents one pole of a motor equipped with shading coils. As the flux produced by the motor wind-

FIGURE 18-8 *Cutaway view of a totally enclosed two-value capacitor motor. (MagneTek, St. Louis, MO.)*

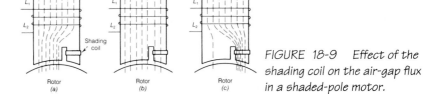

FIGURE 18-9 Effect of the shading coil on the air-gap flux in a shaded-pole motor.

ing begins its increase at the beginning of each cycle, an emf is induced in the short-circuited coil, which causes a current to flow in it. The field produced by the flow of current in the shading coil is in a direction that opposes the change in the main field flux. Thus the increase in flux is less rapid through the shading coil than in the main part of the pole, as shown in Fig. 18-9*a*. Figure 18-9*b* represents the same pole later in the cycle at the instant that the flux has reached its maximum value and is not changing in value. The flux is distributed uniformly across the pole face, since no current is flowing in the shading coil. As the flux decreases in the latter part of the cycle, the induced current in the coil is in a direction such that its magnetic effect opposes the decrease in flux. Consequently, the flux decreases less rapidly through the shading coil than through the other part of the pole, as shown in Fig. 18-9*c*.

The effect of the shading coil is to cause the field flux to shift across the pole face from the unshaded to the shaded portion. This shifting flux, somewhat like a weak rotating field, produces a small starting torque. This method of motor starting is used only on very small motors up to about ¼ hp in size, such as may be used for driving small fans or blowers.

18-9 SINGLE-PHASE SYNCHRONOUS MOTORS

There are three general types of single-phase synchronous motors classified by the type of rotor used: the dc-excited type, the reluctance type, and the hysteresis type. Stator windings used with any of these different motors may be similar to those of single-phase induction motors; that is, they may be split-phase resistance, capacitor, or shaded-pole windings.

The *single-phase dc-excited synchronous motor,* like the larger synchronous motors described in Chap. 17, requires a dc supply to energize the rotor winding. Rotor poles lock into step with the stator rotating magnetic field, and the rotor turns at synchronous speed. Because of its high cost and the need for a separate dc field excitation supply, this motor is seldom used except for a few precision drive applications.

Reluctance-type single-phase synchronous motors have modified squirrel-cage rotors. The laminated steel part of the rotor is cut away at evenly spaced areas, forming salient poles on the rotor surface. The number of poles formed corresponds to the number of poles for which the stator is wound. Rotor bars of conducting material are spaced around the entire rotor circumference as in the induction-motor rotor.

When the stator winding of the reluctance motor is energized, a magnetic field is established as in the single-phase induction motor; currents are induced in the rotor cage winding, and the rotor accelerates to near synchronous speed. Since the salient poles on the rotor provide the lowest reluctance path for the stator flux, they are attracted to the poles of the stator magnetic field and the rotor pulls into step with the rotating stator field.

The amount of torque that a reluctance motor, or any synchronous motor, is able to exert to pull the rotor and driven load into synchronism is called the pull-in torque. Load torque is produced by an angular shift between stator and rotor poles as described for the three-phase synchronous motor in Chap. 17. If load is added to the motor beyond its capability, the rotor is pulled out of synchronism. The torque developed at this point is called the pull-out torque.

Because of their inherently low pull-out torque, low power factor, and poor efficiency, reluctance motors are used only in small fractional sizes where constant speed is essential.

The rotor of the *hysteresis motor* is made of permanent magnet steel. In one construction used, the rotor is made in the form of a hollow, thin-shelled cylinder supported by nonmagnetic material so that its axis coincides with the center line of the rotor shaft. Contrasted with the reluctance rotor, there are no well-defined poles on the rotor surface, nor is there a separate cage winding.

When the stator winding of the hysteresis motor is energized, eddy currents are induced in the rotor surface. These eddy currents, flowing in complex paths, induce poles on the rotor surface. The hardened steel rotor, in effect, acts as a squirrel cage to accelerate the rotor. At synchronous speed, the rotor poles formed by the eddy currents take up fixed positions and the rotor locks into step with the stator magnetic field.

Load torque of the hysteresis motor at synchronous speed is produced by an angular shift between stator and rotor poles as in any synchronous motor. The pull-out torque of this type of motor is a measure of the ability of the rotor steel to retain its magnetism or, in other words, the amount of hysteresis effect present in the rotor steel. Hence, the name *hysteresis motor*.

Because the torque of a hysteresis motor is constant, very smooth, and uniform, this motor was formerly used as a drive for phonographs and tape-reel drives. But it has been largely replaced by transistor-driven permanent-magnet dc motor drives. Likewise, it was universally used as a clock drive but now has been largely displaced by solid-state circuits with digital readout.

18-10 SINGLE-PHASE MOTOR CONTROL

Solid-state power devices such as diodes, diacs, SCRs, and triacs figure prominently in the drive and control circuits for small single-phase motors. Two motor drive and control systems are described in the following paragraphs.

A widely used variable-speed drive for a universal motor using an SCR is shown in Fig. 18-10. As described in Sec. 18-1, the series ac-dc, or universal motor, is used in small domestic appliances and portable tools. This motor has a high torque for high-armature currents, giving it a good starting torque. However, it has a widely varying speed from a very high speed at no load to a very low speed at full load. Speed control in the circuit shown in Fig. 18-10 is accomplished by varying the triggering time or angle of the SCR, which is in series with the motor. The angle at which the SCR fires is controlled by the cathode-to-gate voltage. This voltage is determined at any time both by the setting of the variable resistance R and by the counter emf of the motor, which is proportional to motor speed. For example, after the motor has reached a stable operating speed for a given setting of the resistor R, if load is added to the motor, the rotor speed momentarily decreases, which reduces the counter emf. The resulting reduced cathode-to-gate voltage causes the gate of the SCR to fire earlier, which applies more voltage to the motor, which in turn causes its speed to increase. The counter emf of the motor thus provides a speed-regulative action that results in stable operation of the motor.

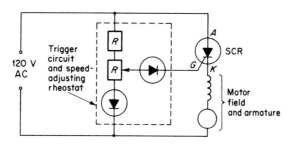

FIGURE 18-10   Circuit diagram of a universal motor control circuit.

The circuit shown in Fig. 18-11 is a basic triac phase-control circuit in which the triac is triggered with a diac. The circuit shown is suitable for controlling single-phase shaded-pole or two-value capacitor fan-drive motors. With minor modifications, the circuit can be used for other purposes such as lamp dimming or heat control.

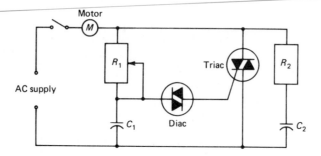

FIGURE 18-11   *Single-phase induction motor control circuit. (RCA.)*

The capacitor C_1, in Fig. 18-11 is charged first in one direction and then the other during each cycle of the applied voltage. As the charge on the capacitor increases in either direction to a value high enough to cause the diac to break over, the capacitor discharges through the diac into the gate of the triac. Either a positive or a negative pulse triggers the triac into conduction, and full line voltage is applied to the motor for the remainder of that half-cycle.

If the resistance of R_1 is increased, the time required for C_1 to charge during each half-cycle is increased. This delays the triggering of the triac, and a smaller percentage of the line voltage is applied to the motor and the motor slows down. Conversely, lowering the resistance of R_1 increases motor speed. The resistor R_2 and capacitor C_2 are connected across the triac to ensure reliable turn-off of the triac when this circuit is used for controlling an induction motor or other highly inductive load.

REVIEW QUESTIONS

1 What are the three general classes of single-phase motors.
2 Name three types of single-phase induction motors.
3 Why is the shunt motor unsatisfactory for ac operation?
4 How does an ac series motor differ in construction from a dc series motor?
5 Is an ac series motor classed as a constant- or variable-speed motor?
6 What is a universal motor?
7 Does a single-phase induction motor develop torque when the rotor is stationary? Why?
8 What causes the rotating magnetic field to be established in a single-phase induction motor? What determines its direction of rotation?
9 What is the split-phase method of motor starting? Name two methods of phase splitting.
10 Why is the starting winding of a split-phase resistance-start motor disconnected after the motor has been started?

11 Which has the better starting characteristics, the resistance-start or the capacitor-start induction motor?

12 What is a permanent-split capacitor motor?

13 What is an application of the permanent-split capacitor motor?

14 What kind of loads might require the use of a two-value capacitor motor?

15 What is the effect of the shading coil when it is used in starting induction motors? Motors using this method of starting are built in sizes up to about what horsepower?

16 Name three types of single-phase synchronous motors.

17 What is a typical application for the hysteresis motor?

18 Compare the speed-regulative action of the motor shown in Fig. 18-10 with that of a dc motor.

19 What is the function of the diac in the motor control circuit shown in Fig. 18-11?

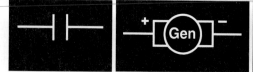

19

Circuit-Protective and Switching Equipment

Protective and switching equipment play very important parts in the operation of a modern electrical system. Since a modern economy is so dependent on the availability of electric energy, it is the goal of those who design, operate, and maintain electrical supply systems to provide as nearly as possible a continuous supply of electric energy to all consumers at a reasonable price. To provide this service, large investments must be made in generating, transmitting, and distributing equipment. Protective and switching equipment is necessary to protect this investment, to provide system operating flexibility, and to ensure a dependable supply of electricity to all consumers. Some of this equipment and its function in modern power systems is described in this chapter.

19-1 SHORT CIRCUITS AND INTERRUPTING RATING

Short circuits are the greatest hazard to the continuity of service. Since protective and switching equipment must either isolate or withstand the effects of short circuits, it is desirable to understand the nature of short circuits, what causes them, and their effects.

All circuits consist of conductors that are isolated from other conductors by insulating materials. Conductors are necessary to carry the current, and insulators are necessary to confine the current to the conductor. Copper and aluminum are the most common conductor materials. Insulating materials may be air, rubber, porcelain, thermoplastic materials, and other similar materials.

Conductors used indoors are usually insulated with rubber, varnished cambric, or thermoplastic materials coated directly over the conductor material so the conductor may be placed in a confined area such as a conduit or raceway. Conductors used on outdoor overhead lines are very often bare and are supported on glass, porcelain, fiberglass, or plastic insulators. Such circuits depend upon the physical separation in air for their insulation, air being the insulator.

Regardless of the construction of an electric circuit, the conductor and the insulator are present. However, regardless of the care used in constructing an electric circuit, the circuit is subject to damage such that the insulator between two or more conductors is destroyed and a *short circuit* or a *fault* is said to exist. A short circuit is then merely an insulation failure in an electric circuit; this failure in insulation provides a low-resistance current path between conductors.

Insulation failures or short circuits in an electric circuit may be caused by inherent defects or by external means. Inherent defects result from improper design or installation of equipment. Insulation may fail because of external causes such as lightning surges, switching surges, mechanical damage, or operating errors. Many short circuits are the result of overloads. Overloads cause a deterioration of insulation and, if sustained, may cause a complete failure of insulation resulting in a short circuit.

The amount of current flowing in a circuit is determined by the applied voltage and the impedance of the circuit. This is true in a normal circuit as well as in a faulted circuit. In the circuit shown in Fig. 19-1, the normal load current is the phase voltage divided by the phase impedance, or

$$I_{norm} = \frac{(480/1.73)}{2} = 139 \, A$$

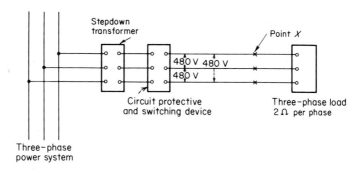

FIGURE 19-1 *Circuit to illustrate magnitudes of short-circuit currents.*

However, if a three-phase short circuit should occur at the point X, the flow of current in the circuit will be much larger since the only impedance to limit the flow of current is then the impedance of the system itself. The total system impedance for the circuit shown in Fig. 19-1 includes the impedance of the step-down transformer and the impedance of the three-phase power system back to the generators. The impedance of the power system is normally very small relative to that of the step-down transformer.

When a transformer impedance is known in ohms, then the short-circuit current for a fault on the secondary side of the transformer may be calculated by dividing the transformer phase voltage by the transformer phase impedance. However, usually the transformer impedance is known in percent as indicated on the transformer nameplate (see Sec. 12-9). When the transformer percent impedance is known and the system impedance is so small that it may be neglected, as it often is, transformer secondary short-circuit current I_{SC} may be computed by the following:

$$I_{SC} = \frac{\text{kVA}_T \times 100}{\% Z \times 1.73 \times \text{kV}} \tag{19-1}$$

where kVA_T = transformer rating, kilovoltamperes
$\% Z$ = transformer percent impedance
kV = transformer secondary line-to-line voltage, kilovolts

Thus, for the circuit shown in Fig. 19-1, if the transformer is rated 500 kVA and has an impedance of 5 percent, the short-circuit current for a fault at point X will be

$$I_{SC} = \frac{500 \times 100}{5 \times 1.73 \times 0.48}$$

$$= 12{,}000 \text{ A}$$

Larger supply systems and supply transformers normally produce higher fault currents. If, for example, the 500-kVA transformer in Fig. 19-1 were replaced by a 2500-kVA 5 percent impedance transformer, then the short-circuit current for a fault at point X, again neglecting the system impedance, would be

$$I_{SC} = \frac{2500 \times 100}{5 \times 1.73 \times 0.48}$$

$$= 60{,}000 \text{ A}$$

As may be seen from the foregoing examples, insulation failures cause abnormally high currents to flow. Such high values of current may damage equipment or property unless the faulty circuit or equipment is removed promptly (deenergized) from the supply system. It is further desirable to remove only the faulty circuit or equipment from the system so that service may be maintained to the remainder of the system.

A circuit-protective and switching device such as the one shown in Fig. 19-1 has two purposes:

1. It carries the normal load current and is used to open and close the cicuit when necessary.
2. It is used to open the circuit to interrupt the flow of abnormal or short-circuit current.

A basic purpose of an overcurrent protective device is to interrupt the flow of short-circuit current before the protected circuit conductors or other components have been damaged by the high levels of energy that are released by the flow of the short-circuit current.

Circuit-protective and switching devices in general have two ratings: a *continuous* rating and an *interrupting* rating. The continuous rating is usually in amperes and is the amount of current that a device can carry continuously without overheating. The interrupting rating of a device is based on the ability of that device to interrupt the flow of short-circuit current. Interrupting ratings of low-voltage devices are usually in amperes and of high-voltage devices either in amperes or in kilovoltamperes at a given voltage.

19-2 ELECTRICAL DISTRIBUTION SYSTEMS

The distribution of electric energy is accomplished by dividing and subdividing large-capacity feeders into feeders of smaller and smaller capacity. At each change in circuit capacity on an electrical system there is a distribution center where circuit-protective and switching equipment is grouped. In general, this is true throughout a system from the generators to the loads.

The system represented in one-line-diagram form in Fig. 19-2 is a simple radial system in which all circuits radiate from the generating station. Typical groups of protective and switching equipment are shown in the diagram. An assembly of metal-clad switchgear and a step-up outdoor substation are shown at the generating station. Another section of metal-clad switchgear is shown at the 34,500- to 4160-V step-down substation. The load-center unit substation, low-voltage metal-enclosed switchgear (which is part of the unit substation), and lighting and power panelboards at the load are shown. This equipment has been grouped in this chapter into two main groups: low-voltage equipment for circuits operating at 600 V and below, and high-voltage equipment for circuits operating at voltages above 600 V.

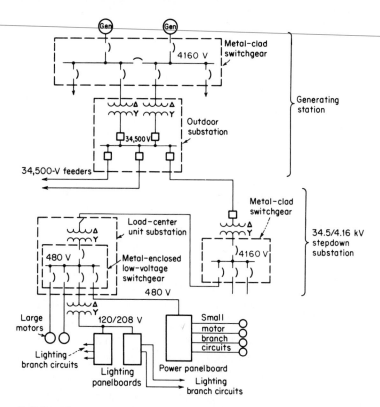

FIGURE 19-2 *Electrical system one-line diagram to illustrate the use of circuit-protective and switching equipment.*

LOW-VOLTAGE PROTECTIVE AND SWITCHING EQUIPMENT

The classification of low-voltage protective and switching equipment as used here applies to that equipment used on circuits rated at 600 V or less. This voltage classification includes a large percentage of the utilization systems such as in homes, commercial establishments, and industries since most of these systems utilize power at 120, 208, 240, 415, 480, or 550 V. Protective and switching equipment used on such systems is used for normal-load switching, for protection of circuits and equipment against overload, and for protection against short circuits.

19-3 LOW-VOLTAGE FUSES   A fuse is an overcurrent protective device that has a circuit-opening fusible member directly heated and destroyed by the passage of overcurrent through it. As indicated by the definition, a fuse contains a current-carrying element so sized that the heat created by the flow of normal current through it is not sufficient to fuse or melt the element. However, when short-circuit or overload currents flow through the fuse, the fusible element melts and opens the circuit. The fusible element of an indoor fuse is enclosed in a protective housing to prevent the heat generated during the clearing of a short circuit from damaging adjacent equipment.

The *plug fuse* is used on low-capacity circuits rated 125 V or less to earth. This fuse consists of a zinc alloy strip fusible element enclosed in a porcelain or glass housing fitted with a screw base. The maximum continuous current-carrying capacity of plug fuses is 30 A, and the commonly used standard sizes are 10, 15, 20, 25, and 30 A. These fuses do not have published interrupting capacities since they are ordinarily used on circuits that have relatively low values of available short-circuit current.

Cartridge fuses are used on circuits with voltage ratings up to 600 V, with common voltage ratings of 250 and 600 V. Cartridge fuses are constructed with a zinc alloy fusible element enclosed in a cylindrical tube or cartridge. The ends of the fusible element are attached to metallic contact pieces at the ends of the tube. The tube is filled with an insulating arc-quenching powder that surrounds the fusible element. On overloads or short circuits, the fusible element is heated to a high temperature, causing it to vaporize. The powder in the fuse cartridge cools and condenses the vapor and quenches the arc, thereby interrupting the flow of current.

Cartridge fuses both in the 250- and 600-V ratings are made to fit standardized fuse-clip sizes. These are 30-, 60-, 100-, 200-, 400-, and 600-A sizes. There are several continuous current ratings for each fuse-clip size. For example, the 100-A size is available in the continuous ratings of 70, 80, 90, and 100 A. Many general-purpose cartridge fuses do not have published interrupting ratings. However, for application on systems having high values of available short-circuit current, certain classes of fuses are available that have rated interrupting ratings of 50,000, 100,000, or 200,000 A.

Time-lag fuses are made in both the plug and cartridge types. These fuses are constructed to have a longer operating time than ordinary fuses, especially for overload currents. They do operate, however, to clear short-circuit currents in about the same time as do the standard fuses. Time-lag fuses have two parts, a thermal-cutout part and a fuse link. The thermal cutout with its long time lag operates on overload currents up to about 500 percent of normal current. Currents above this value are interrupted by the fuse link. Time-lag fuses find their greatest application in motor circuits where it is desirable for the fuse to provide protection for the circuit and yet not operate due to a momentary high current during the starting period of the motor.

Current-limiting fuses are designed to interrupt the flow of fault current in less than one-half cycle (0.008 s on a 60-Hz system). These fuses interrupt the fault current before it can build up to its full peak value.

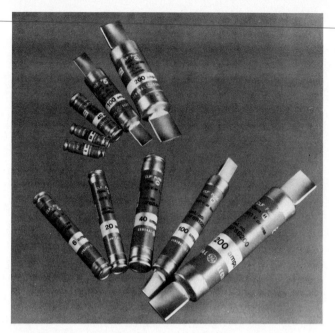

FIGURE 19-3   Current-limiting fuses. (General Electric Company.)

Several types of fuses are classified by electrical standards as current-limiting fuses. Ratings for these fuses, in addition to their current-limiting capability, include a voltage, a continuous-current rating, and an interrupting rating. Several current-limiting fuses are shown in Fig. 19-3.

Since a fuse is a device that is destroyed in the process of interrupting the flow of abnormal current, it must be easily replaceable. Plug fuses have a screw-type base and can be inserted into sockets designed for their use. Cartridge fuses are inserted into fuse clips that make contact with the two ends of the fuse. Fuse sockets and clips usually are assembled into a supporting insulating base provided with means for connecting the fuse into the circuit. The assemblies of fuse, fuse clips, and supports are manufactured and used in many different ways. Two devices in which fuses are generally used are *safety switches* and *panelboards*.

19-4 SAFETY SWITCHES   A switch is a device for isolating parts of an electric circuit or for changing connections in a circuit or system. When a switch is mounted in a metal enclosure and is operable by means of an external handle, it is called a *safety switch*. The switch itself is not designed for interrupting the flow of short-circuit currents. However, switches and fuses are often incorporated into a single device called a *fusible safety switch.*

Safety switches are made in two-, three-, four-, or five-pole assemblies, either fused or unfused. They are made in single-throw and double-throw units, and, depending upon their use, they have a variety of constructional features. One type has a quick-make, quick-break mechanism so arranged that regardless of the speed at which the operating handle is moved, a spring-loaded arrangement causes the contacts to open or close with a quick motion. This type of switch also has a door interlock to prevent the opening of the enclosure door when the switch is closed. A typical heavy-duty three-pole fusible safety switch with the door open and the fuses removed is shown in Fig. 19-4.

Enclosed switches, either fusible or nonfusible, are used as disconnecting devices for main services into buildings, for feeder and branch circuit-protective and switching devices, and for motor protection and switching.

Safety switches are available in two voltage ratings: 250 and 600 V ac. Current ratings are the same as the standard fuse-clip sizes.

FIGURE 19-4   Three-pole safety switch. (Square D Company.)

A circuit breaker is a device designed for interrupting a circuit between separable contacts without injury to itself. Air circuit breakers are circuit breakers in which the circuit interruption occurs in air. The circuit breakers to be discussed here are the so-called low-voltage circuit breakers suitable for application on circuits rated 600 V or lower. Circuit breakers intended for application on high-voltage circuits are discussed in Sec. 19-10.

One commonly used low-voltage air circuit breaker is the *molded-case breaker.* The breaker is assembled as an integral unit in a housing of insulating material. The external appearance of a three-pole molded-case breaker is shown in Fig. 19-5, and a cutaway view of this type of breaker is shown in Fig. 19-6.

An electric circuit is completed or interrupted manually with a circuit breaker by moving the operating handle to the ON or OFF position. In all except the very small breakers, the linkage between the operating handle and the contacts is arranged for a quick-make, quick-break contact action regardless of the speed at which the handle is moved. The handle is also trip-free, which means that the contacts cannot be held closed against a short circuit or overload.

The breaker automatically opens or "trips" when the current through it exceeds a certain value. In the lower current ratings, automatic tripping is accomplished by a thermal tripping device. The thermal trip consists of a bimetallic element so calibrated that the heat from normal current through

FIGURE 19-5   Three-pole molded-case air circuit breaker. (Square D Company.)

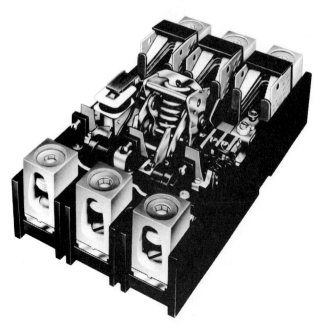

FIGURE 19-6   Cutaway view of a three-pole molded-case circuit
breaker. (Square D Company.)

it does not cause it to deflect. However, an abnormally high current caused
either by an overload or by a short circuit causes the element to deflect and
trip the linkage holding the circuit-breaker contacts closed. The breaker
contacts are opened by spring action.

The bimetallic tripping element, being responsive to heat from the
current flowing through it, has an *inverse-time characteristic*. Inverse
time means that it operates quickly for heavy currents such as those resulting from short circuits and operates more slowly for moderate overload currents. This is a desirable characteristic for an overcurrent protective device
that is used for protecting circuit conductors against overheating. The current-carrying ability of a conductor is limited by the temperature at which
its insulation may be operated safely. The operating temperature of the conductor is the sum of the temperature of the air in which the conductor is
operating and the temperature rise due to the I^2R loss in the conductor. A
circuit breaker that uses a thermal element for tripping depends upon these
same two sources of heat for its operation and, therefore, affords good conductor protection. Thus, when circuit-breaker sizes are properly selected,
they have approximately the same response to temperature as do the conductors that they protect, and they will act to trip the circuit before dangerous overheating can occur in the conductors.

In the larger current ratings of molded-case breakers, the thermal trip element is supplemented by a *magnetic trip element.* The magnetic unit uses the magnetic force surrounding the conductor to operate the breaker-tripping linkage. Breakers with the combination thermal-magnetic trip have inverse-time tripping characteristic for overcurrents up to about 10 times the nameplate rating of the breaker and instantaneous tripping for currents above that value. Stated another way, these breakers have inverse-time thermal tripping for overload currents and instantaneous magnetic tripping for short-circuit currents.

When the contacts part, the current flow causes an arc to form between the contacts. Arc interruption in an air circuit breaker takes place between the separable contacts. Many variations in design and arrangement of contacts and the surrounding chamber are used by manufacturers. One commonly used design is the placing of the contacts in an arc chute made in such a way that the arc is drawn out into the arc chute. As the arc is drawn into the chute, it is divided into small segments and quenched.

Molded-case breakers are manufactured in a wide range of sizes and ratings. Frame sizes range from 50 to 4000 A, each with a standard range of continuous-current ratings. The physical size, rating of the contacts, and interrupting rating are the same for all breakers of a given frame size. The continuous-current rating of any breaker is determined by the rating of its trip elements. Voltage ratings available range from 120 to 600 V, and interrupting capacities range as high as 100,000 A.

Large air circuit breakers are used in industrial and large commercial distribution systems since these breakers are usually available in higher continuous-current and interrupting ratings than are the molded-case breakers. Circuit breakers of this type are available with continuous-current ratings as high as 4000 A and interrupting ratings as high as 130,000 A.

Most modern, large air circuit breakers use a closing device called a *stored energy mechanism* to ensure a fast, positive closing action. In a commonly used breaker-closing mechanism, energy is stored by compressing powerful coil springs that are linked through a latching mechanism to the contact assembly of the breaker. After the springs have been compressed, the latch may be operated to release the springs to provide a quick closing action of the breaker contacts.

Breaker-closing springs may be compressed manually by means of a hand crank or they may be compressed by a small electric motor. Thus, the breaker may be classified as either a manually or an electrically operated breaker. In either type, however, the speed and force of closing the breaker contacts are independent of the action of a human operator.

When a large air circuit breaker is closed, the operating mechanism is latched closed. As the breaker is closed, a set of tripping springs is compressed. The breaker may then be tripped by operating a trip latch manually at the breaker or by electrically energizing a solenoid trip coil by means of a remote-control circuit. When the trip latch is operated either manually or electrically, the breaker contacts are opened quickly by the tripping springs.

Manually operated breakers have a lower initial purchase cost than do electrically operated breakers and are used when infrequent operation is required. Electrically operated breakers are used when breakers need to be operated frequently or when remote control is required. Electrically operated breakers may be closed and tripped from a push-button or control switch mounted in a convenient location remote from the breaker.

As soon as an electrically operated stored energy breaker is tripped, the spring-charging motor is started automatically and the closing springs are compressed so the breaker is ready for the next closing operation. The closing springs of a manually operated breaker are normally compressed by the hand-cranking operation just prior to breaker closing.

The breaker shown in Fig. 19-7 is a manually operated, stored energy, large air circuit breaker. The closing springs in the breaker are compressed by a downward stroke of the large operating handle on the front of the breaker, after which the breaker is closed by manual operation of the small closing lever. The breaker is tripped by the operation of the trip lever on the front of the breaker.

Automatic tripping of large air circuit breakers is accomplished on some breakers by a direct-acting series overcurrent tripping device. The operating coil of the tripping device is connected in series with the power circuit in which the breaker is installed. The flow of an abnormally high current in the circuit causes a plunger of the tripping unit to come in contact with the latch of the breaker-operating mechanism to open the breaker. One other type of tripping system is used with large air circuit breakers in which the tripping function is performed by a static overcurrent trip unit. This tripping unit is supplied energy from the power circuit through current transformers mounted in the breaker. When the tripping unit detects the flow of an overcurrent, a trip coil is energized to trip the breaker. The breaker shown in Fig. 19-7 is equipped with a static (solid state electronic) trip device, which may be seen on the lower right-hand side of the breaker.

Overcurrent tripping devices for large air circuit breakers may be designed to have different tripping characteristics to fit different application requirements. The tripping device used on many breakers is a dual

FIGURE 19-7   Three-pole large air circuit breaker. (Siemens Energy & Automation, Inc.)

device having a long-time-delay element for moderate overload currents and an instantaneous element for short-circuit and severe overload currents. The tripping characteristics of these devices are similar to the tripping devices used with molded-case breakers. Another often used tripping device has both a long-time-delay and a short-time-delay element. Moderate sustained overloads cause the breaker to trip after a relatively long time delay. Short circuits cause the breaker to trip after an intentional short time delay. This device is used in selective protection systems.

By the proper selection of ratings and types of tripping units, a completely selective system may be designed. On such a system, the breaker nearest the fault is caused to trip and remove the faulty circuit or equipment from service while the other breakers on the system, closer to the source, remain closed to carry normal load current to the unfaulted parts of the system.

Arc extinction in large air circuit breakers during the interruption of short circuits is accomplished by different methods by different manufacturers. The method used by one manufacturer is the deionization or physical removal of the conduction particles from the arc path. In breakers using this principle, the arc is forced upward into an arc chute by strong magnetic fields so that heated gas blasts carry the ionized air particles out of the arc path.

Large air circuit breakers are manufactured in several physical sizes as determined by their continuous-current and interrupting ratings. Standard frame sizes are 800, 1600, 2000, 3200, and 4000 A. Each frame size has a short-circuit or interrupting rating at 240, 480, and 600 V. Furthermore, each frame size is available in several continuous-current ratings, the continuous-current rating being determined by the rating of the series trip device used on the breaker.

In general, overcurrent protection is required at every point in a distribution system where wire sizes are reduced. As a consequence, circuit breakers are used at service entrances, on feeder circuits, and on branch circuits for the protection and switching of the various system components.

Circuit breakers used for the protection and switching of motor circuits are selected from breaker-application tables prepared by the circuit-breaker manufacturers. In general, such breakers are selected to have a continuous-current rating not to exceed approximately 250 percent of the full-load current of the motor. For circuits other than motor circuits, circuit breakers are selected to have the same rating as the continuous rating of the conductor used in the circuit they are protecting.

Large air circuit breakers are often assembled in groups together with other control and metering equipment into *metal-enclosed low-voltage air-circuit-breaker switchgear.* This type of switchgear is discussed in Sec. 19-7.

19-6 PANELBOARDS

A *panelboard* consists of a group of overcurrent-protective devices, with or without switches, including buses, assembled into a cabinet that is accessible only from the front. As shown in Fig. 19-2, panelboards are located between the secondary feeders and the branch circuits that supply the utilization equipment. So located, they are the means of control and protection for lighting and power circuits and the equipment supplied by them.

Panelboard overcurrent-protective devices may be fuses or circuit breakers. Fusible panelboards used for lighting and small-appliance loads often have a switch in series with each of the branch-circuit fuses and are called *switch-and-fuse panelboards.*

The panel assembly itself consists of a set of bus bars to which the several branch circuits are connected through the overcurrent or switching devices. The main-supply conductors for the panelboard may connect directly to lugs on the bus bars or may connect to the bus bars through a main circuit breaker, main fuses, or a fused disconnect switch. The panel assembly is enclosed in a sheet-steel cabinet of the dead-front type so that no live parts are exposed when the cabinet door is open.

Fusible and circuit-breaker lighting panelboards are available in several voltage ratings. Two of the more common ratings are the 120/240-V three-wire single-phase grounded neutral and the 120/208-V four-wire three-phase grounded neutral panelboards. Panelboards with these ratings are supplied from four-wire delta- or wye-connected transformers, as described in Sec. 12-17. A partially assembled circuit-breaker lighting panelboard is shown in Fig. 19-8.

Both fusible and circuit-breaker-type panelboards are available for distribution or power circuits with voltage ratings up to 600 V. One commonly used panelboard of this type is the convertible distribution panelboard. Convertible panelboards are of the sectionalized type manufactured in multiples of standardized circuit-breaker dimensions so that it is possible to interchange circuit breakers with different ratings or with different numbers of poles. With this type of panelboard it is possible to provide

FIGURE 19-8   Circuit-breaker lighting panelboard assembly. (Square D Company.)

space in the panelboard cabinet for future additional circuits. These panelboards will accommodate circuit breakers in frame sizes up to 600 A.

19-7 METAL-ENCLOSED LOW-VOLTAGE AIR-CIRCUIT-BREAKER SWITCHGEAR

A metal-enclosed low-voltage air-circuit-breaker switchgear, as the name indicates, is a metal-enclosed assembly containing air circuit breakers, together with bare buses, connections, control wiring, and accessory devices. Air circuit breakers are mounted in individual metal compartments and may be either the stationary or drawout type.

The stationary-type switchgear contains air circuit breakers mounted rigidly on bases with no special arrangement for quick removal from the structure.

Drawout-type switchgear contains breakers arranged so that they may be easily withdrawn from or inserted into their housing. As breakers are removed or inserted, they are automatically disconnected from or connected to their power and control circuits by means of self-aligning contacts. Interlocking is provided so breakers cannot be inserted or removed when they are closed. Drawout-type switchgear is more flexible and more easily maintained than the stationary type. Service interruptions are minimized since a breaker in need of maintenance can be replaced by a spare breaker in a very short time.

A section of low-voltage metal-enclosed switchgear containing several drawout air circuit breakers is shown at the right-hand end of the load-center unit substation in Fig. 19-9. The switchgear housing includes the

FIGURE 19-9   Load-center unit substation. (Siemens Energy & Automation, Inc.)

breaker compartments, breaker supporting rails, buses, connections, and instrument transformers. This switchgear is dead front; that is, no live parts are exposed at any time. Interlocks are provided to prevent inadvertent access to the live parts.

Both manually and electrically operated circuit breakers are used in metal-enclosed low-voltage switchgear. The application of circuit breakers in switchgear is the same as when breakers are used individually as described in Sec. 19-5. The rating of switchgear is dependent upon the continuous current rating of its buses and circuit breakers and the interrupting rating of the circuit breakers used in the switchgear.

19-8 LOAD-CENTER UNIT SUBSTATIONS

Load-center unit substations are combinations of transformers and metal-enclosed switchgear assembled as a unit. These substations may have switchgear on the high-voltage or low-voltage side, or on both sides of the transformer. The unit substation shown in Fig. 19-9 has a high-voltage incoming-line section on the left consisting of an air interrupter switch, a ventilated dry-type transformer section in the center, and a low-voltage metal-enclosed switchgear section on the right. The usual voltage ratings of unit substations are 15,000 V or less for the high-voltage section and 600 V or less for the low-voltage section. Transformers are usually rated 2000 kVA or less.

Modern practices in power-system design include the extensive use of the load-center system of power distribution with distribution voltages of from 2400 to 13,800 V. In this distribution system, power is provided to the point of utilization at the distribution voltage, where it is stepped down to utilization voltage by the load-center unit substation and distributed by short low-voltage feeders. In this way, good system voltage regulation is obtained with relatively small conductor sizes. The use of several small transformers at the load centers rather than one large transformer centrally located on the system also allows the use of smaller circuit breakers at the utilization-voltage level. Breakers are smaller for the load-center type of system because less load current is carried by the breakers, and the available short-circuit current is less because of the higher impedance of the small transformers. Another advantage of the load-center system of distribution is its adaptability to changes in loads. Since the unit substation is factory-assembled as a compact unit, it is readily moved if load centers change.

Load-center unit substations find their greatest application in manufacturing plants and other large, industrial, power-distribution systems.

19-9 HIGH-VOLTAGE FUSES

High-voltage fuses are used both indoors and outdoors for the protection of circuits and equipment with voltage ratings above 600 V. Some of the more generally used fuses and mountings are described briefly in the following paragraphs.

Expulsion fuses consist of a fusible element mounted in a fuse tube and depend upon the vaporization of the fuse element and the fuse-tube liner to expel conducting vapors and metals from the fuse tube, extinguishing the arc formed when current is interrupted. Another type of fuse, called the *liquid fuse,* depends on a spring mechanism to separate quickly the ends of the melted fuse element in a nonflammable liquid to extinguish the arc. Still another type of fuse is the *solid-material fuse,* in which the arc is extinguished in a hole in a solid material. In one type of solid-material fuse, a spring mechanism similar to that of the liquid fuse is used to separate the arcing terminals when the fuse blows. In this fuse, overload and low fault currents are interrupted in a small cylindrical chamber in the solid arc-extinguishing material, and large fault currents are interrupted in a larger chamber in the same fuse holder.

High-voltage fuses are often mounted in the same enclosure with disconnect switches to provide short-circuit protection and switching facilities for circuits and equipment. Typical equipment of this type removed from its enclosure is shown in Fig. 19-10. The equipment shown is for use on 13,800-V circuits and consists of a three-pole load-interrupter switch above and three solid-material fuses below.

Outdoor high-voltage fuses for low-capacity overhead lines are mounted in distribution fuse *cutouts.* Cutouts consist of a fuse support and fuse holder in which the fuse link is installed. An often-used type of cutout is the open cutout illustrated in Fig. 19-11. The cutout shown is designed to be attached to an overhead pole or structure by means of the mounting insert that protrudes from the center of the insulator. The fuse link is enclosed in the fuse tube, which when in the closed position shown completes the protected circuit between the upper and lower contacts of the cutout. When the fuse link is melted by fault current, a spring-loaded "flipper" quickly expels the severed link from the tube. This action also causes the upper end of the tube to swing out and down to an open position, providing a visual indication of the blown fuse. The cutout is placed back in service by inserting a new fuse in the tube and moving the tube back into the closed position.

FIGURE 19-10 Combination load-interrupter switch and solid-material fuses. (S & C Electric Company.)

19-10 HIGH-VOLTAGE CIRCUIT BREAKERS

The term *high-voltage circuit breaker* as used here applies to circuit breakers intended for service on circuits with voltage ratings higher than 600 V. High-voltage circuit breakers have standard voltage ratings of 4160 to 765,000 V and interrupting ratings as high as 63,000 A. Breakers with even higher ratings are being developed.

During the early development of electrical systems, the vast majority of high-voltage breakers used were oil circuit breakers. More recently, however, magnetic air, compressed air, vacuum, and sulfur hexafluoride

FIGURE 19-11 Distribution fuse cutout, open type. (S & C Electric Company.)

breakers have been developed and are now used instead of oil circuit breakers.

The *magnetic air circuit breaker* is available in interrupting ratings up to and including 750,000 kVA at 13,800 V. In this type of breaker, the current is interrupted between separable contacts in air with the aid of magnetic blowout coils. As the main current-carrying contacts part during the interruption of a fault, the arc is drawn out in a horizontal direction and transferred to arcing contacts. At the same time, the blowout coil is connected into the circuit to provide a magnetic field to draw the arc upward into arc chutes. The arc accelerates upward, aided by the magnetic field and natural thermal effects, into the arc chutes where it is elongated and divided into small segments. The arc resistance increases as it elongates and cools until, as the current passes through zero, the arc is broken; after this it does not reestablish itself.

The general construction of the magnetic power circuit breaker is somewhat similar to the large air circuit breaker used on low-voltage circuits except that they are all electrically operated. These breakers are used extensively in metal-clad switchgear assemblies in industrial plants, steel mills, and power plants.

Compressed-air breakers (sometimes called *air-blast breakers*) depend upon a stream of compressed air directed toward the interrupting contacts of the breaker to interrupt the arc formed when current is interrupted. This type of breaker was introduced to the American market in 1940 and has become universally accepted for use in heavy-duty indoor applications. More recently, air-blast breakers have been developed for use in extra-high-voltage outdoor stations with standard ratings up to 765,000 V.

Oil-circuit-breaker contacts are immersed in oil so the current interruption takes place under oil, which by its cooling effect helps quench the arc. Since oil is an insulator, the live parts of oil circuit breakers may be placed closer together than they could be in air. The poles of small oil circuit breakers are all placed in one oil tank, but in the large high-voltage breakers each pole is in a separate oil tank. Tanks of small breakers are suspended from a framework so that the tanks may be lowered for inspection of the contacts. The tanks of very large oil circuit breakers rest directly on a foundation and have handholes for access to the contact assembly.

The oil tanks of oil circuit breakers are sealed. The electric connections between the external circuit and the contacts in the tank are made through porcelain bushings. The breaker contacts are opened and closed by means of insulated lift rods on which the movable contacts are mounted. The lift rods are connected to the operating mechanism by means of a mechanical linkage, so that the contacts of all poles of the breaker are opened and closed together.

Only the very small oil circuit breakers are manually operated. The larger oil-circuit-breaker mechanisms are either pneumatically operated or spring-operated. Pneumatic operators obtain the closing and tripping energy from compressed air provided by a small, automatically controlled air compressor that maintains enough compressed air for several operations of the breaker in an air receiver. Spring operators derive their energy from a spring that is compressed by a small electric motor.

Sulfur hexafluoride breakers use the gas sulfur hexafluoride (SF_6) as the insulating and arc-quenching medium. This gas has been found to have excellent insulating properties as well as being nonflammable, nontoxic, and odorless. A three-pole outdoor frame-mounted SF_6 breaker designed for use on circuits rated 72.5 kV or less is shown in Fig. 19-12. The

FIGURE 19-12   Outdoor SF_6 circuit breaker.
(Siemens Energy & Automation, Inc.)

breaker shown has three cylindrical interrupting modules mounted on top of the cabinet that houses the breaker operating mechanism. These modules are filled with SF_6 gas maintained under a pressure of 75 lb/in^2 (517 kPa). Each gas-filled interrupting module contains the main contacts and a "puffer" mechanism, which momentarily increases the gas pressure in the space immediately surrounding the breaker main contacts during arc interruption. As the contacts part during a fault interruption, the compressed gas flows along the arc, sweeping the hot gases formed by the arc from between the parting contacts, which extinguishes the arc. After the arc is quenched, the SF_6 gas reverts to the original lower pressure in the module, ready for the next operation.

The operating mechanism of the SF_6 breaker is pneumatically operated. As the breaker contacts are closed by the pneumatic mechanism, a spring is compressed, which provides the energy for opening the breaker.

Puffer SF_6 breakers have been used in voltage ratings as high as 765 kV and with interrupting ratings as high as 63,000 A. To obtain the required interrupting capability at voltages higher than 362 kV, two or more interrupter modules are connected in series in each phase of the breaker as is done in other types of high-voltage breakers.

Vacuum breakers interrupt the flow of short-circuit current by separating their contacts in a vacuum chamber. Vacuum interrupters used in these breakers are enclosed in ceramic and steel envelopes or cylinders. Two butt contacts, one stationary and one movable, are used in each phase. A flexible stainless steel bellows is brazed to the stem of the movable contact. This allows the contact to move during operation of the interrupter without breaking the vacuum seal.

Since the current interruption takes place in a vacuum in the vacuum breaker, there is no ionization of the arc products to sustain the arc as is the case when an arc is formed in the air. Thus the separation of the contacts in the interrupter can be very small.

Specialized manufacturing processes are required to ensure proper sealing of the vacuum interrupter. One manufacturer performs the entire operation of evacuating, brazing, and sealing in a vacuum furnace.

The vacuum interrupter has been used in both indoor and outdoor 5- and 15-kV-class breakers. To obtain the insulation levels and interrupting capacity required at higher voltage levels, several vacuum interrupter contacts are used in series in each phase of high-voltage breakers.

19-11 PROTECTIVE RELAYS   Low-voltage air circuit breakers ordinarily have self-contained series trip coils of either the instantaneous or time-delay types. The tripping energy is supplied by the flow of the short-circuit current through the trip coil. Power circuit breakers seldom use series trip coils but are equipped with trip coils designed to operate from a storage battery or a reliable source of alternating current. Auxiliary devices called *protective relays,* designed to detect the presence of short circuits on a system, are used to connect the breaker trip coils to the source of tripping power and trip the breaker.

Protective relays are said to be selective when they trip only the circuit breakers directly supplying the defective part of the system and no other circuit breakers. When relays and circuit breakers are selective, short circuits are removed from a system with a minimum of service interruption. Of course, it is also desirable to isolate the defective system element as quickly as possible. To this end, relays and circuit breakers have been developed that will clear a short circuit in less than 0.1 s. However, selectivity being more important than speed, the tripping of some circuit breakers on a system is delayed intentionally to gain selectivity in clearing faults at certain locations on a system.

Protective-relay operating elements are connected to high-voltage circuits by means of current and potential transformers so that they in effect "measure" the quantities of current, voltage, or phase angle or combinations of these quantities to determine whether or not an abnormal condition exists in the circuit being protected. Relays with widely different operating characteristics have been developed for specific applications in protective schemes for rotating machines, transformers, buses, and lines.

Protective relays have traditionally been electromechanical relays. They have electrical connections to operating coil(s) that magnetically move induction disks to which contacts are attached. The contacts may open or close circuits that operate the protective devices. These electromechanical relays are being replaced by electronic or solid-state relays and, in some instances, directly by computers. Regardless of construction, protective relays perform the same functions. Two of the more frequently used protective relay types are *overcurrent* and *differential* relays.

Overcurrent relays, as the name indicates, operate to close their contacts when current through the relay operating coil exceeds a predetermined value. Overcurrent relays are connected to the protected circuit through current transformers so the current-transformer secondary current through the relay is proportional to the current in the primary circuit.

The most generally used electromechanical overcurrent relay is the induction overcurrent relay. This relay is similar in principle to the induction watthour meter and has a movable disk driven by an electromagnet. However, the induction overcurrent relay disk is not free to revolve continuously as in the watthour meter since the operating torque is opposed by a restraining spring. Furthermore, a movable contact is attached to the disk shaft so that after a partial revolution the movable contact makes contact with the stationary contact and further rotation is impossible.

The pickup current of an overcurrent relay is the current that will just cause the contacts to close. The operating coil of the relay is tapped to provide a means for adjusting the relay for different values of pickup current. Several tap ranges are available for induction relays, a commonly used range being from 4 to 16 A.

Induction overcurrent relays have inverse-time characteristics; that is, the contact closing is fast for high operating-coil currents and slow for low operating-coil currents. Moreover, the operating time for a given current is adjustable, the time of operation being proportional to the distance that the disk is allowed to travel before the contacts close.

With both the pickup current and the operating time being adjustable, induction overcurrent relays may be applied in selective systems with relays close to the generator adjusted for long operating times and with progressively shorter time settings for relays farther away from the generator. Induction overcurrent relays are relatively inexpensive, reliable, and easily calibrated and maintained. Consequently, they are widely used for the protection of transmission and distribution lines and equipment. An induction overcurrent relay withdrawn from its case is shown in Fig. 19-13. This relay

FIGURE 19-13   Induction overcurrent relay removed from drawout case. (General Electric Company.)

has an instantaneous attachment in addition to the basic inverse-time section. The instantaneous element responds only to extremely high fault currents.

Solid-state overcurrent relays (see Fig. 19-14) have a different appearance but perform the same function.

Electromechanical differential relays, like overcurrent relays, have an induction disk assembly on which is mounted a movable contact. However, the differential relay has, in addition to its operating electromagnet, a restraining electromagnet so connected that the flow of current through it provides a restraining torque or a contact-opening torque on the relay during the normal functioning of the equipment being protected by the relay.

Differential relays are used for the protection of rotating machines, transformers, and buses. Differential relays compare the currents entering and leaving each phase winding of the apparatus being protected. Normally the currents entering and leaving the apparatus are equal or are directly proportional to each other. However, on the occurrence of an internal fault in the apparatus, the two currents are no longer equal or proportional, since short-circuit current is fed into the apparatus from one or both sides. Under this condition, the difference of the two currents passes through the operating coil of the relay, causing the relay to close its contacts and trip the necessary circuit breakers to isolate the apparatus from the system. An

FIGURE 19-14 Solid-state overcurrent relays have a different appearance than the induction version; they do not have coils or induction disks. Connections are similar, as are functions. (Basler Electric.)

elementary diagram showing typical connections for a transformer differential relay is shown in Fig. 19-15.

A complete installation of three-phase transformer differential relays consists of three relays, one for each phase. For simplicity the diagram in Fig. 19-15 shows the connections for only one relay. The flow of normal currents is shown in the diagram, the current on the 69,000-V side of the transformer being 100 A and on the 6900-V side being 1000 A. With current-transformer ratios as shown, the secondary currents are balanced, each being 5 A. The 5 A then circulates in the series circuit consisting of the two current transformers and the two relay-restraining windings R and R'. For this normal condition, there is no current in the relay-operating winding 0, and the current in the two restraining windings creates a torque that holds the relay-operating contacts open. However, when a fault develops within the transformer, short-circuit current is fed into the transformer from one or both sides, depending upon the system arrangement. This causes the current in one restraining winding either to reverse or to increase out of proportion to the current in the other restraining winding. The restraining winding currents are no longer balanced, and the unbalanced current flows through the operating winding of the relay, developing a contact-closing torque that causes the relay contacts to close.

Differential relays for the protection of generators and motors are similar to transformer differential relays except that they are ordinarily more sensitive and will operate on a smaller unbalanced current than will transformer relays. A high-speed electromechanical differential relay designed for the protection of large ac generators is shown in Fig. 19-16. Bus differential relays are also similar to transformer relays except that the current transformer secondaries of all incoming lines to the bus are paralleled and connected to one relay-restraining winding, and the current

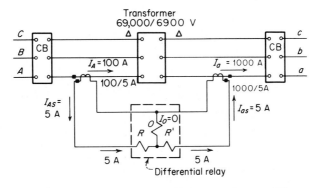

FIGURE 19-15   Connections for a transformer differential relay.

FIGURE 19-16   Electromechanical differential relay in drawout case. (General Electric Company.)

transformer secondaries of the outgoing lines are paralleled and connected to the other restraining winding. A solid-state transformer differential relay is shown in Fig. 19-17.

19-12 METAL-CLAD SWITCHGEAR

Metal-clad switchgear is a type of switchgear assembly in which all parts are completely enclosed in grounded metal enclosures. Circuit breakers used in metal-clad switchgear are air, oil, or vacuum breakers of the removable type, equipped with self-coupling primary and secondary disconnecting devices. All buses, connections, and joints are insulated and are completely isolated from the secondary wiring and control devices. Instruments, relays, and control switches are mounted on the front panels of the switchgear.

Metal-clad switchgear is available in both indoor and outdoor types for use on circuits with voltage ratings below 15,000 V, the standard voltage ratings being 4160, 7200, and 13,800 V. Interrupting and continuous-current ratings are based on the ratings of the circuit breakers used in the switchgear.

A typical assembly of metal-clad switchgear is shown in Fig. 19-18. Instruments, control switches, and protective relays required for control and monitoring the circuit breakers are mounted on the compartment doors above the drawout circuit-breaker compartments.

19-13 OUTDOOR SUBSTATIONS

Outdoor substations consist of structural frameworks on which switches, fuses, buses, and surge arresters are mounted and from which connections are made to circuit breakers, transformers, and other heavy equipment

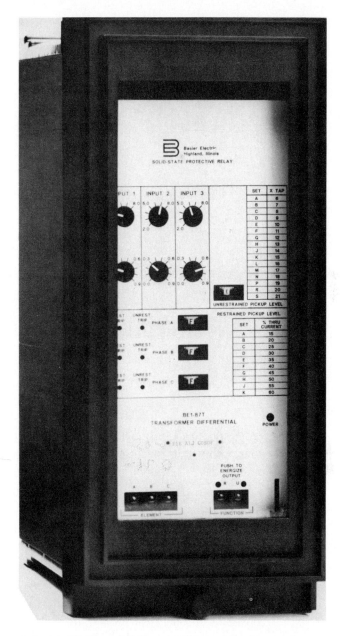

FIGURE 19-17   Solid-state differential relays perform
the same functions as their induction version. This relay is
designed for transformer protection. (Basler Electric.)

installed on the ground under the structure. Structures are usually constructed of galvanized structural steel, but some older low-voltage structures are of wood. Buses, jumpers, and connections are supported on insulators attached to the structural framework.

FIGURE 19-18   Metal-clad switchgear assembly. (Siemens Energy & Automation, Inc.)

There are no limitations to electrical capacity in outdoor substations since equipment of almost any rating or capacity can be installed in stations of this type. Spacings of buses and equipment and arrangement of station are dependent upon the voltage of the station. Arrangement, of course, is also influenced by the function for which the station is intended. Substations are usually designed to fit individual requirements, although some standardization is possible in certain types of substations.

Outdoor substations may be of the step-up type used to connect generating plants to transmission systems or they may be step-down stations for supplying loads from a high-voltage transmission system. Step-down stations for supplying loads or distribution systems at 15,000 V or less often consist of an open-type steel framework for terminating the high-voltage circuit, a three-phase transformer, and an assembly of outdoor metal-clad switchgear for the protection and switching of the low-voltage circuits.

Voltage transformations are not always involved in outdoor substations. Switching stations in which only circuit breakers and other switching equipment are installed are used at junctions and taps in transmission systems.

A typical outdoor substation is shown in Fig. 19-19. This is a step-down substation with a high-voltage rating of 138 kV and a low-voltage rating of 69 kV. The bus, overhead lines, and circuit breakers are shown.

FIGURE 19-19   Outdoor substation showing 138 kV high-voltage and 69 kV low-voltage bus arrangement. (Stanley Consultants.)

19-14 SURGE ARRESTERS   Transient overvoltage surges on an electrical system may be caused by lightning, by switching operations, or by other disturbances. If these overvoltages exceed the insulation level of the system, flashovers occur; these can destroy the insulation of transformers or other apparatus connected to the system. It is the function of surge arresters to provide a path by which these surges are conducted to earth before system flashovers can occur.

The *valve-type surge arrester* was a commonly used type up until a few years ago. These arresters may be likened to a normally open circuit breaker connected from the protected circuit to earth, which closes when a high-voltage surge appears and then reopens when the surge disappears. This arrester consists of two elements, a series gap and a valve assembly, both enclosed in a porcelain housing. The spacing of the series gap is such that it will withstand normal circuit voltage. However, an overvoltage such as may be caused by a lightning surge causes the gap to break down, with a resulting flow of current through the valve assembly to ground. The valve assembly is constructed of a ceramic compound that has the peculiar property of offering a high resistance to current flow when normal system voltage is applied but a low resistance to the flow of high-surge currents. Thus it operates as a current valve and allows the surge current to pass but assumes its former high resistance to stop the flow of current as soon as the voltage drops to its normal value. An installation of heavy-duty station-type valve arresters is shown in Fig. 19-20. These arresters have been largely replaced by metal oxide-based varistors, which have the same physical appearance. They perform the same function but have no parts to burn or wear out. Their principle of operation is described in Sec. 13-11.

FIGURE 19-20   *Installation of valve-type surge arresters.*
(Westinghouse Electric Corporation.)

REVIEW
QUESTIONS

1   What is a short circuit?

2   What are two commonly used automatic circuit-protective devices?

3   In general, where is circuit-protective and switching equipment grouped on an electrical system?

4   What is a fuse? What are two types of low-voltage fuses?

5   What are the standard fuse-clip sizes for cartridge fuses?

6   Describe a time-lag fuse.

7   Where might a current-limiting fuse be used?

8   What is a safety switch? Where is it used?

9   Describe the operation of a molded-case circuit breaker.

10   What is meant when it is said that a circuit breaker has a thermal-magnetic trip element?

11   When is electrical operation of a circuit breaker desirable?

12   What type of automatic tripping is used on large low-voltage air circuit breakers?

13   How do the ratings of large low-voltage air circuit breakers differ from the ratings of molded-case breakers?

14   What is a panelboard? What types of overcurrent devices are used in panelboards?

15   What are two types of construction used in metal-enclosed low-voltage air-circuit-breaker switchgear? What are some advantages of drawout switchgear?

16 What is a load-center unit substation?

17 Name three types of high-voltage fuses.

18 What is a fuse cutout? Where is it used?

19 Name several types of high-voltage circuit breakers.

20 What are some different types of operating mechanisms used with high-voltage circuit breakers?

21 What is the function of protective relays?

22 Describe the principle of operation of overcurrent and differential relays.

23 What is metal-clad switchgear? What are the maximum standard current, voltage, and interrupting ratings?

24 What are the essential features of outdoor substations?

25 Describe briefly the operation of the valve-type surge arrester.

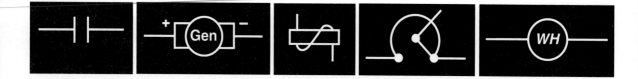

20
Electrical Instruments and Measurements

Electrical measurements play an important role in the design, development, manufacturing, and maintenance of modern industrial equipment. The devices that perform the measurements and indicate values electrically have evolved from very simple meters that move a pointer in response to currents in coils to highly complex, sophisticated, multifunction, solid-state electronic equipment. (See Fig. 20-1.) The indicating parts have changed from pointers and dials to light emitting diodes to miniature video display terminals. Regardless of construction, a basic knowledge of the principles of operation of the devices is essential to the understanding and interpretation of their readings.

This chapter discusses the basics of electric measurements by describing the operation of many of the electromagnetic-based devices. The functions of the electronic types are the same.

20-1 GALVANOMETERS

As was learned from the study of electric motors, when a current-carrying conductor is placed in a magnetic field, a force is developed on the conductor that tends to move the conductor at right angles to the field. This principle is used in current-detecting instruments. A sensitive current-detecting instrument called a *galvanometer,* operating on this principle, is shown diagrammatically in Fig. 20-2.

A coil of very fine insulated wire, usually wound on an aluminum frame or bobbin, is suspended as shown between the poles of a permanent magnet. The coil is suspended by a phosphor-bronze filament that acts as

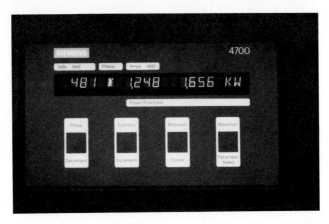

FIGURE 20-1 Multifunction electronic meter with touch selection and programming. (Siemens Energy & Automation, Inc.)

one lead-in wire for the coil. The other lead-in wire is a very flexible spiral wire at the bottom of the coil. When current flows through the coil, a deflecting force proportional to the flux density, the current, and the dimensions of the coil rotates the coil on its vertical axis. The deflecting force is opposed by the restraining force of the suspension filament so that the coil does not continue to rotate as in a motor but turns until the deflecting force is balanced by the restraining force of the suspension filament. Since the deflecting force is directly proportional to the current flowing in the coil, the amount of angular rotation may be used as an indication of the value of the current flowing in the coil.

The amount of deflection and hence the amount of current flowing through the coil may be indicated by a pointer, which is attached to the moving element and moves over a calibrated scale.

On the more sensitive galvanometers a mirror is attached to the moving coil as shown in Fig. 20-2. A light beam is reflected from the mirror onto a ground-glass scale. As the coil is deflected, the light beam moves over the scale.

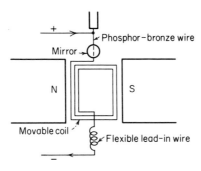

FIGURE 20-2 Diagram of the essentials of a galvanometer. The deflection of the movable coil is proportional to the current flowing.

The restoring force of the suspension filament acts to return the coil and indicating system to the normal or zero position when the current flow through the coil is interrupted.

After a reading is taken and the restoring force acts to return the coil to its normal position, the coil tends to oscillate about the normal position for some time before coming to rest. To prevent excessive oscillation of the coil, a system of damping must be employed. The aluminum frame upon which the coil is wound provides the damping force in the galvanometer just described. Whenever the frame moves in the magnetic field, induced currents are produced that flow around the closed circuit formed by the frame. The induced currents oppose the motion that produces them, bringing the moving element immediately to a standstill.

20-2 THE PERMANENT-MAGNET MOVING-COIL INSTRUMENT

The permanent-magnet moving-coil instrument is universally used for the measurement of direct currents and voltages. It is essentially a galvanometer made for switchboard mounting or for portable use. The moving coil is supported in the magnetic structure in two different ways: by the pivot-and-jewel system or by the taut-band suspension system.

In the pivot-and-jewel construction, the moving coil is wound on a light aluminum frame and the frame turns on steel pivots that rotate in jeweled bearings at each end of the coil. A pointer is attached rigidly to the frame so that as the coil moves, the pointer moves over a graduated scale. Current is brought into the coil through spiral springs, one at each end of the coil. The springs also provide the restoring force for the instrument.

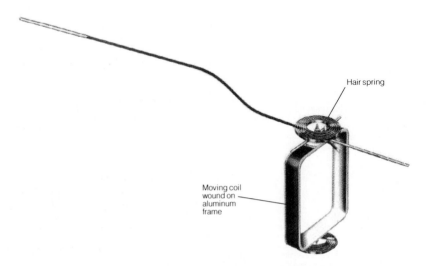

Hair spring

Moving coil wound on aluminum frame

FIGURE 20-3   Moving element of a permanent-magnet moving-coil instrument, pivot-and-jewel construction. (Weston Instruments.)

The movable coil, springs, and pointer of a permanent-magnet moving-coil instrument with the pivot-and-jewel construction are shown in Fig. 20-3.

The magnetic circuit consists of a permanent horseshoe magnet to which are attached soft-iron pole pieces. To form a uniform air gap through which the coil may turn, a cylinder of soft iron is supported between the pole pieces. The method of assembly of the magnetic circuit and moving element is shown in Fig. 20-4. It will be noted that a uniform air gap is provided for the coil for a wide range of deflection of the coil. The use of the uniform air gap results in a deflection of the pointer that is almost directly proportional to the current flowing in the coil. Thus, the scale used with this type of instrument may be made very uniform. The aluminum frame upon which the coil is wound provides the damping force.

The taut-band suspension system for a permanent-magnet moving-coil instrument is illustrated in Fig. 20-5. As shown, the moving coil and pointer are suspended by a taut band at either end of the moving mechanism. The taut bands serve three functions:

1. They support the moving element.
2. They provide the restoring force for the moving element.
3. They carry the current to the coil.

The permanent magnet for the taut-band suspension instrument is a core-type magnet. This is a cylindrical magnet with a hollow center, and the coil thus surrounds the magnet. The return path for the flux is provided

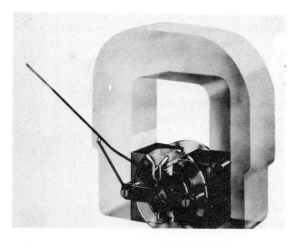

FIGURE 20-4   Permanent-magnet moving-coil mechanism. (Weston Instruments.)

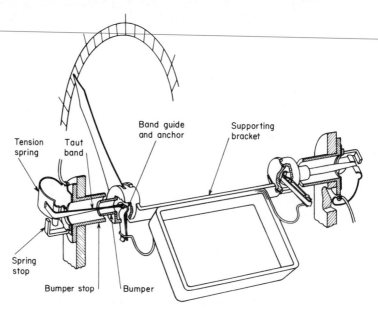

FIGURE 20-5   Moving element of a taut-band suspension
instrument. (Westinghouse Electric Corporation.)

by a tubular piece of soft iron placed concentrically with respect to the
magnet and forming an air gap through which the coil moves.

Since no bearings are used in the taut-band suspension instrument,
the system has no friction and the instrument is highly shock- and vibra-
tion-resistant. The instrument is capable of retaining its calibration and
accuracy over long periods of time with little maintenance.

20-3 DC VOLTMETERS   A voltmeter must be designed to be connected *across* the line. Since the coil
of a permanent-magnet moving-coil instrument has a low resistance, it
must have in series with it a high value of resistance when it is used as a
voltmeter. The value of the series resistance must be high enough that the
rated current of the coil is not exceeded. In portable testing voltmeters, several voltage ranges may be provided by using different values of series resistance for each range. The series resistance is usually contained in the
instrument case itself.

So that the voltmeter will read upscale, the positive (+) terminal of
the voltmeter must be connected to the positive line.

20-4 DC AMMETERS   An ammeter is connected in *series* with the line in which current is to be
measured. Since the moving coil and the spiral springs used as coil connections can be designed for a maximum of only about 50 mA (0.05 A), the
permanent-magnet moving-coil instrument when used for the measurement of current higher than this value must be equipped with a *shunt*. A

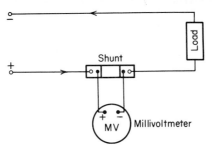

FIGURE 20-6   Diagram showing the method of connecting a shunt and a millivoltmeter for measuring current.

shunt is merely a low resistance that is placed in parallel with the coil circuit. The greater part of the current in the main circuit is then diverted around the coil through the shunt.

Portable testing ammeters are equipped with self-contained shunts in ranges up to about 50 A. For measurement of higher values of current, millivoltmeters are used with external shunts. The millivoltmeter together with the shunt then constitute the ammeter. Shunts used with millivoltmeters are usually designed so that when rated current flows through them the voltage drop across the shunt is 50 mV (0.05 V). Millivoltmeters used with 50-mV shunts are designed to give a full-scale reading at 50 mV. Switchboard-type instruments or portable instruments always used with the same shunt are generally scaled to read directly in amperes. The method of connecting a shunt and millivoltmeter for measuring current is shown in Fig. 20-6. Note the polarity marking of the millivoltmeter with respect to the line polarity.

20-5 MOVING-IRON VANE INSTRUMENTS

When two soft-iron bars are placed parallel to each other inside a current-carrying coil, the bars become magnetized, with adjacent ends of the bars behaving like magnetic poles. The bars being similarly magnetized repel each other, the amount of repulsion being dependent on the amount of magnetizing force of the coil. Repulsion occurs between the bars for either direction of current through the coil, and thus they are repelled when an alternating current is passed through the coil.

The moving-iron vane instrument, widely used for the measurement of alternating currents and voltages, operates on the above-described principle. Two iron vanes, one stationary and one securely attached to the central shaft of the moving element, are placed inside the coil through which the current to be measured is passed. In one type of moving-iron vane instrument, the vanes are concentric, the movable vane being placed inside the stationary vane. This type of construction is shown in Fig. 20-7. Another version of this type of instrument uses flat vanes, one attached to the shaft like a flag, the other one fixed. This

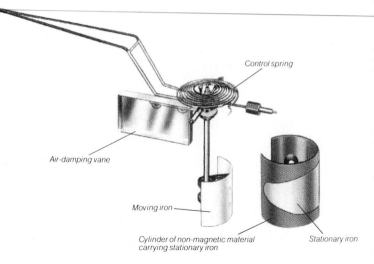

Control spring

Air-damping vane

Moving iron

Cylinder of non-magnetic material
carrying stationary iron

Stationary iron

FIGURE 20-7   Moving element of the moving-iron vane instrument with the
stationary vane shown at the right. (Weston Instruments.)

so-called *radial vane* type of instrument has certain advantages in sensitivity and scale distribution.

When current flows in the coil surrounding the vanes, the movable vane is repelled by the stationary vane and moves the entire moving element clockwise, the amount of movement being dependent on the amount of current flowing in the coil. Restraining action is provided by the spiral spring at the top of the moving element. It should be noted that in this instrument the springs do not carry current. The damping device is an aluminum vane attached to the moving element, as shown in Fig. 20-7, which moves in a closed chamber. The motion of the vane against the air in the chamber prevents excessive oscillation of the pointer. Figure 20-8 shows a phantom view of an assembled moving-iron vane mechanism.

When used as a voltmeter, the moving-iron vane instrument is used with a series resistance. Portable voltmeters are made with self-contained resistors in ranges up to 750 V. Higher ranges may be obtained by using external multipliers or, as is usually the case, voltage transformers (Sec. 12-12). Since the standard low-voltage rating of voltage transformers is 115 or 120 V, voltmeters when used with voltage transformers generally have a full-scale value of 150 V.

When used as an ammeter, the moving-iron vane instrument is wound with wire heavy enough to carry the rated current. Usual practice, however, is to use a 5-A ammeter in conjunction with a current transformer for measuring values of current larger than 5 A.

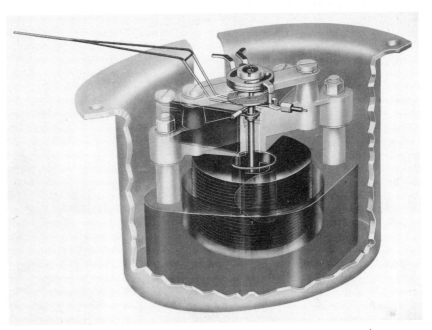

FIGURE 20-8   Moving-iron vane mechanism. (Weston Instruments.)

20-6 ELECTRODYNAMOMETER VOLTMETERS AND AMMETERS

The electrodynamometer instrument operates on the same principle as the permanent-magnet moving-coil instrument in that deflection of the moving element is obtained from the interaction of two magnetic fields. However, in the electrodynamometer instrument, two stationary coils are used in place of the permanent magnet. The movable coil is attached to the central shaft so that it may rotate inside the two stationary coils as shown in Fig. 20-9. Connections are made to the movable coil through the two spiral springs at the top of the mechanism. The springs also provide the restoring force for the moving element. Air damping is obtained by means of the aluminum vanes that move in the enclosed chambers shown at the bottom of the mechanism.

When used as a voltmeter, or as an ammeter for measuring small values of current, the stationary and movable coils are connected in series. Since the polarity of the fields produced by both the stationary and movable coils is reversed by a reversal of current, the deflection of the moving element is always in the same direction, regardless of the direction of current through the coils. For this reason, electrodynamometer instruments may be used with practically equal precision for either dc or ac measurements.

Since the movable coil and its flexible connections can be made to carry only a small amount of current, the electrodynamometer-type instrument is seldom used as an ammeter except for lower current ranges. The permanent-

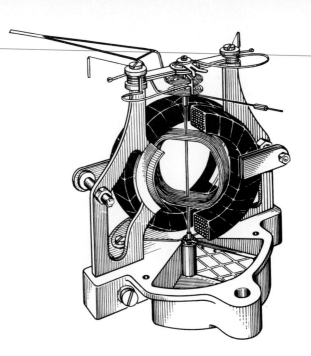

FIGURE 20-9   Electrodynamometer mechanism.
(Weston Instruments.)

magnet moving-coil and the moving-iron vane instruments are more wide-ly used for the measurement of direct and alternating currents, respectively.

20-7 POWER MEASUREMENTS

Power is measured by means of the wattmeter. Since power is a function of both current and voltage, two elements are necessary in a wattmeter. For this reason, wattmeters are generally of the electrodynamometer type. The movable coil with a series resistance forms the potential element, while the stationary coils are used as the current element. Connection is as shown in Fig. 20-10, with the potential and current elements in parallel and series, respectively, with the circuit in which the power is to be measured.

Electrodynamometer wattmeters may be used for the measurement of either ac or dc power. Since power in a dc circuit is always a product of current and voltage, however, it is seldom necessary to use a wattmeter in a dc circuit. The power can be calculated easily from the voltmeter and ammeter readings.

Power in an ac circuit depends on the power factor of the circuit as well as the current and the voltage. At unity power factor, the current in both the current and potential coils of a wattmeter reverses at the same time, resulting in a deflecting force on the moving element that is always in the same direction. However, at power factors less than unity, the current in one element reverses before the current reverses in the other element, resulting in a reverse torque during the time that the two currents are in

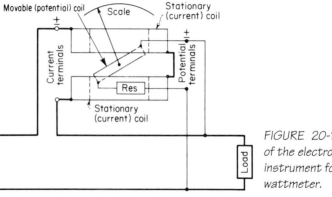

FIGURE 20-10 Connection of the electrodynamometer instrument for use as a wattmeter.

opposite directions. The inertia of the moving element prevents its following the torque reversals, however, so that the resulting indication of the instrument is a resultant of the two torques. For a phase angle of 90°, the two torques are equal and the wattmeter indicates zero power. Thus the wattmeter indicates the true or actual power of an ac circuit.

Power in a three-phase three-wire circuit may be measured by means of two wattmeters W_1 and W_2 connected as shown in Fig. 20-11. The algebraic sum of the readings of the two wattmeters will be the total three-phase power of the circuit for either balanced or unbalanced loads or voltages. For balanced loads at unity power factor, the readings of the two wattmeters will be identical. When the load power factor is 50 percent, W_2 will read zero and W_1 will read the total three-phase power. At power factors between 50 and 100 percent, W_1 will read higher than W_2. At power factors lower than 50 percent, the reading of W_2 will be negative and the total three-phase power will be $W_1 - W_2$. At zero power factor, the wattmeters will have identical readings but of opposite signs, indicating zero power. Thus there is a definite ratio of W_1 to W_2 for each value of circuit power factor.

The two-element or polyphase wattmeter may be used in place of the two single-phase wattmeters shown in Fig. 20-11. Two complete elements

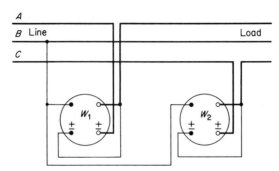

FIGURE 20-11 Diagram of connections of two wattmeters for measuring power in a three-wire three-phase circuit.

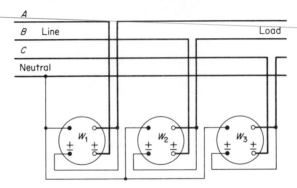

FIGURE 20-12   Connection of three wattmeters
to measure power in a four-wire three-phase circuit.

are mounted on the same shaft and connected in the same manner as two single-phase wattmeters. The deflection of the pointer indicates total three-phase power directly.

In a three-phase four-wire system, three wattmeters or the three elements of a polyphase wattmeter connected as shown in Fig. 20-12 are necessary for a true indication of power under all conditions of unbalance. Total three-phase power is the sum of the readings of the three wattmeters or is indicated directly by the three-element wattmeter. In general, power may be measured in any polyphase system with the number of wattmeters or elements one less than the number of wires in the system.

20-8 THE VARMETER

A standard wattmeter can be used to measure reactive voltamperes, or vars, by applying a voltage to it that is displaced 90° in time phase from the voltage that would normally be used for the measurement of watts. The phase shifting required for measuring vars in a single-phase circuit is accomplished by connecting a combination of resistors and capacitors in series with the potential coil of the meter. For three-phase var measurement, a phase-shifting transformer can be used to shift the phase of the voltages applied to each of the elements of the meter.

The connections of a two-element varmeter and its auxiliary phase-shifting autotransformers are shown in Fig. 20-13. Analysis of the circuit will show that the voltages V_{45} and V_{67} lag the voltages V_{12} and V_{32}, respectively, by 90°.

Varmeter scales are usually the zero-center type and are marked IN and OUT at their left- and right-hand ends, respectively. When the meter reads vars OUT it indicates by convention that magnetizing vars are flowing from the supply to the load. This would be the case when a generator is supplying an inductive load such as an induction motor. Likewise, a var IN

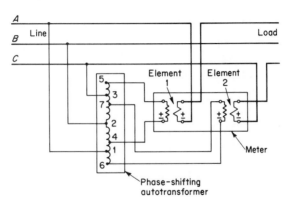

FIGURE 20-13 Connection of a three-phase varmeter.

reading would indicate that the load was capacitive in nature and that the load was causing magnetizing vars to flow back to the supply.

20-9 ENERGY MEASUREMENTS IN AC CIRCUITS

Energy is measured by means of the watthour meter as indicated in Sec. 4-8. Since energy is a product of power and time, both quantities must be taken into account in energy measurements. Induction watthour meters have both potential and current elements and in addition have a rotating element, the speed of which at any time is proportional to the circuit power at that time. The rotating element drives a gear train in a register to record the number of kilowatthours used in a given time. The total number of revolutions made by the rotating element in a given time is then a measure of the energy consumption in that time. A single-phase induction watthour meter is shown in Fig. 20-14.

Single-phase watthour meter connections are the same as single-phase wattmeter connections in that the potential element is connected in parallel and the current element in series with the circuit in which energy is being measured.

Polyphase energy measurements are similar to polyphase power measurements; that is, the number of watthour meters required for correct energy measurements is one less than the number of wires in the system. Multielement watthour meters are generally used, however, with the required number of elements mounted to drive a single register to record the total polyphase energy.

Several schemes of metering polyphase energy are in use in which the number of elements is less than required for absolutely correct measurement. For reasonably balanced loads and voltages, these meters are sufficiently accurate for commercial measurements.

FIGURE 20-14   Single-phase induction watthour meter. (General Electric Company.)

Watthour meters can be equipped with special all solid-state, microprocessor-based registers to record not only the total monthly kilowatthours but also kilowatthours for certain periods of time (such as off-peak times). These registers can also be programmed to record kilowatt demand and the time that the demand occurred during each month. As a further step in the trend toward solid-state measuring equipment, many induction-type watthour meters are being replaced with electronic-type meters that are completely solid state.

20-10 POWER-FACTOR METER

The electrodynamometer-type instrument when made with two movable coils set at right angles to each other may be used to measure power factor. The method of connection of this type of power-factor meter in a three-phase circuit is shown in Fig. 20-15. The two stationary coils S and S' are connected in series in line B. Mounted on the central shaft and

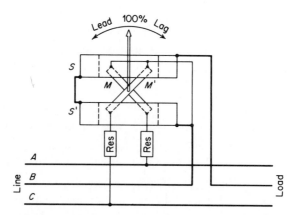

FIGURE 20-15   Connection diagram of a three-phase power-factor meter of the electrodynamometer type.

free to move without restraining or control springs are coils M and M'. Coils M and M' are connected with their series resistors from lines B to A and B to C, respectively. At unity power factor, one potential-coil current leads and one lags the current in line B by 30° so that the coils are balanced in the position shown in Fig. 20-15. A change in power factor causes the current of one potential coil to become more in phase and the other to be more out of phase with the current in line B so that the moving element and pointer take a new position of balance to indicate the new power factor.

The power-factor meter shown in Fig. 20-15, when used with a special reactance-resistance network auxiliary, may be used to measure power factor in a single-phase circuit.

20-11 PHASE-ANGLE METER

The phase-angle meter is a test instrument that is almost indispensable in power circuit testing work, particularly in the testing of metering and protective relaying circuits. It is used to indicate the phase angle between an alternating current and voltage, between two voltages, or between two currents. Since many meters, relays, and other control devices depend upon the proper phase relationships between two or more electrical quantities for their proper operation, the phase-angle meter may be used for installation or maintenance testing of such circuits and equipment. A portable phase-angle meter is shown in Fig. 20-16.

The movable element of the meter shown consists of iron vanes, but unlike the moving-iron instrument described in Sec. 20-5, the vanes are magnetized by a coil on the moving element. Furthermore, the moving element is free to turn a full 360°. When the stationary and moving coils are energized from two sources of the same frequency, the moving element takes a position where the pole strength of the vanes is greatest. This position of greatest pole strength is determined by the phase angle between the sources energizing the two coils. Therefore, the meter may be calibrated so that the phase angle between the two sources is indicated directly upon the meter dial.

The electrodynamometer-type instrument may also be used as a phase-angle meter, and one commercially available meter uses this principle.

20-12 MEASUREMENT OF RESISTANCE; VOLTMETER-AMMETER METHOD

A very simple method of measuring resistance is by the voltmeter-ammeter method. A direct current is passed through the unknown resistance, the current being measured by an ammeter and the voltage drop across the resistance being measured by a voltmeter. By Ohm's law, the resistance is

$$R = \frac{V}{I}$$

FIGURE 20-16   Phase-angle meter.
(Westinghouse Electric Corporation.)

Although this method of resistance measurement is relatively simple, care should be taken so that the currents taken by the instruments themselves do not introduce errors in the results of the measurements. When low values of resistance, such as an armature-circuit resistance, are being measured, the connections for the least amount of error are as shown in Fig. 20-17. To ensure that the voltage drop across the armature is large enough to be read on the voltmeter, a large value of current must be used, care being taken, of course, not to exceed the current rating of the armature. For the connection shown in Fig. 20-17, the ammeter reads the sum of the currents through the armature and the voltmeter. However, the current taken by the

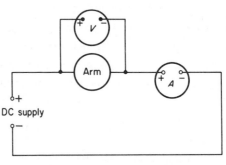

FIGURE 20-17   Connection of a voltmeter and ammeter for measuring a low value of resistance.

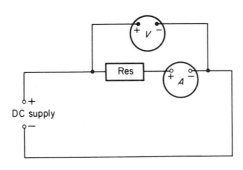

FIGURE 20-18 *Connection of a voltmeter and ammeter for measuring a high value of resistance.*

voltmeter is so small compared with the current through the armature that the error introduced is practically negligible.

When higher values of resistance are being measured and where the value of test current is quite small, the instrument connections should be as shown in Fig. 20-18. Note that the voltmeter is connected across both the ammeter and the unknown resistance. The ammeter reads the true value of the current through the resistance, but the voltmeter reading includes the *IR* drop in the ammeter. Since the test current required is quite low, the ammeter *IR* drop is very low, resulting in an almost negligible error in the measurements.

For precise measurements with either of the two above methods of resistance determination, corrections should be made for the currents taken by the instruments. This is not necessary in most ordinary commercial measurements, however.

20-13 WHEATSTONE BRIDGE The most commonly used device for accurate resistance measurements in the range of 1 to 100,000 Ω is the Wheatstone bridge. As shown in Fig. 20-19, it is composed of resistances R_1, R_2, and R_3, which are accurately known adjustable resistances. The unknown resistance R_x is connected between points D and B. The resistors are arranged in two parallel circuits through which current from the battery can flow. A galvanometer is bridged across the two circuits between the points C and D.

Balancing the bridge consists of closing the battery switch S_1 and adjusting R_1, R_2, and R_3 to such values that when the galvanometer switch S_2 is closed the galvanometer shows no deflection.

With the bridge balanced, the points C and D are at the same potential. This means that the voltage drops from A to C and A to D must be equal and that the drops from C to B and D to B are also equal. Hence

$$I_1R_1 = I_2R_2 \tag{20-1}$$

and $$I_1R_3 = I_2R_x \tag{20-2}$$

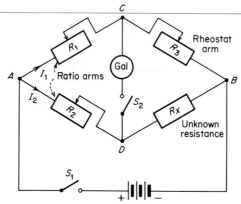

FIGURE 20-19   Schematic diagram of connections of the Wheatstone bridge.

Dividing Eq. (20-1) by Eq. (20-2) and solving for R_x results in

$$R_x = \frac{R_2}{R_1} R_3 \qquad\qquad (20\text{-}3)$$

EXAMPLE 20-1

The Wheatstone bridge represented in Fig. 20-19 when used for determining the value of an unknown resistance R_x is balanced when R_3 is 124 Ω, R_1 is 10 Ω, R_2 is 100 Ω. What is the value of R_x?

$$R_x = \frac{R_2}{R_1} R_3$$

$$= \frac{100}{10} \times 124 = 1240 \ \Omega$$

Resistances R_1 and R_2 are called the ratio arms of the bridge and are usually made to equal a definite ratio such as 1 to 1, 10 to 1, or 100 to 1. Resistance R_3 is called the rheostat arm and is made continuously variable from 1 to 1000 or from 1 to 10,000 Ω. When the bridge is balanced, the value of the unknown resistance is equal to the value of R_3 multiplied by the ratio of R_2 to R_1.

On commercial "dial"-type bridges, the ratio of R_2 to R_1 is selected by means of a rotary switch, the setting being indicated on a dial. The rheostat arm R_3 is adjusted by four or more rotary switches, one of which varies the resistance in steps of 1 Ω, another in steps of 10 Ω, another in steps of 100 Ω, and so on. The value of the rheostat arm R_3 then may be read directly

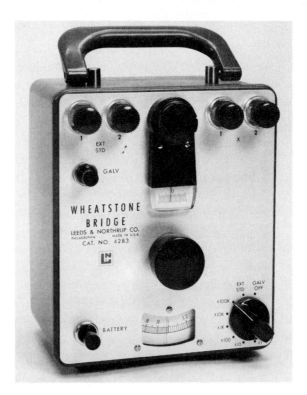

FIGURE 20-20   Portable-type Wheatstone bridge.
(Leeds & Northrup Company.)

from the setting of the several dials on the rotary switches when the bridge is balanced.

In the portable bridge shown in Fig. 20-20, an adjustable slide wire controlled by the knob in the center of the panel forms the ratio arms. The rheostat arm consists of six fixed standard resistors, one of which may be selected by means of the rotary range switch located in the lower right-hand corner of the bridge. The unknown resistance is connected to the two terminals in the upper right-hand corner. The bridge is balanced by selecting the proper fixed resistor with the rotary selector switch and then adjusting the ratio arm slide wire until the galvanometer reads zero. The value of the unknown resistance is then the setting of the slide wire as indicated on its scale, multiplied by the multiplier as set on the rotary selector switch.

**20-14
INSULATION
RESISTANCE**
Insulating materials are materials that offer a high resistance to the flow of an electric current. The materials and the dimensions of the materials used for insulating electric circuits must be such that a very high resistance, usually in the order of megohms, is offered to the flow of current. (One

megohm [MΩ] is equal to 1,000,000 Ω.) For example, the armature coils of a generator must have sufficient insulation to prevent the leakage of current to adjacent coils or to the iron armature core. The operation of electric equipment such as transformers, motors, generators, and cables depends on the maintenance of the proper value of insulation resistance for the circuits involved.

In many cases it is desirable to measure periodically the insulation resistance of electric circuits, so that defective insulation may be found before it causes failure of the equipment. Defective insulation may occur from the effects of heat, the presence of moisture, mechanical injuries, chemical action, or other causes. Of the several known methods for measuring insulation resistance, the voltmeter method and the Megger method are the most commonly used and are described in the following sections.

20-15 MEASUREMENT OF RESISTANCE; VOLTMETER METHOD

A voltmeter may be used for measuring insulation resistance or other high resistances. A high-resistance voltmeter is preferable, and the voltmeter resistance R_v must be known. A reading is first taken with the voltmeter across a constant source of voltage V_s. The voltmeter and the unknown resistance are then connected in series to the same source of voltage V_s, as in Fig. 20-21, and a second reading is taken. From these readings, the value of the unknown resistance may be calculated as shown below.

In Fig. 20-21 the current is

$$I = \frac{V_s}{R_v + R_x} \tag{20-4}$$

The voltmeter reading V when the voltmeter is connected in the circuit of Fig. 20-21 is the voltage drop in its own resistance, or

$$V = IR_v \tag{20-5}$$

Substituting Eq. (20-4) in Eq. (20-5) and solving for R_x results in

$$R_x = R_v\left(\frac{V_s}{V} - 1\right) \tag{20-6}$$

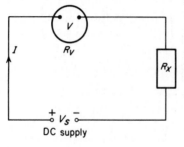

FIGURE 20-21   Voltmeter method of measuring resistance.

The value of the unknown resistance R_x may thus be calculated by substituting the two voltmeter readings in Eq. (20-6).

It is desired to measure the resistance of the insulation between a motor winding and the motor frame. A 300-V, 30,000-Ω voltmeter is used in making the measurements. The voltmeter when connected across the source reads 230 V and when connected in series with the insulation reads 5 V. Find the insulation resistance.

$$R_X = R_V \left(\frac{V_S}{V} - 1 \right)$$

$$= 30,000 \left(\frac{230}{5} - 1 \right) = 30,000 \times 45$$

$$= 1,350,000 \ \Omega, \text{ or } 1.35 \text{ M}\Omega$$

20-16 THE MEGGER INSULATION TESTER

The Megger insulation tester is extensively used for measuring insulation resistance. One model of this line of insulation testers is shown in Fig. 20-22.

Essentially the Megger insulation tester consists of a hand-driven generator and a direct-reading true ohmmeter. A simplified diagram of the

FIGURE 20-22   Megger insulation tester. (James G. Biddle Co.)

electric connections of the instrument is shown in Fig. 20-23. A permanent magnet provides the field for the ohmmeter.

The moving element of the ohmmeter consists of two coils, *A* and *B,* which are mounted rigidly to a pivoted central shaft and which are free to rotate over a stationary C-shaped iron core. The coils are connected to the circuit by means of flexible leads that exert no restoring force on the moving element. Hence the moving element may stand in any position over the scale when the generator is not in operation. Coil *B* is connected in series with the resistance *R* across the generator terminals, while coil *A* is connected in series with the resistance R' between one generator terminal and the test terminal marked LINE.

When current flows from the generator through coil *B,* the coil tends to set itself at right angles to the field of the permanent magnet. With the test terminals open, corresponding to infinite resistance, no current flows in coil *A.* Coil *B* thus governs the motion of the rotating element, causing it to move to its extreme counterclockwise position. The point on the scale indicated by the pointer under this condition is marked infinite resistance.

Coil *A* is wound to produce a clockwise torque on the moving element. With the test terminals marked LINE and EARTH short-circuited, corresponding to zero external resistance, the current flowing through coil *A* is large enough to produce enough torque to overcome the counterclockwise torque of coil *B.* This moves the pointer to its extreme clockwise position, and this point is marked zero resistance on the scale. The resistance R' protects coil *A* from the flow of excessive current when the test terminals are short-circuited.

When an unknown value of resistance is connected between the test terminals LINE and EARTH, the opposing torques of the coil balance each

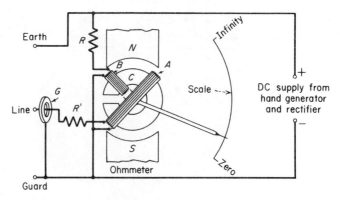

FIGURE 20-23   *Simplified circuit diagram of the Megger insulation tester.* (James G. Biddle Co.)

other so that the pointer comes to rest at some intermediate point on the scale. The scale is calibrated so that the pointer indicates directly the value of the resistance being measured.

The purpose of the guard ring G is to prevent errors in reading due to leakage currents between the LINE and EARTH terminals of the instrument. The GUARD terminal is provided so that unwanted leakage currents in the apparatus under test may also be eliminated.

20-17 THE OHMMETER

The series ohmmeter makes use of the principle of using a voltmeter for measuring resistance as described in Sec. 20-15. Figure 20-24 shows the circuit of the series ohmmeter. A battery, a current-detecting instrument G, and a series resistance R are connected in series between the two terminals A and B. With terminals A and B open (infinite resistance), the pointer of the instrument is at rest at the left-hand end of the scale. The value of the series resistance R is chosen so that with the test terminals A and B short-circuited (zero resistance) the pointer comes to rest at the right-hand end of the scale. Thus an inverse scale is used, zero resistance being indicated on the right-hand end with increasing values of resistance marked on the scale from right to left.

Before a resistance measurement is made, the test terminals are short-circuited and the instrument pointer is adjusted to the zero position by means of the variable-shunt resistance R_s. This is necessary since the battery voltage may vary from time to time and thus affect the readings. The resistance R_x to be measured is then connected between terminals A and B, the value of the resistance being read directly from the scale.

Ohmmeters are very often included as a part of a multifunction, multirange test instrument.

20-18 ELECTRONIC VOLTMETERS

Electric instruments or meters may be classified as being either passive or active. The permanent-magnet moving-coil, moving-iron vane, and electrodynamometer instruments described earlier in this chapter are examples

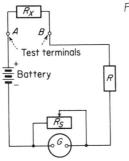

FIGURE 20-24   Connections of the ohmmeter.

of the passive type. Instruments of this type must extract energy from the circuit under test to deflect the meter movement. This loading effect of the meter is of no great consequence in normal 50- or 60-Hz power circuits since the current drawn by the meter is usually very small relative to the current in the circuit being tested. However, in sensitive high-impedance or high-frequency circuits the use of such meters may affect appreciably the operation of the circuit being tested.

Active instruments have their own power supplies, and the energy required to operate the instrument display is obtained from this internal power supply. The loading effect of such instruments therefore is minimal, resulting in very little disturbance of the circuits in which measurements are being made. Electronic voltmeters are active instruments and may be either the *analog* or the *digital* type.

In the *analog electronic voltmeter,* tiny currents flow from an internal power source through two transistors. With the instrument energized and with zero voltage at its input terminals, these two currents are balanced by the operation of a zero-adjustment knob on the instrument so that the pointer reads exactly zero on the meter scale. When a voltage to be measured is then applied to the input terminals of the meter, an unbalance is caused in the flow of current in the two transistors. This causes a deflection of the meter pointer that is proportional to the unbalance in the currents and to the magnitude of the voltage being measured.

The *digital electronic voltmeter* displays the measured voltage on a multidigit numeric display as lighted figures or digits. This is in contrast to a pointer moving over a fixed scale as in the analog voltmeter. However, since voltage is an analog quantity—that is, it can vary from zero to its maximum value with an infinite number of values—its form must be changed to be displayed by a digital voltmeter. This is accomplished by the use of an analog-to-digital converter built into the voltmeter.

A digital voltmeter consists essentially of a signal conditioner or buffer, an analog-to-digital converter, and a display system. A built-in power supply is included also to provide a reference voltage for the analog-to-digital converter and to power the electronic circuits. A voltage to be measured is applied to the signal conditioner, where it is amplified, reduced in value, or passed unchanged to the analog-to-digital converter. The converter changes the input analog voltage to digital form by the use of one of several standardized techniques. The signal is then decoded and displayed in its digital form by the voltmeter's digital circuits. A range switch is provided so that different full-scale values of voltage may be selected by moving the displayed decimal point. Thus, full-scale values may be, for example, 9.999, 99.99, or 999.9 V.

20-19
MULTIMETERS

Voltmeters, ammeters, and ohmmeters are often combined into multifunction, multirange test instruments called *multimeters*. Multimeters may be equipped with either an analog or a digital display system.

Analog multimeters may be either the passive type or the electronic type. The display system for either is the pointer-type movement, usually the permanent-magnet moving-coil type. A typical passive-type combination volt-ohm-milliammeter (usually referred to as a VOM), which has been in use for a considerable time, is shown in Fig. 20-25. This portable meter has self-contained dry cells to provide the source of emf for the ohmmeter. Both dc and ac currents and voltages, as well as resistance, can be measured with this meter.

Analog electronic multimeters have features similar to those of the passive analog VOM except that lower voltage ranges are provided on the electronic type. These meters are made portable by the use of self-contained battery supplies for the ohmmeter and the electronic circuits.

Digital multimeters (DMMs) are active electronic meters that offer some advantages over the analog types for certain applications. The use of the numeric display makes reading easier and eliminates interpretation errors that can occur with the use of multiscale analog meters. Digital multimeters are available in many styles, ratings, and levels of accuracy. A typical DMM is shown in Fig. 20-26. This meter is powered by rechargeable nickel-cadmium batteries, making it suitable for portable use. Measurements of dc and ac currents and voltages as well as resistance all in several ranges can be made with this compact instrument.

FIGURE 20-25   A portable volt-ohm-milliammeter, or VOM. (Simpson Electric Company.)

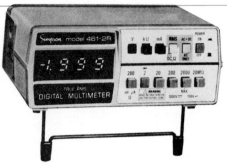

FIGURE 20-26   A digital multi-meter, or DMM. (Simpson Electric Company.)

20-20 THE CATHODE-RAY OSCILLOSCOPE

The cathode-ray oscilloscope is a very useful general-purpose electronic measuring instrument. It is used for studying waveshapes of alternating currents and voltages as well as for measuring voltage, current, power, and frequency, in fact, almost any quantity that involves amplitude and waveform.

The cathode-ray oscilloscope is built around the cathode-ray tube. (Note that the cathode-ray tube is the primary component in a video display terminal also.) This is a high-vacuum glass tube containing a cathode, a cathode heater, a control grid, an anode, and two pairs of deflecting plates, all as shown in simplified form in Fig. 20-27. The cathode is fabricated with a material that, when heated, liberates or emits electrons.

When the cathode of the cathode-ray tube is heated to its emitting temperature and a voltage of the proper magnitude and polarity is applied between the anode and the cathode, the emitted electrons are drawn toward the anode at a high velocity. The anode has an opening in it so that some of the electrons pass on through it. These electrons, because of the velocity attained by their movement from the cathode to the anode, continue to move through the highly evacuated tube until they strike the end wall of the tube. This part of the tube is covered with a fluorescent material that glows when it is struck by the stream of electrons and forms the viewing

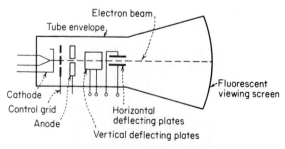

FIGURE 20-27   Diagram of a cathode-ray tube.

screen of the tube. A stream of electrons passing from the cathode through the opening in the anode and in a straight line to the screen appears as a stationary spot on the screen.

Between the anode and the viewing screen and adjacent to the electron stream are placed two pairs of deflecting plates, one horizontal pair and one vertical pair. By applying a voltage to either pair of plates, the electron stream may be deflected from its straight-line path, the amount of deflection being in proportion to the voltage applied. If, for example, an alternating voltage is applied across the two vertical deflecting plates, the electron stream is deflected back and forth along a horizontal line on the viewing screen as the voltage changes in magnitude and direction. Unless the frequency of the applied voltage is very low, the back-and-forth movement of the electron stream appears as a stationary horizontal line on the screen. Likewise, an alternating voltage applied to the horizontal plates appears as a stationary vertical line.

When the oscilloscope is being used to study the waveshape of, say, an alternating voltage, the voltage under test is applied to one set of plates while a second voltage, called a sweep voltage, is applied to the other set of plates. The sweep voltage is of special waveform and is generated in an oscillator that is usually a part of the oscilloscope. The sweep voltage is adjusted until it is synchronized with the voltage under test. With the two voltages properly synchronized, a stationary wave of the exact shape of the voltage under test appears in the form of a graph on the viewing screen. Thus both the waveshape and magnitude of the test voltage may be observed on the screen. The brilliance of the image on the screen may be controlled by means of a control grid placed between the cathode and the deflecting plates.

Because of its versatility in high-frequency measurements, the cathode-ray oscilloscope is almost indispensable in communications research work and in the servicing of communications equipment. It is used in the study of transient phenomena due to lightning and switching surges on power systems. It is also used in the studies of electric arcs and commutation; for the measurements of impedance, transformer ratios, and phase angle; for mechanical measurements; and in many other applications.

REVIEW QUESTIONS

1 What is a galvanometer? What is the fundamental principle upon which it operates?
2 How is the restoring or restraining action produced in a galvanometer? How is damping accomplished?
3 Describe the construction of a permanent-magnet moving-coil instrument. For what kind of measurements is it used?

4 How are voltmeters and ammeters connected to a circuit?

5 How may the range of a voltmeter be extended so that high values of voltage may be measured?

6 How are high values of current measured with a permanent-magnet moving-coil instrument?

7 Describe the principle of the moving-iron vane instrument. How is damping provided? For what kind of measurements is it used?

8 What is an electrodynamometer instrument? Why can this instrument be used for either ac or dc measurements while the permanent-magnet moving-coil instrument is suitable only for dc measurements?

9 How is the electrodynamometer instrument connected when used as a wattmeter? Will this instrument correctly measure power in a circuit in which the current and voltage are out of phase?

10 How many single-phase wattmeters are necessary for accurate measurements of power in a three-wire three-phase circuit? How are the wattmeters connected?

11 How is driving torque obtained in the induction watthour meter? How is retarding torque provided?

12 How do watthour-meter connections compare with wattmeter connections?

13 What type of instrument is used for power-factor measurement?

14 Show how an ammeter and voltmeter should be connected for determining the resistance of a 115-V 10-W lamp. How does this connection differ from that used for measuring a motor armature resistance?

15 Draw a sketch showing the essential circuits of a Wheatstone bridge. Show the relation between the four resistances when the bridge is balanced.

16 What is meant by insulation resistance? Why is it important to know the value of insulation resistance?

17 Describe the method of measuring resistance with a voltmeter.

18 What are the essential parts of the Megger insulation tester? Describe its operation.

19 How is a series ohmmeter adjusted for a zero reading? Why is this adjustment necessary?

20 How do passive test instruments differ from active ones?

21 Why is an analog-to-digital converter a necessary part of a digital electronic voltmeter?

22 What is the basic function of an oscilloscope?

20-1 A 150-V dc voltmeter has a resistance of 15,000 Ω. Find the current flowing through the voltmeter when it is connected to a 120-V line. What will be the voltmeter current for full-scale (150-V) deflection?

20-2 What must be the value of an external resistance used with the voltmeter in Prob. 20-1 if it is desired to use the voltmeter for measuring a maximum of 300 V?

20-3 What is the resistance of a 50-A shunt if the voltage drop across it is 50 mV at rated shunt current?

20-4 A 50-mV millivoltmeter is used with a 100-A 50-mV shunt for measuring the current taken by a dc motor. What is the motor current when the instrument indicates 30 mV? By what factor must any reading of the above millivoltmeter be multiplied to give the motor current in amperes?

20-5 If a 100-mV instrument is used with the shunt in Prob. 20-4, what will be the scale-multiplying factor?

20-6 A millivoltmeter used with a 200-A 50-mV shunt has a 0 to 200 scale marked to read directly in amperes. If the 200-A shunt is replaced by a 100-A 50-mV shunt, by what factor must the instrument readings be multiplied to obtain true values of current?

20-7 A permanent-magnet moving-coil instrument has a coil resistance of 2 Ω and requires a current of 0.025 A for full-scale deflection. The scale reads 0 to 300. Find the value of a series resistance required to make the instrument suitable for use as a 300-V voltmeter.

20-8 Three impedances having resistances of 5 Ω and inductive reactances of 8.66 Ω are wye-connected to a 480-V three-phase three-wire system. Three ammeters and two single-phase wattmeters are connected to read currents and power in the circuit. What will be the readings of each of the ammeters and of each wattmeter?

20-9 A 15,000-Ω voltmeter is to be used for an insulation-resistance measurement. The supply voltage is 120 V. When connected in series with the supply and the insulation under test, the voltmeter reads 10 V. What is the resistance of the insulation?

20-10 With the ratio arms of a Wheatstone bridge set at a 10-to-1 ratio, the bridge balances with the rheostat arm set at 23 Ω. What is the value of the unknown resistance?

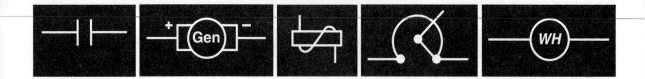

21

Industrial Control

The control of industrial equipment has changed radically from just a few years ago. The control of facilities used pneumatic, hydraulic, or electromechanical relays up until about 1970. That year marked the first time that a computer-based control system was utilized in an industrial environment. These computer-based systems are called *programmable logic controllers,* or PLCs. A PLC is defined by the National Electrical Manufacturers Association (NEMA) as *a digital electronic apparatus with a programmable memory for storing instructions to implement specific functions, such as logic, sequencing, timing, counting, and arithmetic, to control machines and processes.* Essentially, solid-state electronic devices or hardware are used to store instructions that are repeatedly executed in the same order, as with relay systems. However, PLCs, unlike electromechanical relay logic, can alter control strategy based on data collected and computations performed.

This chapter will discuss types of control, what is meant by control logic, the components of PLCs, and their applications.

21-1 CONTROL DESCRIPTION

The control of processes or machines means the execution of a specific set of instructions in response to an input or condition detected by sensors located within the process or machine. Control may be as simple as an on or off operation. A switch is a simple control device when it is used to start and stop equipment. Alternatively, control may be continuous, as in varying a light's output using a dimmer.

The switch is an example of a *digital* device that responds to individual, discrete inputs. The dimmer is an example of an *analog* control device that responds to a continuous input.

The emphasis in this chapter is on digital control since it is the most common type. Digital, or switching-type control, functions in a *binary* manner. A binary system allows devices to exist in only two possible states. A switch is a binary device since it may either be open or closed. Various other designations given to these two states are on-off, yes-no, high-low, or energized-deenergized. The designations most often used, however, particularly in mathematical manipulation of switching or logic functions, are the symbols 1 (one) and 0 (zero).

Digital control has been performed for many years by electromechanical relays, which have moving contacts operated by electromagnets. The same functions are replaced by *static* devices, or devices that perform without moving parts. These electronic devices may take many forms, the most sophisticated of which are PLCs. PLCs have several advantages over electromechanical control including smaller size, greater reliability, and much higher speed of operation.

21-2 BASIC RELAY LOGIC PRINCIPLES

Schematic representations of digital control circuits are called *logic diagrams*. Relay logic diagrams have traditionally utilized the symbols listed in Appendix A and represent actual electrical components such as resistors, capacitors, coils, movable relay contacts, and fuses. The symbols are arranged to show the control and flow of power in a circuit. These same symbols may or may not be used to represent PLC logic as described in Sec. 21-5.

Figure 21-1 contains simple control circuits for a motor control and a lighting circuit. Fig. 21-1a is a simplified standard *motor control circuit.* (Fig. 16-9 shows the more complete circuit.) The START pushbutton is a momentary contact switch held normally open by a spring. The STOP button is held normally closed by a spring. When the START pushbutton is depressed, the contactor coil M is energized. The contact M is connected to contactor M and "follows" (has the same position) as contactor M. The M contact, which is connected in parallel with the START button, closes to "seal in" the contactor. This means that the START button can be released and the coil M remains energized. Depressing the STOP button momentarily breaks the circuit and deenergizes coil M, which opens contact M.

Similarly, Fig. 21-1b illustrates a simple *three-way switching circuit* used for control of a light from two locations. As shown, the light may be switched on or off by either SW 1 or SW 2. If both switches are connected to the same line, the lamp will light. If not, the lamp will be off.

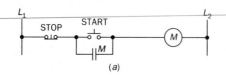

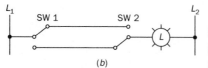

FIGURE 21-1 Relay logic diagram for (a) motor control circuit, and (b) three-way switching circuit.

Complicated control logic may be devised using these simple symbols. Different types of contacts and switches are included in Appendix A, along with types of coils and other input devices.

21-3 BASIC DIGITAL LOGIC PRINCIPLES

Digital logic utilizes static devices to perform control. Static devices do not use moving contacts and coils to accomplish control. Rather, control is accomplished using basic building blocks called *logic gates*. Logic gates are composed of functional groupings of electrical components such as resistors, capacitors, diodes, and transistors. Logic gates can be represented by symbols that are assembled into logic diagrams. However, logic diagrams, unlike those for relay logic, show signal or information circuits, but not necessarily power circuits or connections. Logic diagrams do not represent actual circuit details or components used within the logic gate. The diagrams are nothing more than a shorthand method of representing control circuitry.

There are three basic building blocks or logic gates used in logic diagrams: AND, OR, and NOT gates. When these three basic gates are put together in the right combinations, almost any control or mathematical problem may be solved.

The AND gate has two or more inputs and one output. In order to have an output, all inputs must be present. In accordance with the American National Standards Institute (ANSI) formal definition, *the output of an AND assumes the 1-state if and only if all inputs assume the 1-state.* This, of course, is equivalent to the electromechanical series circuit. The standard symbol, its electromechanical equivalent, and a truth table for the AND are shown in Fig. 21-2.

A *truth table* is a tabular representation of all the possible combinations of inputs and their corresponding resulting outputs for a given device. As shown in the first part of Fig. 21-2, the truth table for a two-input AND gate has three columns, one for each input and one for the output. In the truth table, the 1 and 0 shown are not numbers as used in the usual sense

Function	Symbol	Truth Table		Output	Comments
		Inputs		Output	
AND		A 1 1 0 0	B 1 0 1 0	Y 1 0 0 0	Mechanical equivalent 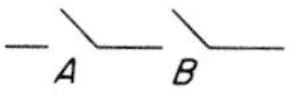
OR		A 1 1 0 0	B 1 0 1 0	Y 1 1 1 0	Mechanical equivalent 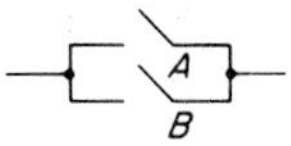
NOT	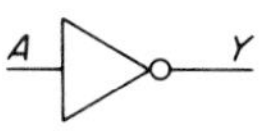	A 1 0		Y 0 1	Inverter
NAND		A 1 1 0 0	B 1 0 1 0	Y 0 1 1 1	Same as AND with output inverted
NOR		A 1 1 0 0	B 1 0 1 0	Y 0 0 0 1	Same as OR with output inverted

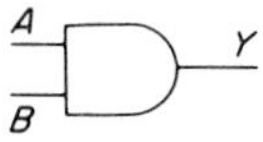

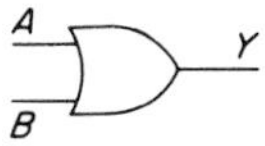

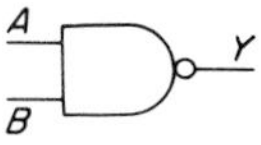

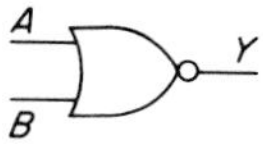

FIGURE 21-2   Basic logic gates.

but are symbols representing *states.* (These states are sometimes referred to as *logic states,* such as *logic 1* or *logic 0.*) Thus, reading across in the AND gate truth table, if inputs A and B have the 1-state, the output state is 1. On the second line of the truth table, if A is 1 and B is 0, the output must be 0 since both inputs must be 1 to result in a 1 output. The same reasoning applies for line 3. Likewise, in line 4, if both A and B are 0, then the output, by definition, must be 0.

The OR gate symbol, truth table, and electromechanical equivalents are shown in the second part of Fig. 21-2. The OR gate, like the AND gate, has two or more inputs and one output. However, the OR gate has an output only when any one of its outputs are present. The ANSI formal definition is *the output of an OR assumes the 1-state if one or more of the inputs assume the 1-state.* This is equivalent to the parallel electromechanical circuit as shown. In the truth table, if A and B are 1, Y is 1. If either A or B is 1 and the other is 0, Y is 1. The output Y is 0 only if both A and B are 0.

The third basic logic gate is the NOT gate, which is usually called an *inverter* because its output is always the opposite of its input. This is defined by ANSI as a logic negation as follows: *the output of a logic negation takes on the 1-state if and only if the input does not take on the 1-state.* The standard symbol and truth table are shown in the third part of Fig. 21-2. As indicated, when A is 1, Y is 0, and when A is 0, Y is 1.

Many logic gates other than the basic three described are used, but these others are mostly derived from the basic three building blocks described. For example, a combination of AND and NOT gates is called a NAND (meaning NOT-AND) and is merely the inverted output of an AND . If both inputs of a NAND are 1s, the output is 1 inverted or 0. Similarly, the NOR (or NOT-OR) is made up of the NOT followed by an OR . Therefore, the input of one or more 1s will result in an output of 0. The NAND and NOR gates are included in Fig. 21-2 for illustration.

The use of logic gates in a motor control circuit is illustrated in Fig. 21-3. As shown, when the START switch is depressed, a 1-state input is provided to input B of the OR. This provides a 1 output from the OR to the A input of the AND. Since the STOP pushbutton is normally closed, it provides the other 1 input for the AND. The 1 output energizes coil *M*. The 1 output is also used to *feedback* or provide a circuit to the A input of the OR. This maintains the 1 output even when the START button is released. When the STOP button is depressed, the B input to the AND becomes 0, causing the AND

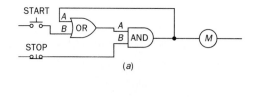

(a)

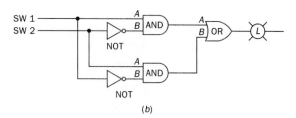

(b)

FIGURE 21-3   The relay logic diagrams of Fig. 21-1 translated to logic gates: (a) motor control, and (b) a three-way switching circuit.

Input		Output
SW 1	SW 2	
1	1	0
0	1	1
1	0	1
0	0	0

FIGURE 21-4   Truth table for Fig. 21-3b.

output to assume the 0-state, thereby deenergizing coil *M*. This logic circuit is sometimes referred to as an *off-return memory logic circuit*.

Similarly, Fig. 21-3b illustrates the three-way lighting circuit. The positions of switches SW 1 and SW 2 are assumed to have a 1-state when they are as shown in Fig. 21-1b. Thus, if both switches provide a 1 input, the A input to each AND gate is 1 and the B input is 0. Therefore, both AND outputs and both OR inputs are 0. Since the resulting OR output is 0, the light is not energized for this position of SW 1 and SW 2. However, if the position of *either* SW 1 or SW 2 is changed, the OR input is changed, resulting in a 1 output to energize the lamp. If the position of both switches is changed, the OR output is again 0 and the lamp is deenergized. The truth table for Fig. 21-3b is included as Fig. 21-4.

21-4 PROGRAMMABLE LOGIC CONTROLLERS (PLCS)

All computers are made up of numerous combinations of gates to perform the computations that form the *logic functions* of the computer. Programmable logic controllers are specialized computers built to withstand the harsher environments of industrial use including dust, dirt, corrosion, and temperature. PLCs have the same basic parts as a normal computer but are more rugged in construction. Figure 21-5 shows the basic components of a PLC.

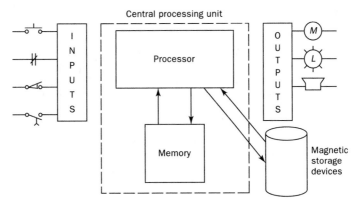

FIGURE 21-5   Major functional parts of a programmable logic controller.

Each of the basic components has specific purposes. The Input devices provide measurement and indicate positions and conditions in the process to be controlled. Inputs include ON-OFF switches, pushbuttons, and relay contacts from other devices. They also may include voltage or current measurements that indicate amount of flow, levels in tanks, speed, or temperature. The type of input utilized depends on the quantity to be measured. As described in Chap. 20, measurements may be performed in a variety of ways.

The Input section also contains input or *signal conditioning* devices, which protect the PLC from damage, change an analog signal to digital, or maintain the input signals within a certain range. Normal current inputs usually range from 4 to 20 ma, while voltage inputs may range from 0 to 5 volts dc.

The Outputs are the responses the PLC has initiated to control or adjust the process. For example, an output may turn on a light to signal the operator that a motor has started or that a valve has closed, or it may turn on a valve to fill a tank or turn on an air pump. All of these outputs are essentially digital, that is, on-off, raise-lower, or open-close. The outputs may also be conditioned; the voltage may be amplified to control a motor, or a relay may be used to control another relay that has higher current carrying capacity contacts.

The *central processing unit* or CPU is the heart of the PLC. It contains the logic circuits and gates that actually execute the programmed instructions. The CPU also contains the *memory,* which actually stores the instructions. Associated with the memory are any auxiliary *storage devices,* such as disk drives or magnetic tape drives that contain the data and programs not continuously utilized by the PLC.

The CPU memory is divided as shown in Fig. 21-6. CPU memory is a

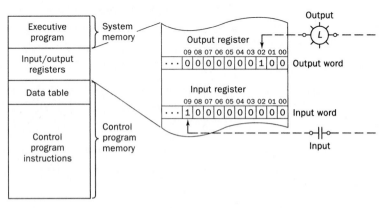

FIGURE 21-6   A simplified diagram or map of a PLC memory with connections for inputs and outputs.

random access memory system that permits the actual storage or retrieval of any instruction or data within the memory to be determined by the machine in any order. All of the areas illustrated in Fig. 21-6 are physical locations in a PLC's memory that have been set aside for specific uses. The following discusses each of these individually.

The *system memory* contains the executive program, which performs the control of the overall operation of the PLC. At least parts of the executive are permanently programmed by the manufacturer into electronic chips for easy access by the PLC. This includes the location of the first instruction for the PLC to execute when the power is switched on, or *booted*. The executive also monitors the condition of the PLC to ensure its proper function electrically, controls access to storage devices, and provides interface to user input and output devices such as keyboards and display terminals.

The inputs and outputs are assigned specific locations in the memory known as *registers*. The input registers are connected to specific input terminals. Each input and output has a specific *bit* associated with it. The *status* (0 or 1) of this bit determines the status of the input or output. Bits are combined to form a *word*. Each bit has a specific location or *address* within a word. One type of word has 16 bits, as shown in Fig. 21-7. Each word also has an address in memory that is accessed or *read* in a prescribed sequence by the program.

The input device condition (open contact, closed contact, pushbutton, etc.) *sets* the status of the input bits. This status cannot be changed by the PLC program. Alternatively, the output bits are totally controlled by the PLC and the "external world" is not allowed to change them.

The remainder of the memory shown in Fig. 21-6 is called the *control program memory* or *application memory* and is divided by function into the *control program instructions* and *data table*. The control program instructions are the actual control functions of the PLC. The data table contains numerical constants, or interim results of mathematical or logic calculations, used by the control program.

The control program will start in a given memory location, but the actual physical length of the program and the location of each program segment will vary by the *application* (process being controlled) and the programming approach utilized by the person entering the instructions.

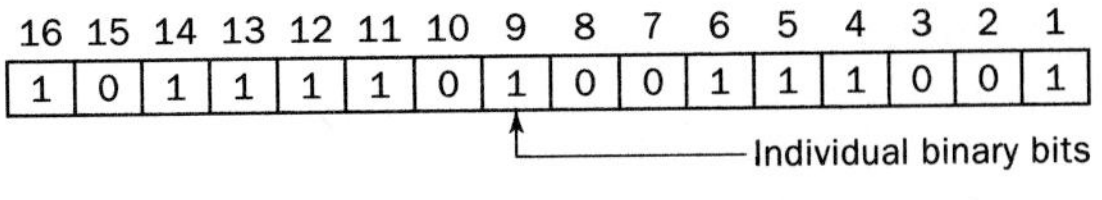

16	15	14	13	12	11	10	9	8	7	6	5	4	3	2	1
1	0	1	1	1	1	0	1	0	0	1	1	1	0	0	1

Individual binary bits

FIGURE 21-7 *A 16-bit word showing locations of individual bits.*

The control program therefore is *software,* as it is created by the user using a given set of standard instructions. The PLC recognizes each individual instruction and executes it. The user or *programmer* decides which instruction to use where and, therefore, determines the control logic or sequence of decisions to be performed by the PLC. The executive decides where the program is to be located in memory.

The control program repeatedly checks for any changes in the input. The time period between checks is known as the *scan rate,* which is usually measured in seconds. The PLC is limited in detecting input changes to the scan rate. Therefore, depending on how fast the process is changing, the scan rate is an extremely important parameter of the PLC. The PLC is not able to react to any changes or "initiate control," that is, begin the decision process to change an output, any faster than it is able to detect a change in the input.

Similarly, outputs are *scanned* to update the outputs based on the logic in the program. The program updates the outputs based on the results of the program logic executed. The program addresses each output terminal in a manner similar to that of the input.

The *data table* is a portion of memory set aside by the executive where constants and other basic data to be used repeatedly by the control program are located. This data may be provided by the programmer or permanently stored in the PLC by the manufacturer.

The size of the memory determines how many inputs and outputs a given PLC is capable of handling as it limits the number of locations a program can address. Also, the amount of *application memory,* memory allocated to the control program and data tables, determines the complexity of control as it limits program physical size that is resident or executable at any one time. The memory size is therefore critical to the correct functioning of the PLC for any specific application.

21-5 SOFTWARE

Software is the means by which the PLC is controlled; it determines the actions to be taken in response to a given set of conditions. The creation of this software is called *programming* or *coding.* Manufacturers supply a standard set of commands, called the programming language, used to program the PLC. There are essentially three types of programming PLC languages:

1. Boolean Languages
2. Ladder Diagrams
3. Higher-Level Languages

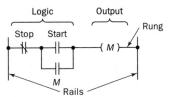

FIGURE 21-8   Simple ladder diagram for circuit shown in Fig. 21-1a.

Boolean languages are those that program by using logic symbols to implement a logical expression. For example, the boolean expression for Fig. 21-3*b* is

$$A\bar{B} + \bar{A}B = 1$$

where A and B are the symbols for SW 1 and SW 2, respectively. The "1" indicates that the statement is "true" or, in this case, the light is turned on. The bar above the symbol indicates that it is the negative or NOT for the symbol.

Although a special set of symbol keys are used to program with logic gates, this is not a very convenient method of programming and can be difficult to *debug*. Debugging is the process used to find and correct mistakes in the control logic.

The most common form of language is the *ladder diagram*. A simple ladder diagram is illustrated in Fig. 21-8, which represents the same functions as Fig. 21-1*a*. It gets its name from its construction in *rungs* and *rails*. The rungs are drawn horizontally and are connected by rails, which are drawn vertically. Each rung contains the logic that controls at least one output. The ladder has been designed to be very similar to relay logic for ease of conversion and universal understanding. The ladder electronically performs the same function as wires do in a hardwired relay-logic scheme.

Higher-level languages are programming languages that use standard written instructions to perform calculations and/or to execute logical statements. The standard instructions are converted to the machine-level language for execution by a special program called a *compiler*. These languages are usually associated with computers and may include BASIC, PASCAL, or C. The languages allow a microcomputer, for example, to interpret inputs and outputs and perform the same functions as a PLC. The program shown later in Fig. 21-14 is an example of using BASIC for control.

**21-6
INPUT/OUTPUT**   There must be devices that allow the user to interface with the PLC to program it and/or access it during actual execution. The PLC may display what is happening to the controlled process or the operator may intervene

and change selected control inputs. These control inputs include *set points* and form the boundaries under which the process may operate. For example, in the water tank problem discussed in Sec. 21-8, the set points could be the maximum and minimum level allowed in the tanks.

Actual programming may be performed using special keyboards whose keys are typewriter style or specialized. The keys may have letters or special ladder diagram symbols that make programming convenient (see Fig. 21-9*a* and Fig. 21-9*b*). The keys may be oversized pushbuttons or may be pressure-sensitive or *touch* keys as shown in Fig. 21-10. Touch keys use a special technology such as *capacitive discharge.* This technique uses the apparent very small change in electrical capacitance at the key to sense the presence of the operator's finger.

The keyboard may be portable and held in the operator's hand or may be attached to a VDT. A VDT, or *video display terminal,* is a cathode ray tube (see Sec. 20-20) that displays both characters and graphic images for

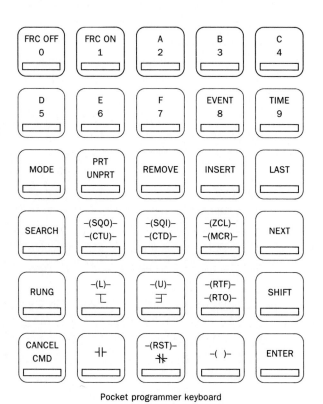

Pocket programmer keyboard

FIGURE 21-9(a)   A portable or "miniprogrammer" device that shows specialized keys for ladder diagram programming. (Allen-Bradley Industrial Control Group.)

Abbreviations and Symbols			
FRC OFF	Force OFF	–(RTF)–	Retentive timer off-delay
FRC ON	Force ON	–(RTO)–	Retentive timer on-delay
PRT	Protect	CANCEL	Cancel command
UNPRT	Not protect	CMD	
–(SQO)–	Sequencer output	–(RST)–	Reset
–(SQI)–	Sequencer input	⊤	Branch open
–(CTU)–	Up counter	⊒	Branch close
–(CTD)–	Down counter		
–(ZCL)–	Zone control last state	⊣⊢	Examine ON
–(MCR)–	Master control reset	⊣⊬	Examine OFF
–(L)–	Latch		Output energize
–(U)–	Unlatch	–( )–	Shift register (use shift key)

FIGURE 21-9(b)   Diagram showing the symbols and abbreviations for the keys shown in Fig. 21-9a. (Allen-Bradley Industrial Control Group.)

the operator. Fig. 21-10 is a VDT. Graphical interfaces such as that shown in Fig. 21-10 are very common as they allow the operator to visualize the actual process during operation.

Programs are stored on a variety of magnetic devices. Normal random access memory was described in Sec. 21-4. Storage must also be provided

FIGURE 21-10   A PLC control panel with pressure-sensitive keys and a VDT. (Allen-Bradley Industrial Control Group.)

outside of the actual PLC for temporary and permanent record of the instructions. It also allows the instructions to be moved from one machine to another. Programs, instructions, and graphics must also be *reloaded* into PLCs after certain occurrences that may erase them from memory. Memory erasures may be caused by power surges, equipment malfunction, or operator error.

Storage devices are all basically magnetic. New devices include optical disks, which are permanent etchings in a plastic disk and are read by small laser devices. The magnetic devices include flexible disks, which are made of material similar to magnetic tape and may be removed from the machine. *Hard drives* may be installed into the machine itself and consist of permanently mounted magnetic platters that are capable of holding large quantities of data.

Other devices include *PROM*s (Programmable Read Only Memory) or *EPROM*s (Erasable Programmable Read Only Memory), which are *electronic chips* similar to static gates that retain their programming after they are deenergized. These devices may store the initial executive program's instructions used to start or boot the PLC as described in Sec. 21-4.

21-7 DATA ENTRY

The PLC may use a variety of devices for input. Usually these devices are electrical measurement devices similar to those described in Chap. 20. These may include the operator's interface console or *thumbswitches* for setting specific values.

Input devices may be discrete, analog, *timer* outputs, or *transducers* among other devices. Timers are solid-state or pneumatic devices that introduce one or more user-defined time delays before closing their contacts. Transducers are devices that receive one type of analog input and produce another type of analog output that is proportional to the input signal. For example, a pressure transducer may have a given range of pressure (force) as input and produce a 4 to 20ma current as its output.

21-8 EXAMPLES OF PLC CONTROL

The use of PLCs can be demonstrated with the following examples. Although these are simple examples that may have been implemented without a PLC, they are useful as illustration.

Motor Control   Several pumps are used to move water from a lower-elevation tank to a tank of higher elevation. The problem is to start one of the pump motors with the pump discharge valve closed. All valves are motor-controlled. Fig. 21-11 is a schematic of the physical system. The sequence to be programmed is as follows:

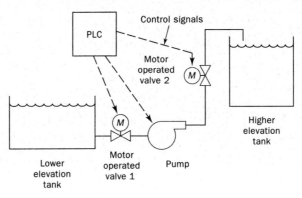

FIGURE 21-11 Physical layout of a system to pump
water from one tank to another.

1. Open the valve for the selected pump.
2. Start the pump and allow it to accelerate to full speed with its discharge valve closed. (The motor is started without a load on it.)
3. Open the discharge valve.

Fig. 21-12 is the ladder diagram to accomplish this. The program is
annotated to explain the logic.

Process Control The water level in a tank is to be maintained
between three and four feet. The problem is to write a higher-level PLC program to accomplish this. A higher level language allows a user to program
in English language-like commands as opposed to using the 1s and 0s that
the computer can understand. The PLC language *interpreter* converts the
standard commands to machine bits and *bytes* (collections of bits) for PLC
use.

Fig. 21-13 illustrates the physical situation. Note that there are float
switches that sense the levels in the tank.

The sequence of events is as follows:

1. Check the level in the tank.
   a. If the level is less than three feet, turn the pump on.
   b. Open valve 1.
   c. Stop pump when the tank level is over four feet.
   d. Close valve 1.
2. If the level is greater than four feet, do nothing.

Figure 21-14 is a simplified listing of a BASIC program to accomplish the
control strategy.

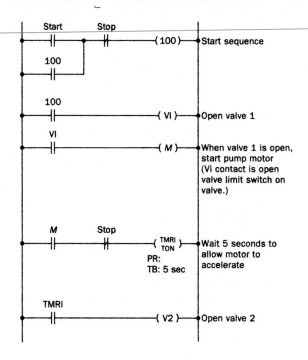

FIGURE 21-12   Ladder diagram for the control of
the physical system shown in Fig. 21-11.

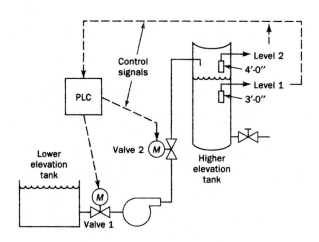

FIGURE 21-13   Physical system for maintaining tank
level by controlling a pump.

```
1        REM A = INPUT FOR 3.0 FT LEVEL SWITCH
2        REM B = INPUT FOR 4.0 FT LEVEL SWITCH
3        REM V1 = OUTPUT FOR VALVE V1
4        REM M1 = MOTOR STARTER OUTPUT FOR PUMP
5        REM
6        REM START PROGRAM
7        REM
10       REM CHECK STATUS OF LEVEL SWITCH
11       IF A = 1 THEN 30
20       GO TO 10
29       REM LEVEL TOO LOW, START PUMP
30       LET V1 = 1
40       LET M1 = 1
50       WAIT SEC = 5
59       REM CHECK STATUS OF HIGH LEVEL SWITCH
60       IF B = 1 THEN 80
70       GO TO 60
79       REM TANK FULL – CLOSE VALVE AND SHUT OFF MOTOR
80       LET V1 = 0
90       LET M1 = 0
100      GO TO 11
110      REM END OF PROGRAM
```

FIGURE 21-14 *Simplified BASIC program to control pump shown in Fig. 21-13.*

REVIEW QUESTIONS

1 What is the basic function of a programmable logic controller (PLC)?

2 In your own words, describe a PLC as it is defined by NEMA.

3 What are some of the differences between a PLC and a relay system?

4 What types of control equipment have been replaced by the PLC?

5 How do PLCs differ basically from electromechanical logic? What advantage does one have over the other?

6 What is the difference between a digital device and an analog device?

7 Define the term *logic gate*. How is a logic gate represented on a drawing?

8 What is shown on a logic diagram?

9 What are the three basic logic gates used in logic diagrams?

10 What is meant by the term *truth table*?

11 Describe an AND gate, an OR gate, and a NOT gate.

12 What is the distinguishing feature of a PLC compared to a computer?

13 What are the three main parts of a PLC?

14 Name several types of input devices for a PLC.

15 Describe the function of the central processing unit (CPU) of a PLC.

16 In a PLC, what is the function of an input register? Of an output register?

17 What is the function of the executive in a PLC memory?

18 What is meant by the scan rate of a PLC?

19 What is contained in a data table of a PLC memory?

20 What is the function of software in a PLC?

21 Name three types of programming languages. Currently, which is the most common form of language?

22 Name some commonly used PLC input devices.

23 What are some magnetic storage devices used with PLCs?

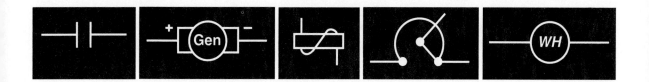

Appendixes

Appendix A

WIRING DIAGRAM SYMBOLS

Device	Standard Symbol	Other Symbols in Use
Battery		
Capacitor		
Circuit breaker, air		
Circuit breaker, oil		
Coil, operating		
Contact, normally closed		
Contact, normally open		
Contact or switch, normally open, time delay closing	TDC	TDC
Contact or switch, normally closed, time delay opening	TDO	TDO
Limit switch, normally open		
Limit switch, normally closed		
Motor-controlled valve		
Pump		
Fuse		
Ground (earth) connection		
Generator		
Galvanometer		

WIRING DIAGRAM SYMBOLS (CONTINUED)

Device	Standard Symbol	Other Symbols in Use
Ammeter		
Voltmeter		
Wattmeter		
Watthour meter		
Lamp (red or green)		
Motor		
Overload relay actuating device		
Pushbutton, spring return, circuit closing		
Pushbutton, spring return, circuit opening		
Resistor, fixed		
Resistor, adjustable		
Switch, single-throw		
Switch, double-throw		
Switch, double-throw, two-poles		
Switch, selector		
Telephone transmitter		
Telephone receiver		

Device	Standard Symbol	Other Symbols in Use
Horn, siren		
Transformer		
Current transformer		
Voltage transformer		
Autotransformer		
Transformer, three-phase, delta-wye connected	$\Delta - Y$	
Winding, motor or generator		
Conductor crossing, no connection		
Conductor junction		
Semiconductor diode		
PNP transistor		
NPN transistor		
Silicon controlled rectified (SCR)	(A) — (K) (G)	(A) — (G) (K)
Triac	(T_2) — (G) (T_1)	
Diac	(T) — (T)	(T) — (T)
Zener diode		(A) — (K)

WIRING DIAGRAM SYMBOLS (CONTINUED)

Device	Standard Symbol	Other Symbols in Use
Photodiode		
Light-emitting diode (LED)		
Phototransistor, PNP	(E) (C) (B)	
Light activated SCR (LASCR)	(A) (G) (K)	

Appendix B

DIMENSIONS, WEIGHTS, AND RESISTANCE OF BARE COPPER WIRE, SOLID, AWG SIZES

Size AWG	Diameter, mils	Area, cmils	Weight, lb per 1000 ft	Resistance, Ω per 1000 ft, 20°C Annealed Wire
0000	460.0	211,600	640.5	0.0490
000	409.6	167,800	507.8	0.0618
00	364.8	133,100	402.8	0.0779
0	324.9	105,600	319.5	0.0983
1	289.3	83,690	253.3	0.124
2	257.6	66,360	200.9	0.156
3	229.4	52,620	159.3	0.197
4	204.3	41,740	126.3	0.249
5	181.9	33,090	100.2	0.313
6	162.0	26,240	79.44	0.395
7	144.3	20,820	63.03	0.498
8	128.5	16,510	49.98	0.628
9	114.4	13,090	39.62	0.793
10	101.9	10,380	31.43	0.999
11	90.7	8,230	24.92	1.26
12	80.8	6,530	19.77	1.59
13	72.0	5,180	15.68	2.00
14	64.1	4,110	12.43	2.52
15	57.1	3,260	9.87	3.18
16	50.8	2,580	7.81	4.02
17	45.3	2,050	6.21	5.05
18	40.3	1,620	4.92	6.39
19	35.9	1,290	3.90	8.05
20	32.0	1,020	3.10	10.1
22	25.3	640	1.94	16.2

called oughts (handwritten annotation next to 0000, 000, 00)

Appendix C

DIMENSIONS, WEIGHTS, AND RESISTANCE OF BARE COPPER WIRE, STRANDED, AWG AND CMIL SIZES

Size, AWG or cmils	Overall Diameter, mils	Number of Strands, Class B Stranding	Weight lb per 1000 ft	Resistance Ω per 1000 ft, 20°C Annealed Wire
2,000,000	1,630	127	6,175	0.00529
1,750,000	1,526	127	5,403	0.00605
1,500,000	1,411	91	4,631	0.00705
1,250,000	1,288	91	3,859	0.00846
1,000,000	1,152	61	3,088	0.0106
900,000	1,094	61	2,779	0.0118
800,000	1,031	61	2,470	0.0132
750,000	998	61	2,316	0.0141
700,000	964	61	2,161	0.0151
600,000	891	61	1,853	0.0176
500,000	813	37	1,544	0.0212
450,000	772	37	1,389	0.0235
400,000	726	37	1,235	0.0265
350,000	679	37	1,081	0.0302
300,000	629	37	925	0.0353
250,000	574	37	772	0.0423
0000	552	19	653	0.0500
000	492	19	518	0.0631
00	414	19	411	0.0795
0	368	19	326	0.100
1	328	19	259	0.126
2	292	7	205	0.159
3	260	7	162	0.201
4	232	7	129	0.253
5	206	7	102	0.320
6	184	7	80.9	0.403
7	164	7	64.2	0.508
8	146	7	51.0	0.641
9	130	7	40.4	0.808
10	116	7	32.1	1.020

Appendix D

DIMENSIONS, WEIGHTS, AND RESISTANCE OF BARE COPPER WIRE, METRIC SIZES

Nominal Area, mm^2	Stranding and Wire Diameter, mm	Approx. Overall Diameter, mm	Nominal Weight, kg/km	Resistance at 20°C,* Ω/km
1000	91/3.73	41.1	9033	0.01765
800	91/3.35	36.9	7284	0.02188
625	91/2.97	32.7	5722	0.02786
500	61/3.25	29.3	4590	0.03472
400	61/2.90	26.1	3642	0.04377
300	61/2.51	22.6	2746	0.05803
240	61/2.26	20.3	2219	0.07181
185	37/2.54	17.8	1699	0.09375
150	37/2.29	16.0	1376	0.1157
120	37/2.06	14.4	1115	0.1429
95	37/1.83	12.8	881	0.1808
70	19/2.18	10.9	645	0.2467
50	19/1.85	9.27	464.8	0.3424
35	19/1.55	7.75	324.5	0.4904
25	19/1.32	6.60	235.8	0.6748
16	7/1.73	5.18	148.3	1.069
10	7/1.37	4.12	93.51	1.695
6	7/0.107	3.21	56.55	2.803
4	7/0.864	2.59	37.08	4.277
2.5	7/0.686	2.06	23.38	6.782
1.5	7/0.533	1.60	14.14	11.21
1.0	1/1.14	1.14	9.122	16.80
0.75	1/0.991	0.99	6.851	22.37
0.50	1/0.813	0.81	4.613	33.23

* The resistance values in this table are based on a conductivity of 100 percent of international annealed copper standard at 20°C.

SOURCE: Based on data published by British Insulated Calendar's Cables, Limited.

Appendix E

SUMMARY OF DC GENERATOR AND MOTOR CHARACTERISTICS

Characteristic	*Generator*	*Motor*
Purpose	To generate an emf	To develop a torque
Capacity rating	Kilowatts	Horsepower (USCS) Kilowatts (SI)
Opposing action	Countertorque	Counter emf
Fundamental equation	$V_t = E - I_a R_a$	$V_t = E + I_a R_a$
Variation in field excitation results in	Change in voltage	Change in speed
Interpoles	Same polarity as following main pole	Opposite polarity from following main pole
Effect of increase in load:		
Shunt machine	Voltage decreases	Small decrease in speed
Series machine	Voltage increases	Large decrease in speed
Cumulative-compound machine	Voltage may remain constant, increase, or decrease, depending on compounding	Large or small decrease in speed depending on compounding

Characteristic	*Generators and Motors*
Generated emf	$E = K\phi n$
Methods of field excitation	Separately excited, shunt, series, compound
Losses	Copper, iron, friction
Efficiency	Percent efficiency $= \dfrac{\text{output}}{\text{input}} \times 100$

Appendix F

FULL-LOAD CURRENTS OF MOTORS

	Three-Phase Motors*						
	Induction-Type Squirrel-Cage and Wound-Rotor, A				Synchronous-Type, Unity Power Factor, A		
Horse-power Rating	200 V	230 V	460 V	575 V	230 V	460 V	575 V
¹/₂	2.3	2	1	0.8			
³/₄	3.2	2.8	1.4	1.1			
1	4.1	3.6	1.8	1.4			
1¹/₂	6.0	5.2	2.6	2.1			
2	7.4	6.4	3.2	2.6			
3	10.4	9.9	4.9	3.8			
5	16.0	14.4	7.2	5.9			
7¹/₂	23.0	22.0	11	8.5			
10	29.0	25.0	12.5	10.4			
15	46	40	20	16			
20	57	50	25	19			
25	80	70	35	26	56	28	23
30		74	37	29	67	34	27
40		98	49	40	88	44	35
50		124	62	50	110	55	44
60		140	70	56	131	66	53
75		174	87	70	162	81	65
100		232	116	95	214	107	86
125			145	116		134	107
150			170	136		160	128
200			230	185		213	171

FULL-LOAD CURRENTS OF MOTORS (CONTINUED)

Direct-Current Motors,* A			Single-Phase Motors,* A		
Horse-power Rating	120 V	240 V	HP Rating	115 V	230 V
$1/4$	2.9	1.5	$1/6$	3.6	1.8
$1/3$	3.6	1.8	$1/4$	5.0	2.5
$1/2$	5.2	2.6	$1/3$	6.4	3.2
$3/4$	7.4	3.7	$1/2$	8.6	4.3
1	9.4	4.7	$3/4$	10.6	5.3
$1 1/2$	13.2	6.6	1	15.4	7.7
2	17	8.5	$1 1/2$	16	8
3	25	12.2	2	27	13
5	40	20	3	34	17
$7 1/2$	58	29	5	56	28
10	76	38	$7 1/2$	80	40
15	. . . .	55	10	100	50
20	. . . .	72			
25	. . . .	89			
30	. . . .	106			
40	. . . .	140			
50	. . . .	173			
60	. . . .	206			
75	. . . .	255			
100	. . . .	341			
125	. . . .	425			
150	. . . .	506			
200	. . . .	675			

* These values of full-load currents are approximate values. Actual currents may be higher or lower than those listed due to variations in torque characteristics, speed, or manufacturer's designs.

Appendix G

TRIGONOMETRIC DEFINITIONS AND FORMULAS

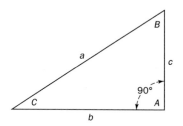

RIGHT TRIANGLES

1. A right triangle is a triangle in which one angle is a right angle (90°).
2. The sum of the three angles (of any triangle) is 180°.
3. The side opposite the right angle is called the hypotenuse. Side a is the hypotenuse in the above figure.
4. The square of the hypotenuse is equal to the sum of the squares of the other two sides. In the above figure,

$$a^2 = b^2 + c^2$$

or

$$a = \sqrt{b^2 + c^2}$$

Also

$$b = \sqrt{a^2 - c^2}$$

and

$$c = \sqrt{a^2 - b^2}$$

Several of the possible ratios between the sides of the right triangle shown in the above figure are listed below. These ratios are called *trigonometric functions*. An abbreviated table of values of the trigonometric functions is given in Appendix H.

5. The ratio of side c to side a is called the sine (sin) of angle C.

$$\sin C = \frac{c}{a} = \frac{\text{opposite side}}{\text{hypotenuse}}$$

6. The ratio of side b to side a is called the cosine (cos) of angle C.

$$\cos C = \frac{b}{a} = \frac{\text{adjacent side}}{\text{hypotenuse}}$$

7. The ratio of side c to side b is called the tangent (tan) of angle C.

$$\tan C = \frac{c}{b} = \frac{\text{opposite side}}{\text{adjacent side}}$$

8. The ratio of side b to side c is called the cotangent (cot) of angle C.

$$\cot C = \frac{b}{c} = \frac{\text{adjacent side}}{\text{opposite side}}$$

The following are examples using the above relationships:

EXAMPLE G-1

If angle B in the triangle shown is $63°$, find angle C.

$\qquad A + B + C = 180°$ (from relation 2)

$\qquad\qquad A = 90°$

Therefore $B + C = 90°$

$\qquad\qquad C = 90 - 63 = 27°$

EXAMPLE G-2

If side c in the figure represents 12 Ω and side b represents 16 Ω, find the angle C.

$$\tan C = \frac{c}{b} \quad \text{(from relation 7)}$$

$$\tan C = \frac{12}{16} = 0.75$$

In the table of tangents (Appendix H), the number nearest 0.75 is 0.7536, which is the tangent of $37°$. Thus the angle C is approximately $37°$ (or by calculator $= 36.87°$).

EXAMPLE G-3

If side a in the figure represents 200 V and angle C is $45°$, find side c.

$$\sin C = \frac{c}{a} \quad \text{(from relation 5)}$$

or $\qquad c = a \sin C$

From the table of sines, $\sin 45 \deg = 0.7071$

Thus $\qquad c = 200 \times 0.707 = 141.4$ V

EXAMPLE G-4   If side a in the figure is 40 units long and angle C is 30°, find sides b and c.

$$b = a \cos C \quad \text{(from relation 6)}$$

$$b = 40 \times \cos 30°$$

$$= 40 \times 0.866 = 34.64$$

$$c = a \sin C \quad \text{(from relation 5)}$$

$$c = 40 \times \sin 30°$$

$$= 40 \times 0.500 = 20$$

Appendix H

TRIGONOMETRIC FUNCTIONS

Angle, degree	Sin	Cos	Tan	Cot	
0	0.0000	1.0000	0.0000	∞	90
1	0.0175	0.9998	0.0175	57.29	89
2	0.0349	0.9994	0.0349	28.64	88
3	0.0523	0.9986	0.0524	19.08	87
4	0.0698	0.9976	0.0699	14.30	86
5	0.0872	0.9962	0.0875	11.43	85
6	0.1045	0.9945	0.1051	9.514	84
7	0.1219	0.9925	0.1228	8.144	83
8	0.1392	0.9903	0.1405	7.115	82
9	0.1564	0.9877	0.1584	6.314	81
10	0.1736	0.9848	0.1763	5.671	80
11	0.1908	0.9816	0.1944	5.145	79
12	0.2079	0.9781	0.2126	4.705	78
13	0.2250	0.9744	0.2309	4.332	77
14	0.2419	0.9703	0.2493	4.011	76
15	0.2588	0.9659	0.2679	3.732	75
16	0.2756	0.9613	0.2867	3.487	74
17	0.2924	0.9563	0.3057	3.271	73
18	0.3090	0.9511	0.3249	3.078	72
19	0.3256	0.9455	0.3433	2.904	71
20	0.3420	0.9397	0.3640	2.748	70
21	0.3584	0.9336	0.3839	2.605	69
22	0.3746	0.9272	0.4040	2.475	68
23	0.3907	0.9205	0.4245	2.356	67
24	0.4067	0.9135	0.4452	2.246	66
25	0.4226	0.9063	0.4663	2.146	65
26	0.4384	0.8988	0.4877	2.050	64
27	0.4540	0.8910	0.5095	1.963	63
28	0.4695	0.8829	0.5317	1.881	62
29	0.4848	0.8746	0.5543	1.804	61
30	0.5000	0.8660	0.5774	1.732	60
31	0.5150	0.8572	0.6009	1.664	59
32	0.5299	0.8480	0.6249	1.600	58
33	0.5446	0.8387	0.6494	1.540	57
34	0.5592	0.8290	0.6745	1.483	56
35	0.5736	0.8192	0.7002	1.428	55
36	0.5878	0.8090	0.7265	1.376	54
37	0.6018	0.7986	0.7536	1.327	53
38	0.6157	0.7880	0.7813	1.280	52
39	0.6293	0.7771	0.8098	1.235	51
40	0.6428	0.7660	0.8391	1.192	50
41	0.6561	0.7547	0.8693	1.150	49
42	0.6691	0.7431	0.9004	1.111	48
43	0.6820	0.7314	0.9325	1.072	47
44	0.6947	0.7193	0.9657	1.036	46
45	0.7071	0.7071	1.0000	1.000	45
	Cos	Sin	Cot	Tan	Angle, degree

For angles over 45°, use titles at foot of table.

Appendix I

Phasors when represented as complex quantities consist of two components, a *real* and a *quadrature* component. Graphically, the real components are represented by points along a horizontal axis, and the quadrature components are represented by points along a vertical axis, as shown in Fig. I-1. The point of intersection of the two axes is called the *origin point O*. Positive values of the real components are represented by points to the right of the origin *O* and negative values to the left of the origin. Positive values of the quadrature components are represented by points above the origin *O* and negative values by points below the origin.

The four sections of the plane as divided by the two axes are called *quadrants* and are numbered as shown. The horizontal line to the right of the origin is called the *reference line*. Starting at the reference line and proceeding in a counterclockwise direction, the first quadrant contains all angles from 0 to 90°; the second quadrant, 90 to 180°; the third quadrant, 180 to 270°; and the fourth, 270 to 360°.

The real and quadrature components of a phasor in any quadrant are the projections of that phasor on the horizontal and vertical axes of the

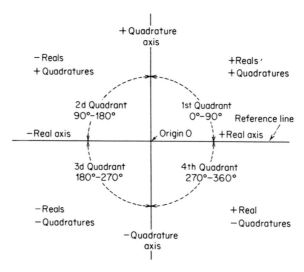

FIGURE I-1   Diagram showing the four quadrants with the positive and negative values of phasor components in each.

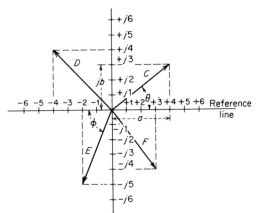

FIGURE I-2   *Diagram to illustrate phasor notation.*

quadrant in which the phasor lies. For example, in Fig. I-2, the phasor C has a real component a (the projection on the horizontal axis) and a quadrature component jb (the projection on the vertical axis). The operator j is used with the quadrature component to indicate that this component is at right angles to the real component.

In accordance with the convention stated, both the real and quadrature components of phasors in the first quadrant have positive values. Mathematically, the phasor C can be written:

$$C = a + jb$$

Using the numerical values shown in Fig. I-2, the phasor C is

$$C = 4 + j3$$

As can be seen in Figs. I-1 and I-2, phasors in the second quadrant have negative real and positive quadrature components, those in the third quadrant have negative real and negative quadrature components, and those in the fourth quadrant have positive real and negative quadrature components. The phasors D, E, and F in Fig. I-2 are expressed numerically as

$$D = -4 + j4$$

$$E = -2 - j5$$

and $$F = +3 - j4$$

The angle that a phasor makes with the horizontal is the angle whose tangent is the quadrature component divided by the real component. For example, the angle θ between the phasor C and the horizontal in Fig. I-2 is

$$\tan \theta = \frac{b}{a} = \frac{3}{4} = 0.75$$

$$\theta = 36.87°$$

Similarly, the angle ϕ for the phasor E is

$$\tan \phi = \frac{5}{2} = 2.5$$

$$\phi = 68.2°$$

The magnitude of a phasor expressed in the $a + jb$ form is $\sqrt{a^2 + b^2}$. For example, in Fig. I-2 the magnitude (length) of each of the four phasors shown is

$$C = \sqrt{4^2 + 3^2} = 5.0$$

$$D = \sqrt{4^2 + 4^2} = 5.66$$

$$E = \sqrt{2^2 + 5^2} = 5.39$$

$$F = \sqrt{3^2 + 4^2} = 5.0$$

Thus the magnitude and the angular position of any phasor are determined fully by the expression $a + jb$, the magnitude being $\sqrt{a^2 + b^2}$, and the angle from the horizontal being the angle whose tangent is b/a. The phasor form $a + jb$ is called the *rectangular form* of phasor notation.

Another way of expressing a phasor is the so-called *polar form* $c\angle\theta°$, where c is the magnitude of the phasor and θ is the angular position of the phasor with respect to the reference line. Using this notation, the phasor C in Fig. I-2 may be represented numerically as $5 \angle 36.87°$ and the phasor E as $5.39 \angle 248.2°$.

Both the rectangular and the polar forms are useful, and many times it is necessary to convert from one form to the other. In Fig. I-2, the two components of the phasor C may be written

$$a = C \cos \theta \qquad \text{and} \qquad b = C \sin \theta$$

Substituting these values in the $a + jb$ form results in

$$C(\cos \theta + j \sin \theta)$$

Again referring to Fig. I-2, if the phasor C is given as $5 \angle 36.87°$, it can be changed to the $a + jb$ form by writing

$$C = 5(\cos 36.87° + j \sin 36.87°)$$
$$= 5(0.8 + j0.6)$$
$$= 4 + j3$$

Furthermore, if the phasor E is represented as $5.39 \angle 248.2°$, it is known that this phasor is in the third quadrant, since the angle is between

180 and 270°. Thus the phasor makes an angle of 248.2° − 180°, or 68.2°, with the horizontal. Since it is in the third quadrant, both the real and the quadrature components will have negative values. Thus

$$E = 5.39(-\cos 68.2° - j \sin 68.2°)$$
$$= 5.39(-0.37 - j0.93)$$
$$= -2 - j5$$

As is shown in the text, currents and voltages are conveniently represented as phasors. Both currents and voltages then may be expressed as complex quantities. Although impedances are not strictly phasor quantities, it is convenient to express impedances as complex quantities. Resistances are represented as real components and reactances as quadrature components. In general, impedance is expressed in the rectangular form as

$$Z = R + j(X_L - X_C)$$

and in the polar form

$$Z\angle\theta°$$

Complex quantities representing phasors in general may be subjected to the usual algebraic operations such as addition, subtraction, multiplication, and division. The $a + jb$ or rectangular form is the more convenient form when adding or subtracting phasors, while the $c\angle\theta°$ or polar form is the more convenient form when multiplying or dividing phasors. The conversion from one form to the other and the algebraic operations using either form are easily accomplished using an inexpensive hand-held electronic calculator.

ADDITION AND SUBTRACTION OF PHASORS To add or subtract phasors expressed as complex quantities, the real and quadrature components are added or subtracted separately, and their sums or differences are combined to form the resultant phasor.

Find the phasor sum of the two voltages $100\angle 0°$ V and $100\angle 30°$ V.   **EXAMPLE I-1**
(Refer to Example 9-4 and Fig. 9-7 in Sec. 9-8 in the text.)
    The two phasors expressed in rectangular form are

$$100\angle 0° = 100(\cos 0° + j \sin 0°)$$
$$= 100(1 + j0)$$
$$= 100 + j0$$

**EXAMPLE I-1
(CONTINUED)**

and $\quad 100\angle 30° = 100(\cos 30° + j \sin 30°)$

$$= 100(0.866 + j0.5)$$

$$= 86.6 + j50$$

Adding the two phasors results in

$$100 + j0$$

$$\underline{86.6 + j50}$$

$$186.6 + j50$$

The magnitude of the resultant voltage is

$$\sqrt{(186.6)^2 + (50)^2} = 193.2 \text{ V}$$

The resultant angle is

$$\tan \theta = \frac{50}{186.6} = 0.268$$

$$\theta = 15°$$

The sum of the voltages is, therefore, $193.2\angle 15°$ V.

EXAMPLE I-2

The total current supplied to the two branches of a single-phase parallel circuit is $40\angle 315°$ A. The current in branch no. 1 is $20\angle 0°$ A. What is the current in branch no. 2? The circuit and phasor diagrams are shown in Fig. I-3.

$$I_t = 40\angle 315° = 40(\cos 45° - j \sin 45°)$$

$$= 40(0.707 - j0.707)$$

$$= 28.28 - j28.28$$

$$I_1 = 20\angle 0° = 20(\cos 0° + j \sin 0°)$$

$$= 20(1 + j0)$$

$$= 20 + j0$$

Subtracting I_1 from I_t results in

$$28.28 \quad - j28.28$$

$$\underline{- (+20 \quad + j0)}$$

$$I_2 = \quad +8.28 \quad - j28.28$$

The magnitude of the current I_2 in the second branch is

EXAMPLE I-2 (CONTINUED)

$$\sqrt{(8.28)^2 + (28.28)^2} = 29.5 \, A$$

and the angle is

$$\tan \theta = \frac{-28.28}{8.28} = -3.416$$

$$\theta = -73.68°$$

Note that the angle θ has a negative value. This indicates that the angular position of I_2 is 73.68° in the *clockwise* direction from the reference line since, by convention, positive angles are measured in a counterclockwise direction from the reference line. A phasor with an angle of 73.68° in the clockwise direction from the reference line is, of course, in the fourth quadrant. This is evident also from the fact that the real component of I_2 is positive and the quadrature component is negative. The angle $-73.68°$ may be expressed as a positive angle by subtracting from 360°:

$$360° - 73.68° = 286.32°$$

The current I_2 then may be expressed in the polar form either as $29.5 \angle{-73.68°}$ or $29.5 \angle{286.32°}$ A.

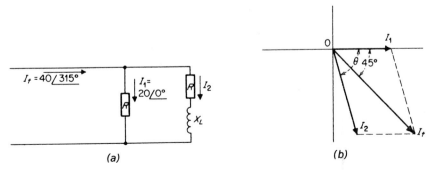

(a) (b)

FIGURE I-3   *Circuit and phasor diagrams for Example I-2.*

MULTIPLICATION OF PHASORS

The product of two phasors $A\angle\theta°$ and $B\angle\phi°$ is $AB\angle\theta° + \phi°$.

EXAMPLE I-3

What is the voltage across an impedance of $40\angle20°\,\Omega$ when a current of $6\angle40°$ A flows through it?

$$V = IZ = 6\angle40° \times 40\angle20°$$

$$= 6 \times 40\angle40° + 20°$$

$$= 240\angle60° \text{ V}$$

DIVISION OF PHASORS

The quotient of two phasors $A\angle\theta°$ and $B\angle\phi°$ is $\dfrac{A}{B}\angle\theta° + \phi°$.

EXAMPLE I-4

What is the current in a circuit that has a resistance of 12 Ω and an inductive reactance of 16 Ω in series when a voltage of $120\angle0°$ V is applied?

$$I = \frac{V}{Z}$$

$$V = 120\angle0°$$

$$Z = 12 + j16 = 20\angle53.13°$$

$$I = \frac{120\angle0°}{20\angle53.13°}$$

$$= \frac{120\angle0-53.13°}{20}$$

$$= 6\angle-53.13° \text{ A}$$

This angle may be expressed as a positive angle by subtracting from 360°:

$$360° - 53.13° = 306.87°$$

Thus $\qquad\qquad I = 6\angle-53.13°$

or $\qquad\qquad I = 6\angle306.87° \text{ A}$

References

Handbooks and Reference Books

Belove, Charles, *Handbook of Modern Electronics and Electrical Engineering*, John Wiley & Sons, Inc., New York, 1986.

Croft, Terrill, and Wilford I. Summers, *American Electrician's Handbook*, 11th ed., McGraw-Hill Book Company, New York, 1987.

Eshbach, Ovid W., and Mott Souders (eds.) *Handbook of Engineering Fundamentals*, 4th ed., John Wiley & Sons, Inc., New York, 1990.

Fink, Donald G., and H. Wayne Beaty (eds.), *Standard Handbook for Electrical Engineers*, 12th ed., McGraw-Hill Book Company, New York, 1987.

Fink, Donald G., and Donald Christensen (eds.), *Electronic Engineer's Handbook*, 2nd ed., McGraw-Hill Book Company, New York, 1982.

IEEE Standard Dictionary of Electrical and Electronic Terms, 4th ed., Institute of Electrical and Electronic Engineers, New York, 1988.

National Electrical Code, National Fire Protection Association, Quincy, MA, 1990.

Weedy, B. M., *Electric Power Systems*, 3rd ed., John Wiley & Sons, Inc., New York, 1987.

Principles and Circuits

Floyd, Thomas L., *Electric Circuit Fundamentals*, 2nd ed., Merrill Publishing, Macmillan Publishing Co., Columbus, OH, 1991.

Grob, Bernard, *Basic Electronics*, 7th ed., Glencoe Division of Macmillan/McGraw-Hill, Columbus, OH, 1992.

Paul, Clayton R., Syed A. Nasar and L. E. Unnewehr, *Introduction to Electrical Engineering*, 2nd ed., McGraw-Hill Book Company, New York, 1992.

Schuler, Charles A., and Richard J. Fowler, *Basic Electricity and Electronics*, Glencoe Division of Macmillan/McGraw-Hill, Columbus, OH, 1988.

Smith, Ralph J., *Circuits, Devices, and Systems*, 4th ed., John Wiley & Sons, Inc., New York, 1984.

Zbar, Paul B., and Gordon Rockmaker, *Basic Electricity: A Text-Lab Manual*, 6th ed., Glencoe Division of Macmillan/McGraw-Hill, Columbus, OH, 1992.

Machinery and Equipment

Anderson, Gunner, *Electric Motors*, 4th ed., Macmillan Publishing Co, Inc., New York, 1983.

Chapman, S. J., *Electric Machinery Fundamentals*, 2nd ed., McGraw-Hill Book Company, New York, 1991.

Gonen, Turan, *Electric Power Distribution System Engineering*, McGraw-Hill Book Company, New York, 1986.

Mason, C. Russell, *The Art and Science of Protective Relaying*, John Wiley & Sons, Inc., New York, 1956.

Nasar, Syed A., *Handbook of Electric Machines*, McGraw-Hill Book Company, New York, 1987.

Ryff, Peter F., *Electric Machinery*, Prentice-Hall, Englewood Cliffs, NJ, 1988.

Veinott, Cyril G., and J. E. Martin, *Fractional and Subfractional Horsepower Electric Motors*, 4th ed., McGraw-Hill Book Company, New York, 1987.

Electrical Measurements

Nachtigal, Chester L., *Instrumentation and Control: Fundamentals and Applications*, John Wiley & Sons, Inc., New York, 1990.

Reissland, Martin U., *Electrical Measurements: Fundamentals, Concepts, Applications*, John Wiley & Sons, New York, 1989.

Semiconductor Devices

Grafham, D. R., and F. B. Golden (eds.), *SCR Manual*, 6th ed., General Electric Company, Auburn, New York, 1979.

Malvino, Albert Paul, *Electronic Principles*, 4th ed., McGraw-Hill Book Company, New York, 1989.

Malvino, Albert Paul, *Semiconductor Circuit Approximations*, 4th ed., Glencoe Division of Macmillan/McGraw-Hill, Columbus, OH, 1985.

Rutkowski, George B., and J. E. Oleksy, *Solid-State Electronics*, 4th. ed., Glencoe Division of Macmillan/McGraw-Hill, Columbus, OH, 1992.

Industrial Control

Chute, George M., and Robert D. Chute, *Electronics in Industry*, 5th ed., Glencoe Division of Macmillan/McGraw-Hill, Columbus, OH, 1979.

Jones, Clarence T., and Luis A. Bryan, *Programmable Controllers-Concepts and Applications*, International Programmable Controls, Inc., Atlanta, GA, 1983.

Answers to Odd-Numbered Problems

Chapter 2	2-1	50 V	2-21	32.67 Ω	2-33	(a) 4 A from A to B
	2-3	4.5 A	2-23	12.0 A; 3 A		2 A from C to F
	2-5	0.5 A	2-25	47.5 Ω		2 A from D to E
	2-7	1600 A	2-27	3.75 Ω; 3.94 Ω		(b) 3.2 V in line AB
	2-9	130 V	2-29	208 V		1.6 V in line CF
	2-11	220 Ω	2-31	(a) 229 V		1.6 V in line DE
	2-13	61 Ω		(b) 235 V		
	2-15	112 V				
	2-17	55 Ω				
	2-19	160 Ω				

Chapter 3	3-1	2,000 mils; 750 mils; 675 mils; 430 mils; 250 mils	3-5	2,750 kg	3-11	0.0033
			3-7	33.6 mm^2	3-13	35.6°C
			3-9	255 Ω		
	3-3	No. 13; No. 10; No. 4; No. 0				

Chapter 4

4-1	20 N		4-13	23.04 Ω
4-3	(a) 320,000 ft-lb per min;	(b) 3.292 kWh or 11.85 MJ	4-15	(a) 406 V;
	(b) 9.7 hp	4-7 480 kWh; 20 kW		(b) 180 W;
4-5	(a) 0.823 kW;	4-9 50.9 A		(c) 288 kWh;
		4-11 7.37 hp		(d) 4.32 kWh

Chapters 5, 6, & 7 have no problems.

Chapter 8

8-1	208.3 A	8-13	0.0392 Ω		(b) 1331 W
8-3	133.3 V	8-15	25 percent	8-25	(a) 1652 W;
8-5	120 V	8-17	2.5 V		(b) 382 W;
8-7	1.6 A	8-19	83.8 percent		(c) 1960 W;
8-9	26.33 A	8-21	91.1 percent		(d) 4810 W
8-11	0.2 Ω	8-23	(a) 501 W;		

Chapter 9

9-1	35.35 A	9-11	10 A	9-21	44.7 Ω
9-3	339.5 V	9-13	0.106 H	9-23	36.1 Ω
9-5	678.9 V; 480 V	9-15	113 Ω; 47.1 Ω	9-25	14.4 Ω
9-7	17.3 A	9-17	66.3 Ω; 33.2 Ω	9-27	6 A
9-9	120 V	9-19	5308 Ω; 0.531 Ω		

Chapter 10

10-1	8320 VA; 8.32 kVA	10-13	(a) 66.5 Ω;	10-23	(a) 13.83 A;
10-3	81 percent		(b) 1.8 A;		(b) 75.9 percent;
10-5	4800 VA; 3360 kW		(c) 60.2 percent		(c) 1260 W
10-7	4800 VA; 83.3 percent		(d) 216 VA;	10-25	(a) 9.23 A; 12 A;
10-9	1118 var; 66.7 percent		(e) 130 W		(b) 4.97 A;
10-11	(a) 20 Ω;	10-15	4.52 A; 4.23 A		(c) 426 W;
	(b) 5.75 A;	10-17	85.4 V		(d) 71.4 percent
	(c) 80 percent;	10-19	0.012 H	10-27	3.3 kvar
	(d) 661 VA;	10-21	(a) 4388 W;	10-29	(a) 12.4 A;
	(e) 529 W		(b) 45.7 A		(b) 40.3 percent

Chapter 11

11-1	208 V	11-11	(a) 1329 V;		(b) 12 A;
11-3	311 V		(b) 597 kVA;		(c) 4320 W
11-5	439 V; 6096 W; 18,288 W		(c) 488 kW	11-21	(a) 208 V;
11-7	7977 V; 3333 kVA; 419 A	11-13	12.96 kW		(b) 36 A;
		11-15	240 V; 92.6 percent		(c) 12,950 W
11-9	86.5 A	11-17	57.36 A	11-23	6 Ω; 4.5 Ω
		11-19	(a) 120 V;		

Chapter 12

12-1	100 turns		44,000 VA;	12-15	5.78 A
12-3	480 V		81.8 percent;	12-17	346 A
12-5	41.7 A; 833.3 A		36,000 W	12-19	100 A; 2080 kVA
12-7	4167 cmil or	12-11	91 A		
	No. 14 AWG;	12-13	(a) 10 kVA;		
	41,667 cmil or No. 4 AWG		(b) 10 kVA		
12-9	2200 V; 20 A;				

Chapters 13 has no problems.

Chapter 14

14-1	36 poles	14-5	1387 V; 250 A;	14-7	(a) 875 kW;
14-3	1500 rpm		600 kVA		(b) 700 kW;
					(c) 438 kW; 211 A

Chapter 15

15-1	33.3 lb-ft		(c) 111.1 V;		(c) 568.5 V;
15-3	2.07 kW		(d) 0 V		(d) 3.16 A
15-5	2.4 A; 77.4 A; 18.57 kW	15-13	3.02 kW	15-21	(a) 85.1 percent;
15-7	1167 rpm	15-15	230 V		(b) 0 percent
15-9	1221 rpm	15-17	5.88 percent	15-23	(a) 112.6 V;
15-11	(a) 2.4 A;	15-19	(a) 146.3 A;		(b) 1.81 hp
	(b) 35.6 A;		(b) 143.1 A;		

Chapter 16

16-1	900 rpm	16-5	3.33 Hz	16-9	9.73 A
16-3	1140 rpm	16-7	354.4 rpm;	16-11	42.3 percent
			54.5 rpm		

Chapter 17

17-1	240 rpm; slow speed	17-5	54 poles; 139 rpm		17-7 5356 kVA
17-3	60 Hz		35.6 rpm;		
			1247 A		

Chapters 18-22 have no problems.

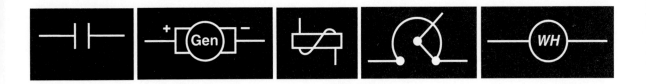

Index

Across-the-line starter, 295
Adjustable-speed drive, for dc motor, 273–74
Alternating current
  advantages of, 117
  components of, 146–47
  defined, 98
  effective value of, 120-21
  frequency of, 120
  instantaneous value of, 120
  maximum value of, 120
  properties of, 130–40
  phase relations of, 122–24
  root-mean-square value of, 121
  in three-wire systems, 29
Alternating emf
  defined, 97, 118
  effective value of, 121–22
  generation of, 118–19
  harmonics with, 230–32
  instantaneous value of, 118
  maximum value of, 119
  in three-phase circuits, 165–66
Alternators. *See* Generators, ac
Aluminum, as conductor material, 36, 41
American Wire Gauge (AWG), 37–38

Ammeters
  defined, 6
  electrodynamometer, 375
  moving-iron vane instrument, 374
  in multimeters, 391
  operation of, 372–73
  permanent-magnet moving coil instrument, 372–73
  placement of, in circuit, 6–7
Amortisseur windings, 247
Ampere, 4, 9
Ampere-hour, 63
Ampere-turn, 77
Amplifier transistor, 217–19
Armature-circuit resistance, in dc motors, 273
Armature reaction
  in ac generators, 118
  in dc generators, 106
Armature windings, in dc generators, 100–102
Arresters, surge, 365
Atom, structure of, 2
Attraction and repulsion
  of electric charges, 3
  of magnetic poles, 74

Autotransformers, 194–95, 201
Autotransformer starters, 297–98

Bar magnets, 72, 73, 74
Batteries
  automobile, 63–63, 65
  charging, 64–65
  chemical process in, 60
  construction of, 61–63
  defined, 55
  lead-acid, 60–65
  maintenance of, 64–65
  nickel-cadmium, 65–69
  rating of, 63–64
  specific gravity of, 61
  See also Cells
Bias, in semiconductor devices, 212, 215
Binary system, 397
Blowout, magnetic, 266
Boolean language (programming), 405
Brakes
  prony, 254
  solenoid-operated, 78–79
Brushes, in dc generators, 94, 100–101, 106

Capacitance, 8, 130, 135, 150, 151
Capacitive reactance, 137–38